Mathematics counts

**Report of the Committee of Inquiry into the Teaching of
Mathematics in Schools under the Chairmanship of
Dr W H Cockcroft**

London: Her Majesty's Stationery Office

© Crown copyright 1982
First published 1982
Fourth impression 1983

ISBN 0 11 270522 7

Foreword

**BY THE SECRETARY OF STATE FOR EDUCATION AND SCIENCE
AND THE SECRETARY OF STATE FOR WALES**

Few subjects in the school curriculum are as important to the future of the nation as mathematics; and few have been the subject of more comment and criticism in recent years. This report tackles that criticism head on. It offers constructive and original proposals for change. It should be read by those responsible for school mathematics at all levels.

The main message is for the education service. The report identifies six agencies whose active response is required. The contribution of all will be necessary if we are to make headway. To the extent that the report calls for extra resources, progress is bound to be conditioned by the continuing need to restrain public expenditure; but many recommendations involve no such call. We hope that there will be widespread discussion of the report's conclusions and that action will follow.

The Committee's terms of reference invited it to consider the teaching of mathematics with particular regard to the mathematics required in further and higher education, employment and adult life generally. The early chapters of the report are concerned with these aspects. They will be of interest to many both within and outside the education service. They reveal the hesitant grasp many adults have of even quite simple mathematical skills. They are particularly valuable in examining closely the mathematics which is in fact needed in different kinds of employment and in everyday life, and relating it to what is taught in the schools. The Committee's findings point to the need for teachers to devote more time to the use of mathematics in applications taken from real life.

This is a first-class report. We are greatly indebted to Dr Cockcroft and the members of his Committee, and commend their work to all those concerned about the quality of mathematics taught in our schools.

Kar Joseph . Nicolas Edwards

January 1982

The Committee of Inquiry into the teaching of Mathematics in primary and secondary schools in England and Wales

MEMBERSHIP OF THE COMMITTEE

Dr W H Cockcroft (Chairman)	Vice Chancellor, New University of Ulster, Coleraine.
Mr A G Ahmed	Head of Mathematics Department, Fairchildes High School, Croydon.
Professor M F Atiyah FRS	Royal Society Research Professor.
Miss K Cross	Assistant Principal and Head of the Faculty of Mathematics and Science, Accrington and Rossendale College, Accrington.
Mr C David	Head teacher, Dyffryn Comprehensive School, Port Talbot.
Mr G Davies	Policy Unit, Prime Minister's Office (resigned April 1980).
Mr K T Dennis	Teacher, Dunmore County Junior School, Abingdon.
Mr T Easingwood	Reader in Mathematical Education, Derby Lonsdale College of Higher Education, Derby.
Mr H R Galleymore	Former Director, Procter and Gamble Limited (appointed August 1979).
Mr R P Harding CBE	Chief Education Officer, Buckinghamshire County Council.
Mr J W Hersee	Executive Director, School Mathematics Project; Chairman, Schools Council Mathematics Committee.
Mrs M Hughes	Head teacher, Yardley Junior School, Birmingham.
Mr A J McIntosh	Principal Adviser in Mathematics, Leicestershire County Council (resigned March 1980).
Mr H Neill	Lecturer in Mathematics, University of Durham.
Mr P Reynolds	Mathematics Adviser, Suffolk County Council (appointed March 1980).
Mr O G Saunders MBE	Welsh Area Secretary, Association of Professional, Executive, Clerical and Computer Staff (appointed March 1979).
Mr H P Scanlon	President, Amalgamated Union of Engineering Workers (resigned January 1979).
Miss H B Shuard	Deputy Principal, Homerton College, Cambridge.
Dr P G Wakely	Chairman and Managing Director, Associated Engineering Developments Limited.
Councillor D Webster	Chairman, Education Committee, Newcastle upon Tyne Borough Council.
Mr L D Wigham	Postgraduate Certificate in Education Student, University of Leeds (appointed November 1978).
Mr P H Halsey (Assessor)	Department of Education and Science.
Mr W J A Mann HMI (Secretary)	Her Majesty's Inspectorate of Schools.
Mr E L Basire (Assistant Secretary)	Department of Education and Science.

The styles, decorations and appointments shown are those held by members at the time of their appointment to the Committee.

10 November 1981

Dear Secretaries of State

On behalf of the Committee of Inquiry into the teaching of mathematics in primary and secondary schools in England and Wales, I have the honour to submit our report to you.

Yours sincerely

W H COCKCROFT

The Rt Hon Sir Keith Joseph Bt MP
Secretary of State for Education and Science

The Rt Hon Nicholas Edwards MP
Secretary of State for Wales

Contents

Introduction

In its report published in July 1977, the Education, Arts and Home Office Sub-Committee of the Parliamentary Expenditure Committee stated that "it is clear from the points which were made over and over again by witnesses that there is a large number of questions about the mathematical attainments of children which need much more careful analysis than we have been able to give during our enquiry. These concern the apparent lack of basic computation skills in many children, the increasing mathematical demands made on adults, the lack of qualified maths teachers, the multiplicity of syllabuses for old, new and mixed maths, the lack of communication between further and higher education, employers and schools about each group's needs and viewpoints, the inadequacy of information on job content or test results over a period of time, and the responsibility of teachers of mathematics and other subjects to equip children with the skills of numeracy". The Committee recommended as "possibly the most important of our recommendations" that the Secretary of State for Education and Science should set up an enquiry into the teaching of mathematics. In their reply presented to Parliament in March 1978, the Government agreed "that issues of the kind listed in the Committee's report need thorough examination" and announced their decision to "establish an Inquiry to consider the teaching of mathematics in primary and secondary schools in England and Wales, with particular regard to its effectiveness and intelligibility and to the match between the mathematical curriculum and the skills required in further education, employment and adult life generally". They further undertook that the Inquiry would examine the suggestion that there should be a full analysis of the mathematical skills required in employment and the problem of the proliferation of mathematics syllabuses at A-level and at 16+.

Terms of reference

Our Committee met for the first time on 25 September 1978 with the following terms of reference:

> To consider the teaching of mathematics in primary and secondary schools in England and Wales, with particular regard to the mathematics required in further and higher education, employment and adult life generally, and to make recommendations.

Meetings and visits

The full Committee has met on 64 days, which have included three residential meetings. Its Working Groups have met on 143 days in all, and there have been less formal discussions on many occasions. 54 schools and 26 companies of various kinds in England and Wales have been visited by members of the Committee and there have been six meetings with groups of teachers in different parts of the country. Small groups of members have visited the Scottish Education Department in Edinburgh, the Institute for the Development of Mathematics Education (IOWO) at Utrecht, Holland, the Institute for the Teaching of Mathematics at the University of Bielefeld, West Germany and the Royal Danish School of Educational Studies in Copenhagen; two members have visited industrial companies in Nuremberg, West Germany. Several members of the Committee were present at the Fourth International Congress on Mathematical Education held at the University of California at Berkeley in August 1980. Individual members of the Committee have been invited to attend the conferences and meetings of a number of professional bodies.

Submissions of evidence

Throughout our work we have been greatly encouraged by the welcome which many people have given to the setting up of the Inquiry and by the helpful response which we have received to our requests for information and written evidence. We have received written submissions, many of them of considerable length, from 930 individuals and bodies of many kinds. 73 individuals and groups have met members of the Committee for discussion. A list of those who have submitted evidence and who have met members of the Committee for discussion is given in Appendix 3.

Research studies

When we started to consider how best we might respond to our terms of reference, we became aware that we needed more detailed information about the mathematical needs of employment and of adult life generally than we were likely either to receive in written evidence or to be able to obtain by our own efforts. We therefore requested the Department of Education and Science (DES) to commission two complementary studies into the mathematical needs of employment and also a small study into the mathematical needs of adult life. One of the studies into the mathematical needs of employment was based at the University of Bath under the direction of Professor D E Bailey, assisted by Mr A Fitzgerald of the University of Birmingham, and the other at the Shell Centre for Mathematical Education, University of Nottingham, under the direction of Mr R L Lindsay. The study into the mathematical needs of adult life was carried out by Mrs B Sewell on behalf both of the Committee and of the Advisory Council for Adult and Continuing Education. The DES also agreed to commission a review of existing research on the teaching and learning of mathematics which was carried out by Dr A Bell of the University of Nottingham and Dr A Bishop of the University of Cambridge. The Steering Groups for all these studies have included members of the Committee, and relevant evidence which has been received has been made available on a confidential basis to those engaged in the studies. The reports which have been produced have proved to be of very considerable help to us; we refer to them and draw on their conclusions in a number of the chapters which follow. At a later stage the DES commissioned a small survey of mathematics teachers in secondary schools who were in their first three years of teaching; this was carried out for us by the National Foundation for Educational Research.

Since we started our work a considerable number of official reports and other publications have been issued which relate wholly or in part to the teaching of mathematics in schools. These include from the DES,

Mathematical development. Primary survey reports Nos 1 and 2 (The APU Primary Surveys)
Mathematical development. Secondary survey report No 1 (The APU Secondary Survey)
Local authority arrangements for the school curriculum
A basis for choice
Proposals for a Certificate of Extended Education (The Keohane Report)
Secondary school examinations: a single system at 16 plus
A framework for the school curriculum
The school curriculum
Examinations 16–18: a consultative paper
Education for 16–19 year olds;

from HM Inspectorate,

Primary education in England (Report of the National Primary Survey)
Aspects of secondary education in England (Report of the National Secondary Survey)
Aspects of secondary education in England: supplementary information on mathematics
Mathematics 5-11: a handbook of suggestions
Developments in the BEd degree course
PGCE in the public sector
Teacher training and the secondary school
A view of the curriculum;

from the Schools Council,

Mathematics and the 10-year old (Schools Council Working Paper 61)
Mathematics in school and employment: a study of liaison activities (Schools Council Working Paper 68)
Statistics in schools 11-16: a review (Schools Council Working Paper 69)
The practical curriculum (Schools Council Working Paper 70);

from other sources,

Engineering our future (Report of the Finniston Committee)
The funding and organisation of courses in higher education (Report of the Education, Science and Arts Committee of the House of Commons)
The PGCE course and the training of specialist teachers for secondary schools (Universities Council for the Education of Teachers)
A minimal core syllabus for A-level mathematics (Standing Conference on University Entrance and Council for National Academic Awards)
Children's understanding of mathematics: 11-16 (Report of the Concepts in Secondary Mathematics and Science Project).

We have studied all these documents and make reference to several of them in the course of this report.

Since we started work, the Government have announced that the present O-level and CSE examinations are to be replaced by a single system of examining at 16+, that GCE A-levels are to be retained, that the Certificate of Extended Education will not be introduced but that there will be a pre-vocational examination at 17+, and that consideration is being given to the introduction of Intermediate levels (I-levels). We have considered the implications of these announcements so far as mathematics is concerned.

Statistical information

As a result of the work which has been carried out for us by the Statistics Branch of the DES and by the Universities Statistical Record we have been able to obtain a considerable amount of information which has not hitherto been available. We refer to this, and to other existing information, from time to time throughout the report. In general we quote this information in rounded terms or present it in diagrammatic form. The detailed tables from which the information is taken are set out in Appendix 1, which also gives in each case the source from which the information has been obtained. This Appendix also contains some tables to which no direct reference is made in the text but which we believe to be of interest. The Appendix discusses, where appropriate, any assumptions which it has been necessary to make in order to prepare the tables and includes brief comments on some of them.

Views from the past

In the light of present day criticism of standards, it is interesting to assemble a collection of quotations from documents of various kinds, some of which date back to the last century, which draw attention to the allegedly poor mathematical standards of the day. We content ourselves with examples from approximately a century, a half-century and a quarter-century ago.

> In arithmetic, I regret to say worse results than ever before have been obtained – this is partly attributable, no doubt, to my having so framed my sums as to require rather more intelligence than before: the failures are almost invariably traceable to radically imperfect teaching.

> The failures in arithmetic are mainly due to the scarcity of good teachers of it.

Those comments are taken from reports by HM Inspectors written in 1876.

> Many who are in a position to criticise the capacity of young people who have passed through the public elementary schools have experienced some uneasiness about the condition of arithmetical knowledge and teaching at the present time. It has been said, for instance, that accuracy in the manipulation of figures does not reach the same standard which was reached twenty years ago. Some employers express surprise and concern at the inability of young persons to perform simple numerical operations involved in business. Some evening school teachers complain that the knowledge of arithmetic shown by their pupils does not reach their expectations. It is sometimes alleged in consequence, though not as a rule with the support of definite evidence, that the teacher no longer prosecutes his attack on this subject with the energy or purposefulness for which his predecessors are given credit.

That extract comes from a Board of Education Report of 1925.

> The standard of mathematical ability of entrants to trade courses is often very low . . . . . Experience shows that a large proportion of entrants have forgotten how to deal with simple vulgar and decimal fractions, have very hazy ideas on some easy

arithmetical processes, and retain no trace of knowledge of algebra, graphs or geometry, if, in fact, they ever did possess any. Some improvements in this position may be expected as a result of the raising of the school leaving age, but there is as yet no evidence of any marked change.

Our final quotation comes from a Mathematical Association Report of 1954; the school leaving age was raised to 15 in 1947.

It is therefore clear that criticism of mathematical education is not new. Indeed, throughout the time for which we have been working we have been conscious that for many years a great deal of advice to teachers about good practice in mathematics teaching has been available in published form from a variety of sources. These include the publications of the DES, of HM Inspectorate, of the Schools Council and of the professional mathematical associations; there have also been references to mathematics teaching in the reports of Committees of Inquiry, for example that of the Newsom Committee. Much of this advice is still relevant today and serves as a background to our own work.

General approach

In writing our report we have tried so far as is possible to avoid the use of technical language and to put forward our views in a way which we hope will be intelligible to mathematician and non-mathematician alike. For this reason we have at times omitted detail which, had we been writing only for those engaged in mathematical education, we would have included. We hope that those who would have wished us to discuss certain matters in greater detail will understand the reason why we have in certain places used a somewhat 'broad brush'. We hope that our attempt to draw attention to those aspects of the teaching of mathematics which we believe to be of fundamental importance will be of use both inside and outside the classroom.

We wish to stress that many of the chapters in our report, and especially those in Part 2, are inter-related. For example, in Chapters 5 and 6 we discuss the elements of mathematics teaching at some length; the fact that we do not repeat this discussion in Chapter 9 (Mathematics in the secondary years), but deal mainly with matters of syllabus content and organisation, does not mean that the teaching approaches we have recommended in earlier chapters are not equally applicable at the secondary stage. We therefore hope that those who read our report will view it as a whole.

In our report we have not considered the needs of pupils with severe learning difficulties. We hope, however, that those who teach pupils of this kind will find that our discussion of mathematics teaching in general, and of the needs of low-attaining pupils in particular, as well as our discussion of the mathematical needs of adult life, will be of assistance.

Acknowledgements

We would like to express our thanks to all those who have written to us and with whom we have talked both formally and informally. We are grateful for the help we have received from the heads, staff and pupils of the schools we have visited, from the teachers whom we have met at meetings in different parts of the country and from those at all levels whom we have met during our visits to commerce and industry. We are grateful, too, to those who were kind enough to arrange these visits and meetings. We wish to thank the Scottish Education Department for arranging our visit to Edinburgh and those whom we visited in Denmark, Holland and West Germany for the help they gave us and the arrangements they made on our behalf.

Our thanks are due to all those who have carried out the various research studies and also to those in DES Statistics Branch and the Universities Statistical Record who have undertaken a great deal of work for us and found ways of answering our questions. We are grateful for the help we have received from many officers in the DES, in particular our Assessor, Mr P H Halsey; and from members of HM Inspectorate, especially Mr T J Fletcher who has, at our invitation, attended many of our meetings.

We wish also to express our thanks for the help and support which we have received from the members of the Committee's secretariat. Mr W M White has taken major responsibility for obtaining, putting in order and interpreting a very great deal of statistical information. Mr E L Basire, our Assistant Secretary, Miss E Kirszberg and Mr R W Le Cheminant, in addition to their other duties, have given much personal help to members of the Committee.

Among those to whom we are indebted, our Secretary, Mr W J A Mann HMI, stands out. We are conscious of the burden we have placed upon his shoulders and of the conscientious way in which he has accepted the load. Most of all, we are grateful to him for the able and efficient way in which he has taken the outcome of our many diffuse and varied deliberations and moulded it into a coherent whole.

Explanatory note

Throughout the report there are certain passages which are printed in heavier type. In selecting these passages, we have chosen those which either relate to matters we consider to be of significance for all our readers or which call for action by those who are outside the classroom. This means that, especially in Chapters 5 to 11 which are concerned in particular with the teaching of mathematics, we have not picked out many of the passages in which we make suggestions relating to classroom practice. As we have pointed out in the Introduction, Chapters 5 to 11 are inter-related and we have not wished to draw attention only to certain passages in them, except in so far as these passages fulfil the purposes we have already set out.

We do not regard the passages printed in heavier type as in any way constituting a summary of the report.

Part 1

1 Why teach mathematics?

1 There can be no doubt that there is general agreement that every child should study mathematics at school; indeed, the study of mathematics, together with that of English, is regarded by most people as being essential. It might therefore be argued that there is no need to answer the question which we have used as our chapter heading. It would be very difficult—perhaps impossible—to live a normal life in very many parts of the world in the twentieth century without making use of mathematics of some kind. This fact in itself could be thought to provide a sufficient reason for teaching mathematics, and in one sense this is undoubtedly true. However, we believe that it is of value to try to provide a more detailed answer.

2 Mathematics is only one of many subjects which are included in the school curriculum, yet there is greater pressure for children to succeed at mathematics than, for example, at history or geography, even though it is generally accepted that these subjects should also form part of the curriculum. This suggests that mathematics is in some way thought to be of especial importance. If we ask why this should be so, one of the reasons which is frequently given is that mathematics is 'useful'; it is clear, too, that this usefulness is in some way seen to be of a different kind from that of many other subjects in the curriculum. The usefulness of mathematics is perceived in different ways. For many it is seen in terms of the arithmetic skills which are needed for use at home or in the office or workshop; some see mathematics as the basis of scientific development and modern technology; some emphasise the increasing use of mathematical techniques as a management tool in commerce and industry.

3 **We believe that all these perceptions of the usefulness of mathematics arise from the fact that mathematics provides a means of communication which is powerful, concise and unambiguous.** Even though many of those who consider mathematics to be useful would probably not express the reason in these terms, we believe that it is the fact that mathematics can be used as a powerful means of communication which provides the principal reason for teaching mathematics to all children.

4 Mathematics can be used to present information in many ways, not only by means of figures and letters but also through the use of tables, charts and diagrams as well as of graphs and geometrical or technical drawings. Furthermore, the figures and other symbols which are used in mathematics can be manipulated and combined in systematic ways so that it is often possible to

deduce further information about the situation to which the mathematics relates. For example, if we are told that a car has travelled for 3 hours at an average speed of 20 miles per hour, we can deduce that it has covered a distance of 60 miles. In order to obtain this result we made use of the fact that:

$$20 \times 3 = 60.$$

However, this mathematical statement also represents the calculation required to find the cost of 20 articles each costing 3p, the area of carpet required to cover a corridor 20 metres long and 3 metres wide and many other things as well. This provides an illustration of the fact that the same mathematical statement can arise from and represent many different situations. This fact has important consequences. Because the same mathematical statement can relate to more than one situation, results which have been obtained in solving a problem arising from one situation can often be seen to apply to a different situation. In this way mathematics can be used not only to explain the outcome of an event which has already occurred but also, and perhaps more importantly, to predict the outcome of an event which has yet to take place. Such a prediction may be simple, for example the amount of petrol which will be needed for a journey, its cost and the time which the journey will take; or it may be complex, such as the path which will be taken by a rocket launched into space or the load which can be supported by a bridge of given design. Indeed, it is the ability of mathematics to predict which has made possible many of the technological advances of recent years.

5 A second important reason for teaching mathematics must be its importance and usefulness in many other fields. It is fundamental to the study of the physical sciences and of engineering of all kinds. It is increasingly being used in medicine and the biological sciences, in geography and economics, in business and management studies. It is essential to the operations of industry and commerce in both office and workshop.

6 It is often suggested that mathematics should be studied in order to develop powers of logical thinking, accuracy and spatial awareness. The study of mathematics can certainly contribute to these ends but the extent to which it does so depends on the way in which mathematics is taught. Nor is its contribution unique; many other activities and the study of a number of other subjects can develop these powers as well. We therefore believe that the need to develop these powers does not in itself constitute a sufficient reason for studying mathematics rather than other things. However, teachers should be aware of the contribution which mathematics can make.

7 The inherent interest of mathematics and the appeal which it can have for many children and adults provide yet another reason for teaching mathematics in schools. The fact that 'puzzle corners' of various kinds appear in so many papers and periodicals testifies to the fact that the appeal of relatively elementary problems and puzzles is widespread; attempts to solve them can both provide enjoyment and also, in many cases, lead to increased mathematical understanding. For some people, too, the appeal of mathematics can be even greater and more intense. For instance:

Anna and I had both seen that maths was more than just working out problems. It was a doorway to magic, mysterious, brain-cracking worlds, worlds where you had to tread carefully, worlds where you made up your own rules, worlds where you had to accept complete responsibility for your actions. But it was exciting and vast beyond understanding.*

*Fynn. *Mister God, this is Anna.* Collins Fount Paperbacks 1974.

Even though it may be given to relatively few to achieve the insight and sense of wonder of 7 year old Anna and of the young man who in later years wrote the book, we believe it to be important that opportunities to do so should not be denied to anyone. Indeed, we hope that all those who learn mathematics will be enabled to become aware of the 'view through the doorway' which many pieces of mathematics can provide and be encouraged to venture through this doorway. However, we have to recognise that there are some who, even though they may glimpse the view from time to time as they become interested in particular activities, see in it no lasting attraction and remain indifferent or in some cases actively hostile to mathematics.

8   There are other reasons for teaching mathematics besides those which we have put forward in this chapter. However, we believe that the reasons which we have given make a more than sufficient case for teaching mathematics to all boys and girls and that foremost among them is the fact that mathematics can be used as a powerful means of communication—to represent, to explain and to predict.

9   It is interesting to note two very different uses which have been made of mathematics in the current Voyager space programme. Not only has the predictive power of mathematics been used to plan the details of the journeys of the two Voyager spacecraft but examples of mathematics have been included in the information about life on Earth which was affixed to each of the spacecraft before they were launched in 1977 to explore the outer Solar System and then to become "emissaries of Earth to the realm of the stars".* The reason for including examples of mathematics is explained in these words:

*Carl Sagan. *Murmurs of Earth.* Hodder and Stoughton 1979

So far as we can tell, mathematical relationships should be valid for all planets, biologies, cultures, philosophies. We can imagine a planet with uranium hexafluoride in the atmosphere or a life form that lives mostly off interstellar dust, even if these are extremely unlikely contingencies. But we cannot imagine a civilization for which one and one does not equal two or for which there is an integer interposed between eight and nine. For this reason, simple mathematical relationships may be even better means of communication between diverse species than references to physics and astronomy. The early part of the pictorial information on the Voyager record is rich in arithmetic, which also provides a kind of dictionary for simple mathematical information contained in later pictures, such as the size of a human being.

10   Mathematics provides a means of communicating information concisely and unambiguously because it makes extensive use of symbolic notation. However, it is the necessity of using and interpreting this notation and of grasping the abstract ideas and concepts which underlie it which proves a stumbling block to many people. Indeed, the symbolic notation which ena-

∴ imp that
(P) understands
basic + − etc notation

bles mathematics to be used as a means of communication and so helps to make it 'useful' can also make mathematics difficult to understand and to use.

11 The problems of learning to use mathematics as a means of communication are not the same as those of learning to use one's native language. Native language provides a means of communication which is in use all the time and which, for the great majority of people, 'comes naturally', even though command of language needs to be developed and extended in the classroom. Furthermore, mistakes of grammar or of spelling do not, in general, render unintelligible the message which is being conveyed. On the other hand, mathematics does not 'come naturally' to most people in the way which is true of native language. It is not constantly being used; it has to be learned and practised; mistakes are of greater consequence. Mathematics also conveys information in a much more precise and concentrated way than is usually the case with the spoken or written word. For these reasons many people take a long time not only to become familiar with mathematical skills and ideas but to develop confidence in making use of them. Those who have been able to develop such confidence with relative ease should not underestimate the difficulties which many others experience, nor the extent of the help which can be required in order to be able to understand and to use mathematics.

hence need for
every day appearance
on curric.

Implications for teachers

12 We conclude this chapter by drawing the attention of those who teach mathematics in schools to what we believe to the implications of the reasons for teaching mathematics which we have discussed. **In our view the mathematics teacher has the task**

- of enabling each pupil to develop, within his capabilities, the mathematical skills and understanding required for adult life, for employment and for further study and training, while remaining aware of the difficulties which some pupils will experience in trying to gain such an appropriate understanding;

- of providing each pupil with such mathematics as may be needed for his study of other subjects;

- of helping each pupil to develop so far as is possible his appreciation and enjoyment of mathematics itself and his realisation of the role which it has played and will continue to play both in the development of science and technology and of our civilisation;

- above all, of making each pupil aware that mathematics provides him with a powerful means of communication.

2   The mathematical needs of adult life

13   There are indeed many adults in Britain who have the greatest difficulty with even such apparently simple matters as adding up money, checking their change in shops or working out the cost of five gallons of petrol. Yet these adults are not just the unintelligent or the uneducated. They come from many walks of life and some are very highly educated indeed, but they are hopeless at arithmetic and they want to do something about it.

Make it count. A study by David Stringer. Independent Broadcasting Authority 1979.

The above quotation comes from the preface to the research study* on the Yorkshire Television series *Make it count*—a series of thirteen programmes for adults broadcast nationally for the first time in 1978. In the conclusion to the study we read

During this investigation the firm impression has built up—in the investigator's mind, at least—that functional innumeracy is far more widespread than anyone has cared to believe.

14 A copy of this study was made available to us soon after we started work and at about the same time the Advisory Council for Adult and Continuing Education (ACACE) drew our attention to the fact that one of the outcomes of the very successful adult literacy campaign of recent years had been an increasing demand for adult numeracy classes. The Council passed on to us some of the experience which had been gained in the course of its work on adult literacy and we are grateful for its willingness to share this with us. In particular the Council urged that, tempting though the approach might seem, we should not set out to try to define the mathematical needs of adult life solely in terms of some kind of 'shopping list' of necessary or desirable skills but should also investigate attitudes towards mathematics and the strategies used by those whose mathematical abilities are limited in their efforts to cope with the mathematics needed in everyday life.

15 Since there appeared to have been very little research carried out to identify the mathematical needs of adults, we decided, as a result of the *Make it count* study and our discussions with ACACE, that we would ask the DES to commission a small study to be carried out on behalf both of ACACE and of our Committee. We suggested that those involved should investigate the mathematical needs of adults in daily life, and, in particular, try to identify the strategies which were used by those whose mathematical skills and understanding were limited. We felt that such an investigation would be of use to both bodies because, although the problems and methods of teaching adults are different from those of teaching children, an understanding of the goals to be achieved should be of value both to those

who teach adults and to teachers in schools. The outcome of this investigation has drawn attention to a number of matters which we believe to be worthy of note, not only by teachers but by many others as well.

The research study

*The results of the study are reported in detail in *Use of mathematics by adults in daily life*: Bridgid Sewell, which may be purchased from the Advisory Council for Adult and Continuing Education. The Advisory Council has also published a summary of the report, together with a summary of the results of a Gallup Poll national survey, in *Adults' mathematical ability and performance*.

16 The study* was carried out in two stages. The first stage consisted of interviews designed to cover four areas:

● a discussion of selected situations, related to shopping and household matters generally, in which mathematics might be involved;

● brief questions on other matters such as the reading of timetables and the use of calculators;

● attitudes to mathematics;

● background information.

These interviews were designed to give some indication of levels of mathematical competence but did not require specific mathematical calculations to be carried out. During the first stage 107 people were interviewed, chosen so far as was possible to reflect the five occupational groups into which the Registrar General divides the population. However, in one sense, very many more people were involved because there proved to be a widespread reluctance to be interviewed about mathematics. In the words of the report:

> Both direct and indirect approaches were tried, the word 'mathematics' was replaced by 'arithmetic' or 'everyday use of numbers' but it was clear that the reason for people's refusal to be interviewed was simply that the subject was mathematics ... Several personal contacts pursued by the enquiry officer were also adamant in their refusals. Evidently there were some painful associations which they feared might be uncovered. This apparently widespread perception amongst adults of mathematics as a daunting subject pervaded a great deal of the sample selection; half of the people approached as being appropriate for inclusion in the sample refused to take part.

17 In the second stage, about half of those who had already been interviewed were interviewed again at greater length. They were invited to answer a series of mathematical questions about a range of everyday situations, some of which were related to topics explored during the first interview; they were not pressed to respond, unless they wished to do so, to questions relating to situations of which they did not have direct experience. Some questions required specific calculations to be carried out, some required an explanation of method but no calculation, some required the interpretation of information presented in mathematical terms. Original documents such as bills, pay-slips and timetables were used whenever appropriate; there were no questions which tested computation by itself unrelated to a real situation. Those being interviewed were free to work out the answers in their head, to use pencil and paper or to use a calculator, as they wished.

18 Because the sample of adults had been small, ACACE decided that it would be desirable to try to validate the findings of the study in some way. The

Advisory Council therefore made arrangements for a selection of questions, of the kind which had been used in the study, to be included as part of a national enquiry undertaken by the Gallup Poll. This enquiry covered a representative sample of the population of Great Britain aged 16 or over; almost 3000 people were interviewed. The results of this enquiry, which are included in the ACACE booklet *Adults' mathematical ability and performance*, suggest that the findings of the original small study are in no way untypical.

Findings of the research study

19   There are, of course, many people who are able to cope confidently and competently with any situation which they may meet in the course of their everyday life which requires them to make use of mathematics. However, the results of the study suggest that there are many others of whom quite the reverse is true.

20   The extent to which the need to undertake even an apparently simple and straightforward piece of mathematics could induce feelings of anxiety, helplessness, fear and even guilt in some of those interviewed was, perhaps, the most striking feature of the study. No connection was found between the extent to which those interviewed used mathematics and the level of their educational qualifications; there were science graduates who claimed to use no arithmetic and others with no qualifications who displayed a high level of arithmetical competence. Nor did there appear to be any connection between mathematical competence and occupational group; people of widely varied mathematical competence were found in each of the five occupational groups. The estimates which those who were interviewed gave of their own mathematical competence did not relate closely to the extent to which they made use of mathematics. There were some who said that they managed very well but who appeared to avoid numbers and others who, although apparently highly competent in the conduct of their everyday affairs, were very hesitant about claiming mathematical skill. There were also some who, while apparently able to perform adequately in the situations which they normally encountered, admitted that they were working at the limit of their mathematical competence and were anxious lest anything more complicated should be required of them.

21   The feelings of guilt to which we referred earlier appeared to be especially marked among those whose academic qualifications were high and who, in consequence of this, felt that they 'ought' to have a confident understanding of mathematics, even though this was not the case. Furthermore, they were aware that others, to whom it was evident that they were well-qualified in general terms, took it for granted that they would be mathematically competent. "People assume you're good at maths if you're good at other things." Those who were not academically well-qualified did not appear to feel guilty in the same way. Some arts graduates who had gained O-level passes in mathematics were nevertheless so aware of a lack of confident understanding of the subject that their career choices were seriously reduced as a result of their determination to avoid mathematics.

22   There was another group consisting of those who, although able to per-

form the calculations which they normally required, felt a sense of inadequacy because they were aware that they did not use what they considered to be the 'proper' method; in other words, they did not make use of the standard methods for setting out written calculations which are normally taught in the classroom. In fact, the study, whilst revealing a very wide variety of approaches to the questions which were asked, also found that many individuals appeared to have only one method of tackling a given problem. If this failed, or if the calculation involved became too cumbersome, they lacked the ability and confidence to attempt a different approach. Nor, in some cases, were they even aware that there might be alternative and possibly more straightforward methods which could be used.

23 Again, just as some felt that there was always a 'proper' method, some felt that there should always be an exact answer to questions involving mathematics and so found themselves in difficulties when it became necessary to approximate or to round off a result. ''I get lost on long sums and never know what to do with the 'leftovers'.'' ''My mind boggles at the arithmetic in estimation.''

24 Failure and consequent dislike of mathematics was often ascribed to a specific cause when young. Such causes included change of teacher or of school, absence through illness, being promoted to a higher class and becoming left behind, having an irascible or unsympathetic teacher who failed to resolve difficulties, or even over-expectation on the part of parents, usually fathers. Criticism by husbands or wives or by other members of the family, especially comment about slowness or the need to use pencil and paper instead of performing a calculation mentally, also eroded confidence and contributed to decreasing use of mathematics. ''I'm afraid I have to write it down. My brother can do it in his head.'' ''My husband says I'm stupid.''

25 The report also refers to ''those who dreaded what they saw as the innate characteristics of learning mathematics such as accuracy and speed, as well as the traditional requirement to show all working neatly. This recalled the long buried anxieties caused by the pupil's arriving at an answer by a mental method and being required to produce a written solution demonstrating a method which had not been used''. This perception of mathematics, and especially arithmetic, as something which is supposed to lead to exact answers by the use of proper methods seemed to be quite common despite the fact that the numbers which arise in everyday life very often need to be rounded or approximated in some way.

26 Another feature revealed by the study was a widespread inability to understand percentages. Many of those interviewed said that they did not understand them or never used them. ''I'm hopeless at percentages really.'' Others who said that they were able to calculate 10 per cent and perhaps, but with greater difficulty, 15 per cent indicated that they would not be able to cope with 8 per cent or 12 per cent. Nor did they seem to have realised that the introduction of a decimalised currency had made it easier to evaluate percentages of sums of money than had previously been the case. It is clear that

politicians, administrators, businessmen, journalists and advertisers all assume that the public at large will understand the many statements which are made which express comparisons in percentage terms. The study suggests that this is very far from being the case and the results of the Gallup Poll enquiry confirm this. Even though those who say that they do not understand percentages probably realise that, for example, a 12 per cent increase is greater than one of 10 per cent, it seems that they would certainly not be able to work out the actual size of the difference in relation to their own salary or wage.

27 A further question revealed that one statistic which is normally expressed in percentage terms, that of rate of inflation, is even more widely misunderstood, with many thinking that a fall in the rate of inflation ought to be associated with an overall drop in the level of prices (even though they often did not think that this was likely to happen) rather than a lessening of the rate at which prices were increasing.

28 The reading of charts and timetables was another area which presented difficulty to many of those interviewed but there was a much higher rate of success on a question which tested ability to read a map and to estimate the distance between two points on it. Understanding of the relative sizes of imperial and metric measures in common use was not widespread.

29 About 70 per cent of those interviewed in the first sample had access to a calculator if they required it but one-third of them said that they never used one. Some of the latter admitted that they did not know how to use a calculator and others expressed doubt and distrust. "I never use it because of the risk of major mistakes." There were also those who maintained that "brains are better" or that "they make you lazy". Some who had tried to use a calculator had been discouraged by the large number of figures which had appeared after the decimal point, for instance when dividing by 3, and had lacked confidence to persevere and to discover how to interpret the answers they had obtained. On the other hand there were some, whose computational skills were weak, for whom the use of a calculator made all the difference. "I know the theory but without the calculator I couldn't do it."

30 Many strategies were encountered for coping with the mathematical demands of everyday life. These included always buying £10 worth of petrol, always paying by cheque, always taking far more money than was likely to be needed when going shopping so as to be certain of being able to pay bills without embarrassment. There was frequent reliance on husbands, wives or children to check and pay bills, to measure or to read timetables; and also reliance on past experience. Sadly, it was also clear that lack of mathematical ability had prevented some people from applying for jobs or from following courses of training which they would otherwise have wished to undertake. In this sense, they had been unable to cope.

The mathematical needs of adult life

31 What, then, are the mathematical needs of adult life? In the first place, it is clear that there is hardly any piece of mathematics which everyone uses. For

example, those who do not travel by bus or train probably have no need to consult timetables; those who do not drive a car have no need to buy petrol; those who do not have meals in hotels or restaurants have no need to be able to calculate a service charge. The study shows that some people appear to use practically no mathematics because they have organised their lives so as to avoid its use or so as to make use of the mathematical skills of others. There are, however, very few people who do not at some time need to be able to read numbers, to count, to tell the time or to undertake a minimal amount of shopping. This, perhaps, represents a minimum list but it is apparent that many of those who possess only this minimum of mathematical skill, as well as some whose attainment is a good deal greater, frequently experience feelings of stress, inadequacy or helplessness, even though they may have found methods of coping with their everyday needs.

32 Therefore, whilst realising that there are some who will not achieve all of them, we would include among the mathematical needs of adult life the ability to read numbers and to count, to tell the time, to pay for purchases and to give change, to weigh and measure, to understand straightforward timetables and simple graphs and charts, and to carry out any necessary calculations associated with these. There are many who, because of the requirements of their employment, their hobbies or their own interest in mathematics, are able to achieve a great deal more than this. Some develop very specialised skills; for example, of the kind which are frequently exhibited by those who play darts or make use of betting shops. **However, we believe that those who teach mathematics in schools should do all that is possible to enable their pupils to include as part of their mathematical knowledge those abilities which we have listed.**

33 We believe too that, as a necessary accompaniment to the list which we have given, it is important to have the feeling for number which permits sensible estimation and approximation—of the kind, for instance, which makes it possible to realise that the cost of 3 items at 95p each will be a little less than £3—and which enables straightforward mental calculation to be accomplished.

34 Most important of all is the need to have sufficient confidence to make effective use of whatever mathematical skill and understanding is possessed, whether this be little or much.

Numeracy

35 The words 'numeracy' and 'numerate' occur in many of the written submissions which we have received. In the light of our discussion in the preceding paragraphs we believe that it is appropriate to ask whether or not an ability to cope confidently with the mathematical needs of adult life, as we have described them, should be thought to be sufficient to constitute 'numeracy'.

*15 to 18. A report of the Central Advisory Council for Education (England). HMSO 1959.

36 The concept of numeracy and the word itself were introduced in the Crowther Report* published in 1959. In a section devoted to the curriculum

of the sixth form, 'numerate' is defined as "a word to represent the mirror image of literacy". Later paragraphs in the report make clear that this definition is intended to imply a quite sophisticated level of mathematical understanding. "On the one hand . . . an understanding of the scientific approach to the study of phenomena—observation, hypothesis, experiment, verification. On the other hand . . . the need in the modern world to think quantitatively, to realise how far our problems are problems of degree even when they appear as problems of kind. Statistical ignorance and statistical fallacies are quite as widespread and quite as dangerous as the logical fallacies which come under the heading of illiteracy." "However able a boy may be . . . if his numeracy has stopped short at the usual fifth form level, he is in danger of relapsing into innumeracy."

37 In none of the submissions which we have received are the words 'numeracy' or 'numerate' used in the sense in which the Crowther Report defines them. Indeed, we are in no doubt that the words, as commonly used, have changed their meaning considerably in the last twenty years. The association with science is no longer present and the level of mathematical understanding to which the words refer is much lower. This change is reflected in the various dictionary definitions of these words. Whereas the Oxford English Dictionary defines 'numerate' to mean "acquainted with the basic principles of mathematics and science", Collins Concise Dictionary gives "able to perform basic arithmetic operations".

38 The second of these definitions reflects the meaning which seems to be intended by most of those who have used the word in submissions to us. However, if we are to equate numeracy with an ability to cope confidently with the mathematical demands of adult life, this definition is too restricted because it refers only to ability to perform basic arithmetic operations and not to ability to make use of them with confidence in practical everyday situations.

39 **We would wish the word 'numerate' to imply the possession of two attributes**. The first of these is an 'at-homeness' with numbers and an ability to make use of mathematical skills which enables an individual to cope with the practical mathematical demands of his everyday life. The second is an ability to have some appreciation and understanding of information which is presented in mathematical terms, for instance in graphs, charts or tables or by reference to percentage increase or decrease. Taken together, these imply that a numerate person should be expected to be able to appreciate and understand some of the ways in which mathematics can be used as a means of communication, as we have described in the previous chapter. We are, in fact, asking for more than is included in the definition in Collins but not as much as is implied by that in the Oxford dictionary—though it will, of course, be the case that anyone who fulfils the latter criteria will be numerate. **Our concern is that those who set out to make their pupils 'numerate' should pay attention to the wider aspects of numeracy and not be content merely to develop the skills of computation.**

3 The mathematical needs of employment

Views expressed before the Committee was set up

*House of Commons. Tenth Report from the Expenditure Committee. *The attainments of the school leaver*. HMSO 1977.

40 It is clear from the report of the Parliamentary Expenditure Committee* to which we referred in the introduction that the volume of complaints which seemed to be coming from employers about lack of mathematical competence on the part of some school leavers was one of the principal reasons for its recommendation that our Inquiry should be set up.

41 We believe that these complaints started to come to the fore in 1973 and 1974 when a number of articles and letters which were highly critical of mathematics teaching in schools, and of 'modern mathematics' in particular, appeared in *Skill*, a news-sheet published by the Engineering Industry Training Board for group training schemes in the engineering industry. These complaints were followed in 1975 and 1976 by articles and letters in *Blueprint*, the newspaper of the Engineering Industry Training Board, which also expressed dissatisfaction with the mathematical attainment of some entrants to the industry. More widespread criticism appeared in newspaper articles and correspondence columns during these years.

42 In his speech made at Ruskin College, Oxford in October 1976, Mr James Callaghan, at that time Prime Minister, said:

> I am concerned on my journeys to find complaints from industry that new recruits from the schools sometimes do not have the basic tools to do the job that is required. ..... There is concern about the standards of numeracy of school leavers. Is there not a case for a professional review of the mathematics needed by industry at different levels? To what extent are these deficiencies the result of insufficient co-ordination between schools and industry? Indeed how much of the criticism about basic skills and attitudes is due to industry's own shortcomings rather than to the educational system?

43 In written evidence to the Parliamentary Expenditure Committee, the Confederation of British Industry (CBI) stated:

> Employers are becoming increasingly concerned that many school leavers, particularly those leaving at the statutory age have not acquired a minimum acceptable standard in the fundamental skills involved in reading, writing, arithmetic and communication. This shows up in the results of nearly every educational enquiry made amongst the CBI membership, and is backed up by continuing evidence from training officers in industry and further education lecturers that young people at 16+ cannot pass simple tests in mathematics and require remedial tuition before training and further education courses can be started.

In oral evidence to the Expenditure Committee a CBI representative stated:

> Mathematics, I think – or arithmetic, which is really the primary concern rather than mathematics themselves – is the one area which is really brought up every time as a problem. It seems that industry's needs are greater in this respect than almost any other. This is the way, certainly, in which shortfall in the education of children makes itself most manifest immediately to an employer.

Written evidence to the Expenditure Committee from the Engineering Industry Training Board (EITB) stated:

> The Engineering Industry Training Board, over the last two years, received from its industry increasing criticism, with supporting evidence, of the level of attainment, particularly in arithmetical skills, of school leavers offering themselves for craft and technician training. ..... In the view of the Engineering Industry Training Board the industry needs a higher level of attainment in basic mathematics among recruits than it is now getting and believes that, with closer co-operation between school and industry, children can while still at school be motivated to achieve this. ..... Mathematics is, however, not simply a question of basic manipulative skills. An understanding of the concepts is also needed and these are better taught by 'innovative methods', which also appear to enhance the ability to acquire planning and diagnostic skills, of great importance to craft and technician employees.

44. With the above quotations in mind — and we could have quoted many more — it has naturally been our concern throughout the time we have been working to investigate complaints about low levels of numeracy among young entrants to employment and the need for improved liaison between schools and industry.

Employers' views expressed to the Committee

45 We have received written evidence, amounting to some 200 submissions in all, from major companies, industry training boards, the Confederation of British Industry, the Trades Union Congress, the Association of British Chambers of Commerce and, with the help of Rotary International in Great Britain and Ireland, from many smaller employers. We have studied the reports of the research studies into the mathematical needs of various types of employment carried out at the Universities of Bath and Nottingham and have also had access to the detailed notes of visits to more than 100 companies on which these reports have been based. Members of the Committee have themselves visited 26 firms chosen to span a wide range of types of employment. During these visits it was possible to talk with young employees as well as with management and training staff, and so to receive their views directly. We have also taken oral evidence from the CBI, from two major training boards and from groups of small employers.

46. The overall picture which has emerged is much more encouraging than the earlier complaints had led us to expect. We have found little real dissatisfaction amongst employers with the mathematical capabilities of those whom they recruit from schools except in respect of entrants to the retail trade and to engineering apprenticeships, both of which involve significant numbers of young people; we discuss these two groups in paragraphs 53 and 54. We have

also found little evidence that employers find difficulty in recruiting young people whose mathematical capabilities are adequate.

47 Because we were concerned that there might nevertheless be a substantial amount of criticism which had not reached us, we made special efforts to invite further evidence. However, amongst other initiatives, an article published in the Autumn 1979 issue of *Skill,* which drew attention to the existence of our Committee and invited its readers to write to us, resulted in only one letter. A further article published in the *Financial Times* in July 1980 produced three letters, only one of which contained major criticisms.

48 Employers have, in general, expressed no more than mild reservation about the fluency in arithmetical skills possessed by the young people whom they recruit; lack of fluency in mental arithmetic has attracted the most comment. However, employers have also told us that, under the motivation which employment can provide, these problems can soon disappear. This situation is consistent with the preliminary findings, published in the Spring of 1980, of the study *Young people and employment,* at present being carried out at Lancaster University under the direction of Professor Gareth Williams. These show that, of a sample of 300 "employing establishments" in England and Wales, only 14 per cent criticised the educational standards of their recruits. More common criticisms found in the Lancaster study were of inability to take work seriously, lack of interest, unwillingness to work, bad time keeping and poor attendance.

49 There appear to be several reasons for this apparent change in the views expressed by employers. In part it may reflect the efforts which we believe have been made by many schools to meet criticisms of the kind which were voiced during the 'Great Debate'. It may also reflect the increasing number of local initiatives which have led to the setting-up of school/industry liaison groups of various kinds. We discuss these groups in paragraph 103.

50 We believe that there is increased understanding on the part of employers of the changes which have occurred in the education system in the post-war years and, in particular, of the fact that, because more young people are going from school to further and higher education, the recruitment of 16-18 year olds to jobs at a given level is taking place from further down the ability spectrum than was the case in earlier years. We may add here that we were much encouraged to hear one of the CBI representatives draw attention in oral evidence to his own company's experience. Although some of the craft apprentices who were currently being recruited were less well qualified academically than some of their predecessors, they were nevertheless being trained successfully to carry out more complex work than their predecessors had been required to undertake. The other CBI representatives agreed with this view.

51 It is possible that a further reason for the present absence of criticism is the current high level of unemployment among young people which allows employers to pick and choose to an extent which was not possible in, for

instance, the mid-sixties and early seventies. It is, of course, not possible to quantify the extent to which those young people who are at present unemployed would, in a time of full employment, exhibit mathematical shortcomings in their work. However, it may be of relevance to consider the pattern of unemployment and notified vacancies among school leavers over the last twenty years which is shown in Figure 1(overleaf).

52 The raising of the school leaving age to 16 in 1973 undoubtedly had some effect on recruitment patterns in 1973-74, but this was also the time at which complaints about lack of numeracy among school leavers came sharply into the public notice, typified by letters and articles in *Skill*. It does seem possible that high levels of employment, and so a smaller pool of applicants from which to choose, may result in greater criticism from employers about the mathematical abilities of their young recruits. We are not aware of the same expression of concern in the mid-sixties, so the connection between high employment and complaint must be treated with caution. Furthermore, industry training boards were only just coming into existence at that time and so there were perhaps not as many vehicles for complaint by employers. However, since it may perhaps be the case that reduction in criticism is in part a result of high levels of unemployment among young people, we are very conscious that this fact must not be allowed to result in complacency about the state of mathematical education at the present time.

53 We have already referred to the fact that most of the criticism of young employees which we have received refers to entrants to the retail trade and to engineering apprenticeships. The former attract the more serious criticism. Those who enter the retail trade on leaving school at 16 commonly have very modest or no mathematical qualifications but are often required from the outset to give change, count stock, fill in stock sheets and calculate discounts. One large employer has complained to us that it is necessary to spend a substantial amount of time teaching newly recruited shop assistants to carry out tasks of this kind.

54 The complaints which we have received relating to engineering apprentices seem to stem largely from the performance of applicants in company selection tests, which are very often tests of computational ability only. However, most of the criticism relates to those applicants who are rejected. Employers have expressed comparatively little dissatisfaction with the mathematical performance of those whom they have taken on as apprentices. Moreover, where difficulties, mainly in arithmetic, do arise, we have been told that they can almost always be overcome relatively quickly during the initial apprenticeship year as apprentices gain practical experience in the workshop and so realise why it is necessary to be able to carry out certain specific calculations.

55 It is important at this stage to note that, while agreeing with the importance of arithmetical attainment, the EITB has stressed to us its view that tests of arithmetical skill play too dominant a role in selection for engineering apprenticeships and that conceptual skills, such as spatial

Figure 1 *Vacancies for young people and numbers registered as unemployed*

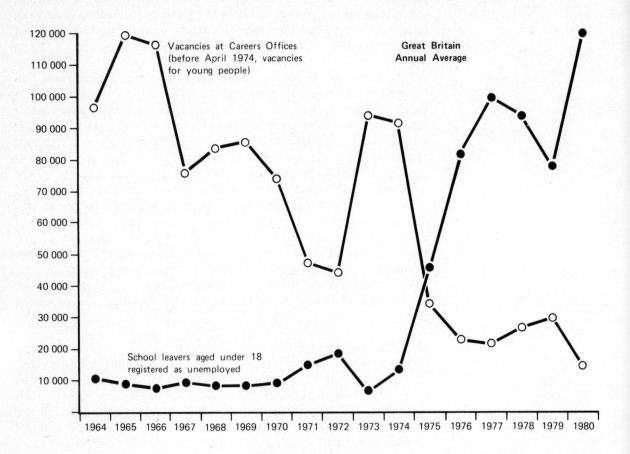

Notes: 1 The lines joining the points have been drawn to "assist the eye"; intermediate points on these lines have no significance.

2 Following the implementation of the Employment and Training Act 1973, the division of responsibilities, at the age of 18, between employment offices and careers offices was abolished. As a result vacancies at careers offices include some which are suitable for adults and vacancies at employment offices include some suitable for young people.

Source: Department of Employment

awareness, the understanding of orders of magnitude, approximation and optimisation, are of equal or greater importance. In the course of oral evidence, one of the EITB representatives expressed concern that some schools, in over-reacting to the perceived criticisms of industry and concentrating excessively on computational work, were failing to give to their pupils the wider mathematical experience which was of value for the intending craftsman.

56   A number of submissions have drawn attention to the need which is experienced by many employers to provide so-called 'remedial' training for young employees; we have referred to these in earlier paragraphs. It is important at the outset to distinguish between three kinds of training for which this term may be used. The first is the need to revise skills which may have become rusty through lack of practice — and all those who work within classrooms know not only how much can apparently be 'forgotten' during the summer holiday, but also how quickly memory can be refreshed. The second is the need to provide opportunity to become familiar with applications, possibly of quite elementary mathematics, which may not have been encountered in the classroom even though they are of frequent use in a particular job, or to teach certain specialised techniques. The third is the need to spend time on training which is remedial in the true sense of the word; in other words, to spend time teaching mathematics which, although it has been included in the school course, has not been understood or is being used incorrectly.

57   We believe that the need to undertake a certain amount of training of the first two kinds, which should only take a short time, is to be expected; though we appreciate that for the small employer or the business which has a rapid staff turnover it will represent some drain on resources. On the other hand, we accept that employers may with reason expect not to have to undertake training of the third kind, since this may properly be considered to come within the responsibility of schools. Inevitably, the line between the two categories is not always easy to draw. In the course of our work we have become aware that a degree of over-expectation exists in many quarters as to the level of attainment in mathematics which some school leavers are able to reach; we discuss this in paragraph 189. This is one of many reasons why good liaison between employers and schools, which we discuss later in this chapter, is of such great importance.

58   There is a further point to which we wish to draw attention. Because of the high level of unemployment which exists among young people at the present time, lack of use of mathematics on their part may well lead to a degree of 'rustiness' which will require sympathetic consideration on the part of employers to whom these young people apply for jobs, and perhaps also the provision of special assistance.

The research studies   59   We turn now to the findings of the two research studies into the mathematical needs of various types of employment which were carried out at the University of Bath and the Shell Centre for Mathematical Education at the University of Nottingham. For the sake of brevity we shall henceforward refer to them as the Bath study and the Nottingham study.

60 The reports which these studies have produced are extensive and wide-ranging. We have space in our own report to draw attention only to the points which we believe to be most significant. We hope that means will be found to make the detailed findings of the studies available widely*; **we believe that they will be of assistance to those who teach in schools and colleges and to those who employ and train young people. We believe, too, that they may well be of considerable interest to a much wider section of the public.**

61 Both studies paid particular attention to the types of employment open to those leaving school at ages from 16 to 18. However, both also included some jobs which, because of legal or other restrictions, were open only to employees over the age of 18 if it was the case that these jobs required no greater experience or qualifications than had been obtained at school.

62 The Bath study was designed to cover as wide and as representative a sample of the various categories of employment as possible. More than 90 companies and other establishments were visited and a considerable amount of time was spent both in observation of work which was going on and in discussion with employees themselves and with managerial and training staff.

63 The Nottingham study was designed to complement the work undertaken at Bath by looking in greater depth at the more specific areas of clerical and retail work, agriculture, and the medical and para-medical occupations within the Health Service which recruit from those who leave school at ages from 16 to 18. It was also able to draw on similar work which had already been undertaken in some other areas, notably the training of engineering apprentices. In both studies stress was laid on obtaining first-hand information. A report was written on each visit and agreed with the company or other establishment concerned.

64 The report of the Bath study draws attention to the fact that it is possible when observing work in progress to describe certain aspects of it in mathematical terms. For instance, the mathematical concept of geometrical symmetry is present within many manufacturing processes; it is possible to describe different methods of stacking and packing in geometrical terms or in terms of sorting and classifying. However, even if the mathematical concepts involved have at some time been encountered in the classroom, the employee will probably not consciously analyse in such terms the operations which are being performed; nor, if he were to do so, would he necessarily be able to do his job any better.

65 On the other hand, many jobs require the employee to make explicit use of mathematics — for instance, to measure, to calculate dimensions from a drawing, to work out costs and discounts. In these cases the job cannot be carried out without recourse to the necessary mathematics and it is this latter use of mathematics to which we refer in the paragraphs which follow. However, even when mathematics is being used, frequent repetition and

*Copies of the Bath Study *Mathematics in employment 16-18* may be purchased from the School of Mathematics, University of Bath.

A series of reports based on the Nottingham Study will be available for purchase from the Shell Centre for Mathematical Education, University of Nottingham.

increasing familiarity with a task may mean that it may cease to be thought of as mathematics and become an almost automatic part of the job. A remark which was overheard — "that's not mathematics, it's common sense" — is an illustration of this and, indeed, indicates an 'at-homeness' with that particular piece of mathematics which we would wish to commend.

66 Both studies draw attention to the diversity of types of employment which exists, to the variety of mathematical demands within each and to the considerable differences which were found to exist even within occupations which might be assumed from their titles to be similar. It is therefore not possible to produce definitive lists of the mathematical topics of which a knowledge will be needed in order to carry out jobs with a particular title. It is, however, possible to describe in general terms the types and levels of mathematics which are likely to be encountered by certain broad categories of employees. These descriptions may be found in paragraphs 120 to 148 at the end of this chapter; those of our readers who have a detailed interest in the mathematical needs of employment may wish to study them before proceeding further.

Some common findings

67 A number of findings emerged from the studies with sufficient regularity to suggest that they are likely to have general validity; it is these which we discuss in the paragraphs which follow.

Range of mathematics required
68 Both studies found that almost all of the mathematics which young people need to use, whatever their job, is included within all the existing O-level and CSE Mode 1* syllabuses. Even the restricted content of the limited grade CSE syllabuses which are offered by some boards includes the mathematics required in most of the jobs which those who take these examinations are likely to have the opportunity to enter. Since most pupils do not know in advance the type of employment which they will follow in later life, it is important that this should be the case.

*Modes of examining vary according to the degree of involvement on the part of the candidate's school. The principal forms are:

Mode 1 — examinations conducted by the examining board on syllabuses set and published by the board;

Mode 2 — examinations conducted by the examining board on syllabuses devised by individual schools or groups of schools and approved by the board;

Mode 3 — examinations set and marked internally by the individual school or groups of schools but moderated by the board.

69 Nevertheless, the studies also identified certain important differences between the ways in which mathematics is used in employment and the ways in which the same mathematics is often encountered in the classroom. We believe that these differences may be among the factors which have contributed to the criticisms which have been voiced in recent years.

Calculation
70 The need to be able to carry out aritthmetical calculations of various kinds appears among the mathematical requirements of almost all the types of employment which we discuss in paragraphs 120 to 148. These calculations are sometimes carried out mentally, sometimes with pencil and paper and sometimes with a calculator. Some jobs specifically require an ability to carry out mental calculations of various kinds. In almost all jobs the ability to carry out some calculations mentally is of value and lack of ability to do this is a frequent cause of complaint by employers. It was also found to be a cause for complaint by some young employees who maintained that an ability to carry

out mental calculations which they had possessed in their primary and early secondary years had been allowed to atrophy as a result of lack of practice in the classroom.

71 Both the Bath and the Nottingham studies found that the methods which are used when at work to carry out calculations with pencil and paper are frequently not those which are traditionally taught in the classroom. Employees use a variety of idiosyncratic and 'back of an envelope' methods, especially for long multiplication and division. Sometimes these methods have been devised by the employees themselves but frequently they have been passed on by their fellow workers. The methods which are used depend very much on the user's confidence in his own mathematical ability. However, at all levels there appears to be a preference for carrying out a calculation in a succession of relatively short stages rather than for making use of a single calculation which is mathematically more sophisticated, and perhaps quicker, but which may be more difficult to use with confidence. For example, we read in the report of the Bath study:

> We met a number of situations where people by-passed the more traditional school methods, eg a boy who had left school before examinations and at work performed quite a complex check (which he was handling with confidence) on whether he was up to quota. He required 7×96, which we might classify as use of the seven times table. However, he proceeded as follows:

$$
\begin{array}{r}
\text{First } 3 \times 96 = 288 \\
\text{then} \quad 288 \\
288 \\
\hline
576 \\
96 \\
\hline
672
\end{array}
$$

> Although this employee did not know his seven times table, his method shows that he knew what "seven times" something meant. This is one of the many examples we found where people resorted to longer methods with which they evidently felt confident, rather than using what a mathematics teacher might have thought to be the obvious direct method.

72 The use of percentages is widespread in offices and laboratories, but is much less frequent in workshops. Percentages occur most often in calculations involving money, for example discount, value-added tax, profit or loss; they are also used in a wide range of other calculations in both offices and laboratories. We are aware of a number of complaints from employers about lack of understanding of percentages. These refer not only to clerical staff, whose difficulties can usually be overcome by making use of a formula or a standard calculator procedure, but also to management trainees and even managers.

Use of calculators

73   Our own enquiries, and the evidence which we have received, lead us to believe that there is an ambivalent attitude to the use of calculators in industry and commerce at the present time. The Bath and Nottingham studies found their use to be widespread in many types of employment. These include a wide range of clerical and administrative jobs such as accounts departments, banks, insurance offices and related areas of employment. Calculators are also used widely by those who work in laboratories and engineering design offices and by those concerned with quality and production control; these are all jobs which require a considerable amount of calculation and analysis of data. In all of these situations calculators are regarded as desirable aids to speed and accuracy.

74   Calculators are also used increasingly by many who work on the shop floor but their use is still viewed with suspicion by some managers and supervisors who were themselves trained to use slide rules or logarithm tables. This seems to be especially true of those who supervise engineering and other technical apprentices and craftsmen of various kinds. We are aware of situations in which new apprentices, who had been issued with logarithm tables by their training supervisor, preferred instead to make use of the personal calculators which they were encouraged to use on the college courses that they attended. There are, of course, many straightforward calculations which a craftsman needs to be able to carry out mentally but this does not seem to be a reason for denying the use of a calculator when it is sensible and time-saving to use one. However, the majority of young employees who were seen to be using calculators at work had not been trained in their use either at school or on the job. In consequence, calculators were frequently not being used in the most effective way.

Fractions

75   Although fractions are still widely used within engineering and some other craft work these are almost always fractions whose denominators are included in the sequence 2, 4, 8, ..., 64. This sequence is visibly present on rules and other measuring instruments and equivalences* are apparent. Addition or subtraction of lengths which involve fractions of this kind can be done directly by making use of the gradations on the rule. When the calculation is carried out with pencil and paper it is never necessary to work out the common denominator which will be required because it is always present already; for example $2\frac{1}{4} + 3\frac{5}{16}$ has the necessary denominator, 16, already visible. Even so, the methods used are not always those which might be expected. The report of the Bath study comments that

> Frequency of use promotes assimilation of equivalents. This is evident from the following method that a craftsman used to add fractions:

$$\frac{3}{16} + \frac{5}{64} = \frac{3}{16} + \frac{1}{16} + \frac{1}{64}$$

$$= \frac{4}{16} + \frac{1}{64}$$

$$= \frac{16}{64} + \frac{1}{64} = \frac{17}{64}$$

This is another instance of someone rewriting numbers in a more convenient

*For example, that $\frac{6}{16}$ and $\frac{3}{8}$ have the same value

form. It could be described as very mathematical but it is not necessarily the method which would be reinforced in schools.

The need to perform operations such as $\frac{2}{5} + \frac{3}{7}$ does not normally arise, and manipulation of fractions of the kind which is commonly practised in the classroom is hardly ever carried out. In the rare instances in which it is necessary to multiply or divide fractions, it is usual to convert each to a decimal before performing the operation, if necessary with the help of a calculator.

76 The notation of fractions appears in some clerical and retail jobs, for instance $4\frac{3}{7}$ to represent 4 weeks and 3 days or $2\frac{5}{D}$ to represent 2 dozens and 5 singles. However, school-type manipulation is rarely found and then only in very simple cases; for instance, the calculation required to find the charge for 3 days based on a weekly rate is division by 7 followed by multiplication by 3.

Algebra

77 One of the more surprising results of the studies is the little explicit use which is made of algebra. Formulae, sometimes using single letters for variables but more often expressed in words or abbreviations, are widely used by technicians, craftsmen, clerical workers and some operatives but all that is usually required is the substitution of numbers in these formulae and perhaps the use of a calculator. The report of the Bath study quotes an example of one such formula which was not even written down, but was remembered by the employee concerned:

> A wages clerk explained: 'To get the rate of pay per hour, we add together the gross pay, the employer's National Insurance, and holiday stamp money, subtract any bonus, and then divide by the hours worked'.

Formulae are also used regularly in nursing. An example quoted in the report of the Nottingham study is:

$$\text{Child's dose} = \frac{\text{Age} \times \text{Adult dose}}{\text{Age} + 12} .$$

It is not normally necessary to transform a formula; any form which is likely to be required will be available or can be looked up. Nor is it necessary to remove brackets, simplify expressions or solve simultaneous or quadratic equations, although algebra of this kind is sometimes encountered on courses at further education colleges. Solution of linear equations is required very occasionally.

Estimation

78 **Industry and commerce rely extensively on the ability to estimate. Two aspects of this are important.** The first is the ability to judge whether the result of a calculation which has been carried out or a measurement which has been taken seems to be reasonable. This enables mistakes to be detected or avoided; examples are the monthly account which is markedly different from its predecessors or the measured dose of medicine which appears unexpec-

tedly large or small. The second is the ability to make subjective judgements about a variety of measures. This is of use in situations in which measurement is difficult or awkward, in which trial and error is possible or in which tolerances are large. Skills of estimation develop on the job but employers often complain that young entrants to industry and commerce lack a 'feel' for both number and measurement, even in terms of the units, whether metric or imperial, which they can be expected to have encountered at school or in everyday life.

Measurement

79 **Although estimation is important, counting and measurement are paramount. A very great deal of the mathematics which is used in employment is concerned with measurement in one or other of a wide variety of forms, by no means all of which are directly concerned with the use of measuring instruments.** Measurements are specified in a variety of ways, for instance in terms of number of items or total of money; of length, weight or volume; of ratio, percentage or rate. Use is made of metric and imperial units as well as of units which are peculiar to a particular industry.

80 There are two different aspects of measurement — the first in which an existing measurement such as length, weight or number in stock has to be determined, the second in which it is necessary to create a desired measurement. In every case it is necessary to be aware of how accurate the measurement needs to be; in some cases it is necessary to be able to choose and use an appropriate measuring instrument. The labourer may well measure in terms of buckets or shovels full. The skilled engineering craftsman may work to different tolerances for different parts of a job and will need in each case to choose the appropriate tool or instrument for creating or checking the measurement he requires. The cashier will need to count the money in the till exactly.

81 There are also those who have to be keenly aware of the meaning of measurements, even though they seldom measure anything for themselves. These include staff who are involved in ordering and costing, in calculating turnover and profit margins. Here again, measurement is seldom exact but needs to operate within specified degrees of accuracy. This concept of measurement as something which is sometimes exact but more often needs to lie within stated limits is one which is very different from that which is commonly encountered in the classroom; it is also one which requires a conceptual understanding which takes time to develop. We discuss this important point in more detail in paragraphs 269 to 272.

Metric and imperial units of measurement

82 We have received a number of submissions which draw attention to the fact that imperial units are still used in many parts of industry. Although there are many companies which use only metric units, others still operate mainly with imperial units and many companies use a mixture of both metric and imperial. There are also certain companies which make use of units which are

peculiar to a particular industry. There are a number of reasons for the continuing use of imperial units. Among the most significant are the expense of re-equipping workshops with new machinery and instrumentation, the need to maintain a supply of spare parts which conform to standard imperial sizes and the needs of customers overseas who still make use of imperial measures. However, where imperial units are still in use, only a limited range is normally encountered on any particular job, for example yards or feet and inches, pounds and ounces. The use of fractions of an inch and of the 'thou' (thousandth of an inch) is also common. Where conversion from imperial to metric units, or vice versa, is necessary it is usual to make use of a conversion table.

Implications for the classroom

83 **It is of fundamental importance — and, we believe, not as self-evident as some might suppose — to appreciate the fact that all the mathematics which is used at work is related directly to specific and often limited tasks which soon become familiar.** The different aspects of mathematics which are used are related to each other by their context so that their application is immediately evident. Increasing experience of carrying out calculations or measurements helps to develop skills of estimation and approximation and an awareness of whether or not a result is sensible.

84 However great the effort which is made, illustrations of the practical applications of mathematics within employment which are given to a group of pupils, whose members will enter many different types of job, cannot provide the immediacy of the actual job itself. Nevertheless, **it is important that the mathematical foundation which has been provided in the classroom should be such as to enable competence in particular applications to develop within a reasonably short time once the necessary employment situation is encountered.**

85 The preceding paragraphs give some indication of the kinds of mathematical skill and understanding which are needed. **We believe that it is possible to summarise a very large part of the mathematical needs of employment as 'a feeling for measurement'.** This implies very much more than an ability to calculate, to estimate and to use measuring instruments, although all of these are part of it. It implies an understanding of the nature and purposes of measurement, of the many different methods of measurement which are used and of the situations in which each is found; it also implies an ability to interpret measurements expressed in a variety of ways.

86 If boys and girls are to leave school equipped in this way, they will need, in the mathematics classroom and elsewhere, to have taken part in a wide variety of appropriate mathematical activities and to have discussed these at length with their teachers and with each other. We consider ways in which this may be achieved in the second part of our report.

Employers' selection tests

87 Some firms, especially those in the engineering industry and some branches of the retail trade, require applicants for jobs to undergo written

selection tests in mathematics. We have received many comments about these tests both in submissions from teachers and from employers themselves. We are grateful to the many employers who have provided us with copies of the selection tests which they use and also, in some cases, with photocopies of scripts completed by applicants for employment.

88 Employers who make use of written selection tests appear to do so for one or more of the following reasons:

- because recruitment takes place before the results of O-level and CSE examinations are available;

- because employers wish to satisfy themselves that those whom they engage are able to carry out specific types of calculation;

- because, even when the results of O-level and CSE examinations are available, there is a feeling that success in the examination, especially if it is at a low grade, does not guarantee competence in the particular mathematical skills which the employer requires;

- because of unwillingness to rely solely on reports provided by schools since, as some employers have explained to us, these reports are often written in terms which are too general and sometimes include predictions of likely examination results which do not prove to be accurate;

- because it is necessary to choose from a large number of applicants or to sort them in some other way.

89 In respect of the last of the above reasons, it has been pointed out to us that, in a time of high unemployment, the need to choose from a large number of applicants does not necessarily imply that those who score most highly will be accepted. It may not be to the advantage either of the applicant or of the firm to employ someone whose skills are too high because of the frustration which may result from doing a job which is below the level of which the employee is capable.

90 Although a few employers are prepared to make offers of employment conditional upon the achievement of specific examination results, this practice does not seem to be widespread and could clearly lead to problems of either under or over recruitment in cases in which the number of vacancies was small. However, some firms use examination results in conjunction with the results of selection tests to discriminate between those who are offered craft apprenticeships and those who are offered technician apprenticeships.

91 The kinds of tests which are used and the level at which they are set vary widely. We are concerned that testing procedures are often in the hands of people who have neither training nor appropriate experience in testing procedures, including the setting and marking of papers. We have been surprised to find that this can be the case even within major companies.

92 Although some tests are specifically related to tasks which the employee will be required to undertake as soon as he starts work, many others are set at a level which is higher than this. In some cases it is claimed that this is because a higher standard of mathematics than the job requires will be needed for success in, and perhaps entry to, some form of training at a college of further education either immediately or in the future. It may also be because a higher standard of mathematics will be required for jobs to which the employee may be promoted in due course.

93 We have also been told that some tests are set at a higher level than will be required on the job because it is considered that success in a harder test will indicate an ability to perform well at more elementary tasks or will indicate a more 'flexible' employee whom it will be possible to employ on a variety of jobs. Some also take the view that ability in mathematics is a good indicator of a more general ability. We are aware, too, of some cases in which a test is set at a higher level merely because a commercially produced test is being used which does not reflect the specific needs of the job.

94 Even in cases in which the requirements of a job have been carefully identified and a test devised which is matched to these, the test may not serve the purpose which is intended and so lead to unjustified criticism of the performance of those who are required to attempt it. There can be many reasons for this. One of the most common is that questions are very often worded ambiguously or set out on the page in a confusing manner. The latter can apply especially to question papers which have been typed and stencilled; fractions, for example, are difficult to set out clearly on a typewriter and the fact that the same key is used for both a decimal point and a full stop can lead to confusion unless great care is taken. Some tests which we have seen use language which is mathematically incorrect; some also use the jargon of the job — 'cast these figures across and down' — which those taking the test may not have met before and so do not understand.

95 We are also concerned at the rigid way in which test papers are very often marked and at the number of errors in marking which we have identified on scripts which have been sent to us. Marking schemes are usually kept very simple, often with just one mark for each answer and frequently no credit is given if, as a result of a single mistake in one part of the question, several subsequent answers are incorrect, even though the mathematical operations involved have been carried out correctly. We have also found instances in which candidates have dealt successfully with one possible interpretation of an ambiguously worded question but have been given no credit for it. In our view, too, the time allowed for completing some tests is far too short so that candidates do not have the opportunity of showing properly what they can do. Furthermore, the stress of working is increased when it is necessary to work against the clock; provided that the mathematics is understood, speed of working will develop with continuing use on the job.

96 **We wish to draw attention to one further aspect of the testing undertaken by employers. There seems to be a very general unwillingness to provide**

applicants with a specimen of the test paper which will be set. Indeed, we have to record with regret that a few employers have not even been willing to make available to the Committee a copy of the test which they use and have devised, despite an assurance that its confidentiality would be preserved. We find it difficult to appreciate the reason for this unwillingness to provide a specimen test paper. Candidates for public examinations of all kinds, whether they are pupils at school, students in further or higher education establishments, or applicants for the membership of professional bodies, are provided both with an examination syllabus and also access to papers set in earlier years. If, as we believe should be the case, one of the purposes of tests set by employers is to discover whether the applicant is able to carry out certain mathematical tasks satisfactorily, it must surely be of value to make clear what these tasks are. This can apply particularly to certain clerical and accounting operations which, although making use only of relatively elementary mathematics which will certainly have been studied at school, often require it to be applied in ways which may not have become familiar in the classroom. If, as a result of being provided with a specimen test paper, the candidate wishes to practise these tasks and improve his ability at them, this can only be of advantage to all concerned. Nor need the provision of a specimen test paper in any way invalidate the confidentiality of the actual test which is used or inhibit its use over a period of years if this is felt to be desirable. We believe the same to be true of the commercially produced tests used by some employers; in our view specimens of these tests should also be made available.

97 In some areas local groups of teachers and employers have discussed the recruitment tests used by the employers, often as part of the work of liaison groups to which we refer in paragraph 103. As a result of such discussions, tests have often been modified so as to enable them to fulfil their purpose better. In a few cases, a further development has been that several employers in the same locality have agreed to use the same test. This avoids the necessity for boys and girls who have applied to several firms to take a succession of tests. On the other hand, a system of this kind may act to an applicant's disadvantage unless it provides opportunity for a second attempt to redeem a poor, and perhaps uncharacteristic, performance at the first attempt.

98 We are aware of an increasing number of schemes which set out to provide evidence of achievement for pupils whose ability is below that for which the CSE examination is intended. Some of these schemes have been developed by a single school or a group of schools, and some by local education authorities. There is also the national SLAPONS test (School Leaver's Attainment Profile of Numerical Skills) which was developed by the Schools and Industry Committee of the Mathematical Association and initially administered by the Shell Centre for Mathematical Education; it is now administered by the Royal Society of Arts. Although these schemes are not, in the strict sense, employers' tests, the intention of all of them is to provide information for employers about the mathematical attainment of school leavers and so render separate recruitment tests unnecessary. We discuss schemes of this kind in Chapter 10.

Liaison

99   Many of those who have submitted evidence to us have drawn attention to the need for better liaison between schools and employers. There is very general agreement that liaison activities are crucial and that greater efforts need to be made to encourage, develop and extend them. We have received encouraging evidence of the way in which liaison of this kind is developing in many areas.

100   **It must be a primary aim of liaison to enable schools and employers to obtain a better understanding of each other's needs and problems and of the way in which each operates.** Such an improved understanding will both assist pupils to make a more informed choice of employment and also ease their transition to working life, whether this takes place immediately on leaving school or after a period of further or higher eduation. Heads and other teachers in schools, local careers officers, staff in colleges of further and higher education and local employers all have a contribution to make. Many organisations participate in careers activities in individual schools and in careers conventions organised by LEA Careers Services. Many companies hold careers evenings on their own premises for pupils and parents. In some cases pupils visit firms to see at first hand the kind of work which is going on and to learn about the career opportunities which are available. In other cases, some pupils may have the opportunity of gaining work experience by spending a period of time with a firm. Joint activities of this kind can provide valuable and we believe necessary opportunities for employers to keep abreast of current educational developments and for schools to be aware of changes in patterns of employment and of opportunities which are available.

101   In the following paragraphs we consider only liaison which contains a mathematical dimension, although it is likely to be the case that such liaison will also contribute to a more general understanding of the school/ employment interface for both teachers and pupils, including the development of a better understanding of the place of industry in the economy.

102   Moves to develop school/employment liaison relating to the mathematics curriculum are by no means new. The Schools and Industry Committee of the Mathematical Association has existed for some twenty years and consists of representatives of industry, commerce and mathematical education. Through its meetings and other related activities it has provided opportunity for much valuable discussion. This discussion has also revealed some of the problems which have to be overcome if liaison is to be effective. These include the difficulties of making both teachers and employers aware of the work which is being done and of persuading teachers who lack direct knowledge of industry that they should nevertheless look for opportunities of introducing examples of the industrial and commercial uses of mathematics into their work in the classroom.

103   A few local groups were in existence before the Schools and Industry Committee was set up. Since that time other groups and projects have come into being, a few of them with national coverage and, especially since the mid-1970s, many others at local level. The latter have usually consisted of

teachers, local employers and the local careers officer. The LEA mathematics adviser and staff from a nearby further education college or college of education have often been involved as well. In 1978 Professor D E Bailey of the University of Bath published a list of some seventy such local groups* and subsequent work has identified some twenty more. In the main, each of these has been concerned with one or more of three broad areas:

*A survey of mathematics projects involving education and employment. University of Bath; also supplement, May 1980.

- the identification of the mathematical needs of various types of employment;

- the preparation of materials for use in the classroom which provide examples of ways in which mathematics is used in industry;

- the discussion of recruitment tests used by employers.

We have referred to the last of these in the previous section; we consider the other two in greater detail in the paragraphs which follow.

Identification of mathematical needs

104   Many local groups and a number of research projects have set out to identify the mathematical needs of various types of employment. The scope of these investigations has varied considerably and, because most of the groups and projects have worked in isolation from each other, it has perhaps been inevitable that certain investigations have been repeated many times. In drawing attention to this fact, we in no way wish to undervalue the work which has been done, the very considerable expenditure of time and effort which it has involved, the valuable experience which has been gained by those who have carried out the work and the increased understanding between employers and schools which has resulted. However we hope that the results of the work which has been carried out at our request at Bath and Nottingham, together with the results of the Schools Council Project *Mathematics and the young entrant to employment*, will provide a firm base from which more detailed, and perhaps local, studies can develop without the necessity of going yet again over the preliminary ground.

*D Bird and M Hiscox. *Mathematics in schools and employment: a study of liaison activities.* Schools Council Working Paper 68. Methuen Educational 1981.

*See note to paragraph 68

105   A not infrequent outcome of the studies which have been made has been some modification of the mathematics courses in local schools. In some cases, too, CSE Mode 3* examination syllabuses have been adopted which have reflected these revised courses. Another outcome has been that the performance of pupils in certain areas of mathematics has been certificated separately by the school or sometimes by the local education authority in cases where several schools have been involved. We are also aware of some local schemes in which experiments are being made with profile reporting, which gives more detailed information about a pupil's mathematical strengths and weaknesses. We suspect, however, that there are relatively few employers for whom such detailed information is of real value.

106   Where modifications to mathematics courses have resulted from local initiatives, these seem to have been directed mainly towards pupils of average or below average attainment. However, an exception to this has been the Mathematics in Education and Industry Project (MEI) which was set up in 1963 specifically to develop new school syllabuses, initially at A-level, which were more in line with the mathematics to be found in industry and in higher education than were those which existed at that time. These syllabuses were

based on the experience of teachers who had spent time on secondment to various large firms. One difficulty which arose was that, while there were plenty of suggestions for new topics which could be included in the curriculum, it was not easy to decide what should be omitted to make room for them. The MEI work has been influential, both through its own A-level (although the size of the entry has been small) and through its influence on other A-level syllabuses. However, its work has not yet had very much influence on the 11-16 curriculum. We have been told that attempts by local MEI groups to work with local employers on ideas suitable for those in the 11-16 age range have not hitherto met with very much success, although there have been recent indications of improvement.

Classroom materials

107   A number of local groups have set out to produce materials for use in the classroom based on the mathematics used in various types of employment. Such materials have most usually been produced as a result of a teacher or teachers visiting local firms to observe the different kinds of work which were going on. Following the visits, worksheets have been produced which are based on examples of the uses of mathematics which have been observed and which explain the situations in which these occur. The preparation of materials of this kind seems to have been most effective when it has been carried out in consultation with a member of staff of the firm concerned, who has sometimes also been able to visit the school to talk with pupils or demonstrate equipment. However, the evidence which is available to us suggests that, although the benefits of such materials can be considerable both to the teacher who has prepared them and to his pupils, **it should not be assumed that these benefits are necessarily easily transferable to other teachers, whether in the same school or in other schools, who have not had the same direct involvement with the firm concerned. It is therefore necessary to provide appropriate in-service training for teachers in the latter category so that they may be able to use the materials effectively.**

108   Preparation of materials for the classroom has also been carried out on a national scale. The School Mathematics Project (SMP), which has from the outset stressed the use of mathematics in application, supported an Industrial Fellow for two years in 1964-66. His task was to gather examples of the uses of mathematics in various kinds of employment so that these could be incorporated into the SMP texts wherever possible. The fact that this initiative achieved only limited success again draws attention to the fact that the preparation of classroom materials related to the world of work is more difficult than might be expected.

109   A more recent initiative on a national scale has been the setting-up in 1975 of the Working Mathematics Group, which developed from the Continuing Mathematics Project and receives non-financial sponsorship from the Council for Educational Technology. The Group is a working liaison of industrialists and teachers which produces instructional materials in modular form designed to show mathematics in action in the world of work. The members of the Group work in pairs, one from industry and one from education, and give time voluntarily to the preparation of materials. The Group

published its first ten instructional units in the autumn of 1980, so it is as yet too early to assess the contribution which these materials will be able to make in the classroom. In its written evidence to us, the Group drew attention to the long time which is required to produce classroom materials — a fact which once again underlines the problems attendant on work of this kind and the time scale which is involved.

Visits

110   Although visits by groups of pupils to companies and other employers can be of considerable value, we have to accept that the number of such visits which can be made by any one school, and by any one group of pupils within the school, will inevitably be small. When it is possible to arrange visits of this kind, **it is essential that there is both adequate preparation of pupils before the visit and also follow-up when the visit has taken place, otherwise much of the value is likely to be lost. Nor should the opportunity for visits be restricted, as sometimes happens, to the academically less able**; pupils of high ability can also benefit from visits of this kind and should have the opportunity to take part.

Attachments for teachers

111   A number of the submissions which we have received urge the provision of greater opportunities for mathematics teachers and mathematics staff in training institutions to undertake a period of attachment to industry. **We believe that such attachments can be of considerable value to those concerned and we commend schemes such as those offered by the Understanding British Industry project** despite the fact that it has been pointed out to us, somewhat ruefully, by the mathematics adviser of one LEA that an unforeseen outcome of three such attachments to industry was the loss to teaching of three good mathematics teachers. We have been told that at least one LEA has appointed full-time mathematics teachers to its supply staff in order to permit mathematics teachers in schools to be released for attachment to industry. Whatever the method adopted, suitable arrangements need to be made to make replacement staffing available so that industrial attachments can take place. However, we have to accept that the number for whom it will be possible to arrange attachment to industry will remain small in comparison with the total number of mathematics teachers. It is therefore of the utmost importance that every effort is made to use to the full the experience which teachers gain during such attachment; and that efforts are made to share with other members of staff the experience which has been gained and to prepare teaching materials based on this experience for use in the classroom, notwithstanding the difficulties to which we referred in paragraph 107.

112   **It is essential that, in any scheme for liaison between schools and employers, the two-way nature of the relationship is accepted by both sides**. Not only is it necessary for teachers to have a better understanding of the needs of industry, but also for those in industry to maintain an up-to-date knowledge of present day practice in schools. We believe that representatives of industry and commerce should visit classrooms to observe teaching in progress and be encouraged to give a talk or series of talks to pupils.

Liaison with further education

113   We end this section by drawing attention to the need for liaison between schools and local further education (FE) establishments. Liaison between

further education and industry has traditionally been relatively strong. A major reason for this is that the part-time and block release courses which FE establishments provide for young employees provide opportunity to establish links between company training officers and college staff; furthermore, many lecturers in FE colleges, especially at craft level, have traditionally been recruited from industry. Training officers and managers are also often members of college governing bodies and advisory committees; and the Business Education Council (BEC) and Technician Education Council (TEC), as part of their course approval and monitoring procedures, require evidence that discussions have taken place between colleges and employers to identify the needs of the young employees concerned.

114 Direct and regular contact between schools and further education establishments seems often to be less effective. 'Link courses', in which staff of both school and college are involved, can provide valuable opportunity for co-operation and also considerable motivation for pupils. However, these courses are almost invariably more expensive than those provided solely in schools. At a time of pressure on resources and cash limits we can understand that LEAs may wish to limit the provision of courses of this kind. **We believe, however, that such limitation is short-sighted and we would wish to see a development of link courses wherever practicable.** In any case, whether link courses are arranged or not, general liaison between schools and colleges needs to be developed.

115 Links between the school and the further education sectors at regional and national levels are also tenuous. For example, further education is generally only weakly represented on GCE and CSE subject panels. Similarly schools are poorly represented on City and Guilds of London Institute (CGLI), BEC and TEC committees and similar bodies. **We believe that much more three-way co-operation between the school, further education and industry sectors should take place in a variety of ways.**

116 It is clear that over the last few years a good deal of effort has been put into liaison activities of various kinds. However, if the improvements achieved by the work of local groups and by other local initiatives are to be maintained, continuing discussion is needed; resources, too, must be maintained and where at all possible improved. It is not the case that discussion followed by the publication of a report, syllabus or test paper will set things right once and for all. All parties must continue to learn from each other. To maintain a continuing discussion is perhaps more demanding than the initial effort required to produce a document of some kind. It also makes a continuing demand on the time of the teachers, industrialists and others who are involved. **Local education authorities, careers advisers, teachers and employers will all need to take responsibility and initiative for the various aspects of liaison.**

The future

117 At a time of rapid technological and social change such as the present, it is particularly difficult to assess the extent to which the mathematical

needs of those entering employment are likely to change. The dramatic reduction in the cost and size of computing and automation equipment made possible by the advent of the silicon chip is already beginning to have a profound effect on our working and social lives and it is evident that this effect will increase in the coming years. Automation has already been introduced into many industries engaged in mass production; leaders in this field have been the process and electronic industries. More recently there have been substantial advances in the automotive and some other industries. However the introduction of automation has not been limited to production only; computer-aided techniques are increasingly being used in design and in various aspects of management information processing. Within manufacturing industry a growing number of examples can be quoted of the integration within one system of many different but inter-related facets of the same business — for instance, design, manufacture, sales and management control. In the wider fields of commerce and of the retail and distribution industries, a similar move towards automation and integration is apparent and it seems likely that few, if any, fields of employment will be unaffected by the time that those children now entering primary school are ready to take up employment.

118 In this situation, and with the possibility of the introduction of factors at present unknown, it is impossible to anticipate future needs in detail. However, we believe that it is possible to put forward some general conclusions. The principal effect of increasing automation will be to reduce the amount of human effort involved in producing a product or providing a service. Thus, for the vast majority of those in employment, it would appear that there could be some reduction in the use of mathematics, at least at the level of arithmetical calculation. This may be considered as an extension of the effect of the introduction of the electronic calculator which, as we have already pointed out, is being used more and more in employment. However, for a limited number of people we expect that there will be an increased need to use mathematics; these will be those who are involved in the maintenance of automated equipment and in the development of computer-aided engineering systems. At still more sophisticated levels we also expect that there will be an increased use of mathematics in conjunction with computers for modelling purposes. At all levels of employment we believe that an understanding of simple mathematical concepts will enable those in employment to take an intelligent interest in their work as it changes under the impact of automation. The need for understanding will apply not only to those operations with which the employee is directly concerned in his work; it will also be important if the employee is to understand the financial performance of his company and participate intelligently in those aspects of its management which he is in a position to influence. **The laying of adequate mathematical foundations at school will, moreover, remain of central importance so as to provide a basis for any further training which career development or change of employment may require.**

119 Bearing in mind the pace of technological development and the unexpected directions which history shows such development may take, we see no reason for supposing that the levels of mathematical need of those who

will enter employment in ten or twenty years time will be less than they are today. Although the specific needs may well be different, a secure grasp of simple mathematical ideas and the ability to apply them will remain as important as ever. For a limited number of people there will be a call for more mathematics than is needed by their counterparts currently in employment; there will also be a need for a certain number of people who are very highly qualified mathematically to be responsible for the development work which very rapidly changing technology will require. **In any event it will be of the utmost importance to maintain and develop liaison between schools and industry.**

The mathematical needs of some areas of employment

120 In the remainder of this chapter we discuss in very general terms the types and levels of mathematics which are likely to be encountered by certain broad categories of employees. Although these categories do not cover all types of employment available to those who leave school at 16, we believe that they include a wide cross-section. In some cases we indicate the recruitment procedures which are most commonly used and the qualifications required. We wish to emphasise that even the general descriptions which we give must not be taken to apply to all those who may seem to be in the categories which we have chosen; it is likely that only some of the mathematics which we mention will be needed by any particular person. The requirements are also likely to vary with the size of company in which an employee works. Those who work in small companies may well undertake a wider range of work, and so make use of a wider range of mathematics, than those who work in a specialised section of a large company.

Manufacturing industry

121 Within manufacturing industry, three levels are commonly identified — operatives, craftsmen and technicians — though the boundaries of each category are by no means clearly defined and the category to which an employee may be assigned varies from industry to industry, and also from firm to firm within the same industry.

Operatives
122 Among those identified as operatives there are very many whose jobs do not appear to require any formal application of mathematics. Their jobs include feeding and removing articles from machines, assembling small components, trimming off surplus material and operating pre-set production machinery. It seems to be extremely rare for selection for these jobs to be based on school qualifications; personal qualities and the way in which applicants present themselves at interview are usually considered to be of greater importance. Selection tests are rarely used, though there may in some cases be tests for specific requirements such as manual dexterity or freedom from colour blindness; training is usually undertaken on the job. However, in cases where operatives in this group obtain promotion, for instance to senior operator or chargehand, arithmetical work may be required and so they will move into a second category, also large in number, consisting of operatives who use a limited range of mathematical skills.

123   Operatives in this second category may need to count articles which are being stacked and to record the result; to recognise, copy and interpret numerals, for instance code numbers; to add, subtract, multiply and sometimes divide whole numbers, perhaps with the help of a calculator; to carry out straightforward mental calculations; to read dials and gauges, though this may merely involve making sure that a needle or other indicator stays within specified limits; to weigh and measure in both metric and imperial units, including measurement involving fractions of an inch; to be familiar with the idea of gross and net weights. Recruitment procedures are usually similar to those for the previous group.

124   A third category is that of operatives who make use either of a wider range of elementary skills or of certain more advanced mathematical techniques; some of the tasks which these operatives carry out are also undertaken by those designated as craftsmen. These tasks include checking dimensions using a micrometer, vernier or other type of gauge; calculating or estimating area; understanding tolerances expressed as $\pm$ or in other ways; reading engineering drawings, often in the form of freehand sketches with some dimensions marked but sometimes also of a more complicated kind. Dimensions may be given in millimetres or inches; the latter may involve both fractions of an inch in the sequence $\frac{1}{2}$, $\frac{1}{4}$, ..., $\frac{1}{64}$ and multiples of these, and also decimals which are frequently expressed in 'thou' (thousandth of an inch). Work of this kind requires the operative to possess the necessary geometrical awareness to interpret a two-dimensional drawing in three-dimensional terms. Associated with the reading of gauges and dials may be a need to understand rates of many kinds, for instance rotational speed, fluid flow or pressure. Some operators will be required to use a calculator-type keyboard, to read pen-trace graphs and to use a variety of reference tables. Others may need to mix substances in a given ratio, perhaps expressed in percentage terms. It may occasionally be necessary to substitute numbers into a simple formula, which will usually be expressed in words. However, it is important to remember that any one operative will hardly ever need to undertake more than a small number of these tasks.

Craftsmen

125   Many industries involve craft-type work and in consequence the mathematical needs vary widely. In many cases the needs will be similar to those listed in the previous paragraph, though within a given industry the craftsman will usually make use of a wider range of mathematical skills than will the operative. He may also need to understand a broader range of geometrical concepts so as to be able, for instance, to estimate or calculate areas and volumes of non-rectilinear shapes in two and three dimensions, measure angles and carry out simple geometrical constructions.

126   Engineering craft apprentices frequently need a still broader mathematical base. Most still need in their work to be fluent with fractions in the sequence $\frac{1}{2}$, $\frac{1}{4}$, ..., $\frac{1}{64}$ and their multiples and to be familiar with the decimal equivalents of these; to be able to select the nearest fractional equivalent in

this sequence to a given decimal; to be able to make conversions between millimetres and inches, probably with the help of a reference table. They need to be able to work within given tolerances and to understand the significance of figures in digital read-out meters. The reading of meters may also require understanding of prefixes such as micro-, milli-, kilo-, mega- and their associated symbols. The need to be able to use a wide range of drawings, for example in plan view, first angle projection and cross-section, requires familiarity with the conventions of technical drawing and a good grasp of geometrical concepts; ability to make use of geometrical instruments may also be necessary. Further geometrical requirements can include the understanding of similar triangles and Pythagoras' theorem and the use of sine and tangent of an angle in right-angled triangles. Craft apprentices have little need of formal algebra but may be required to substitute numbers into a simple formula, very often given in words, to solve simple linear equations and to draw graphs of experimental results with suitable choice of scale and axes. The need to transpose algebraic equations and formulae is usually avoided.

127 Although few young employees are likely to need the breadth of mathematical knowledge implied by the total list in the preceding paragraphs, many will meet most of the topics we have listed during their first year of training. In some cases, one reason for this is that, within the engineering industry, off-the-job training in the first year is very often the same for both craft and technician apprentices (see paragraph 129). Firms adopt widely differing strategies for recruiting craftsmen. A primary requirement is that the applicant should show the motivation and other interests which indicate likely success as a craftsman. In the event, many so recruited come from the CSE grade 4 level (the grade awarded to the average candidate) in appropriate subjects. However, school leavers with qualifications at this level will rarely have confidence in the whole range of mathematical skills and concepts which we have outlined; we believe that it is with this group of young entrants to employment that most concern has been expressed by manufacturing industry.

Technicians
128 Almost all of those who are recruited as engineering technician apprentices also undertake an associated course of further education, commonly a TEC course at certificate or diploma level. These courses are designed for those with at least CSE grade 3 in appropriate subjects, including mathematics. For this reason engineering technician apprentices are normally required to have at least this level of qualification; many have O-level grade C or higher. Many will also have been required to take a company selection test in mathematics.

129 The relationship between craft and technician apprentices is by no means precise and we have already drawn attention to the fact that both may follow the same off-the-job training in the first year. Thereafter, technician apprentices are likely to encounter a wider variety of tasks.

However, some firms make little or no distinction in the first two or three years in so far as in-company work is concerned.

130   The mathematical requirements will be those which we have already listed for craft apprentices. Although by no means all this mathematics will be needed at work during the first year, all is likely to be covered in the associated TEC course. Indeed, the demands of the further education course are likely to be considerably more than the demands of the job itself, a point to which we return when we discuss these courses in the next chapter. We may however note at this point that, because the initial training of craft and technician apprentices is frequently the same, some craft apprentices in particular can be confronted by an unnecessarily taxing programme either in the training centre or at college, or both.

131   Many training centres for craft and technician engineering apprentices include a period of training in 'basic mathematical skills' in the early weeks of training. This training may take from a few hours up to two or three weeks. We have been given evidence of the improvement which is achieved during this period of intensive concentration on number skills and simple mathematical topics. However, we believe that the two main factors at work here are that revision refreshes topics which may not have been used for several months immediately before the start of the training period and that apprentices have a strong motivation to succeed, since it is at this stage that they can begin to see the uses which will be made of mathematics within their own job. There is little evidence either from industrial training centres or from the performance of technicians on TEC programmes to suggest that weakness in mathematics is a general source of problems during the first year of training.

132   The specific mathematical requirements of other groups of technician apprentices vary with the nature of the employment. In general the needs are less broad than for engineering, though a deeper understanding of certain topics may be required. For example, a science laboratory technician may have to plot and interpret graphs of various kinds arising from the results of experiments, the calibration of instruments or the output of computerised testing machines; he may need to determine points of inflexion, to recognise linear and non-linear relationships and to use graphs with logarithmic scales. Some technicians need a sound understanding of numerical processes and orders of magnitude so that they can, for example, weigh and measure very small quantities on digital read-out machines in a variety of units. They may also need to make use of both direct and inverse proportion to scale results obtained from tests which have been carried out on substances diluted for test purposes. Such work also needs a good understanding of the degrees of accuracy which are required in the appropriate practical situation. For work of this kind the use of calculators is widespread.

Clerical work

133   Clerical work covers a wide range of jobs such as accounts, sales, wages and records; the work of many typists, receptionists and secretaries also includes a clerical element. The mathematics required by this large body of

employees, most of whom are female, is predominantly arithmetic, although some tasks may include substitution of numbers into a formula, usually expressed in words, and drawing graphs of sales or production. The total list of possible arithmetical skills which may be needed is large, though any one job is likely to require the use of only a limited range of them. The skills include counting, recording numbers and arranging them in order or in tabulated form; addition, subtraction, multiplication and division with whole numbers, decimals and money; a limited use of fractions, usually in situations involving division by a whole number; percentages applied in a variety of ways; ratio and proportion; rates, for example in the form of wages per hour or cost per tonne. It may be necessary to make use of reference tables and to round up or down, for instance to the nearest penny. The use of calculators for almost all of this work is now widely accepted. Appropriate checking procedures, although perhaps of a different kind from those used formerly, continue to be of crucial importance.

The retail trade

134   Young employees in the retail trade, usually employed as sales assistants or trainee managers, normally need only a limited range of arithmetical skills for their work. Many of their tasks are similar to those listed for clerical workers in the preceding paragraph. The work may entail using a till, ready-reckoner or table of goods and prices. It may be necessary to count goods in dozens and singles and to record the result on stock sheets; to prepare bills, perhaps allowing for discount or value added tax; to work with averages and percentages. Tables of numerical data may require adding both by rows and columns. Although calculators may be used for the more complicated tasks, sales assistants will often be required to carry out simple calculations mentally.

Agriculture

135   The mathematical requirements of those who work in agriculture may appear to be modest, the main skills needed being the ability to count and measure. However, the present high cost of basic materials and advancing technology mean that quantities of such things as animal feed, fertiliser and weed-killer must be calculated and measured with considerable accuracy. Precision drilling requires the farm worker to carry out accurate calibration and provide precise quantities of seed. The feeding of dairy cattle is now finely balanced in respect of such things as starch and protein requirements and these need to be calculated with reference to the characteristics of a range of feeding stuffs. Milk yields have to be recorded.

136   Farm machinery, too, is complex. It is necessary for the farm worker to be able to read a range of dials and gauges and, for instance, to be able to insert the appropriate discs to control the spray rate of a crop sprayer or make the appropriate settings for different types of seed. Although metric quantities are increasingly being used, a knowledge of imperial units is also still required, not least because much of the machinery at present in use has been built to non-metric standards.

137   In addition to being familiar with the aspects which we have already listed, the farm manager needs to be able to deal with farm accounts and to

have planning skills so as to make efficient use of both manpower and machinery. An ability to estimate, especially in respect of irregularly shaped areas, is also desirable.

The construction industry

138 As with other sectors of industry, it is difficult to identify precise mathematical needs for all employees. The size of firms within the industry varies greatly. It is estimated that there are over 40 000 firms employing less than ten people; but 50 large firms account for almost 10 per cent of the construction work which is carried out each year. Within a large organisation, the work of an individual may become very stereotyped and involve much the same routines for most of the time. On the other hand, the self-employed craftsman is likely to have to deal with a wide range of different tasks. For him, as for the small firm in general, estimation of materials is crucial. An over-estimate may result in a residue of materials which cannot be used up elsewhere; an underestimate can lead to delays and loss of time and profit.

139 Many of those who work in the construction industry will need to interpret plans and drawings and to be capable of accurate measurement. The bricklayer, the plasterer, the painter and decorator must become adept at estimating the amount of material they will require for a variety of tasks. Employers in the construction industry often complain that young entrants have initial difficulties in this field and lack any 'feel' for orders of magnitude, especially in respect of length, area and volume.

140 Some tradesmen require specific mathematical techniques. For example, the electrician may need to use the appropriate formula for combining resistances or calculating the current in a circuit. The plumber requires a sound appreciation of the quantities and shapes involved when working in three dimensions, for instance in order to shape the flashings around chimneys or to erect pipework. The joiner may have to carry out quite detailed calculations when constructing stairways or erecting roofing timbers. The foreman who is responsible for laying out a building site will need good planning and measurement skills.

Hotels and catering

141 The main mathematical requirement for work in the hotel and catering industry is an ability to carry out arithmetical calculations accurately. Many workers are concerned with calculations involving money, with weighing and measuring and with counting stock.

142 Those concerned with reception and accounts have to keep records and account books and to calculate service charges and value added tax. They may also be called upon to help visitors with bus or train timetables. Those who prepare food are concerned with weighing and measuring and with calculating the time required to prepare dishes. They require a knowledge of proportion in order to be able to adapt recipes for larger or smaller quantities and may also need to work out the cost per portion. Some of those employed in institutional catering must also be able to calculate specific nutritional requirements and take account of factors such as the provision of a balanced diet; accurate costing is likely to be essential.

143 Waiters and bar staff have to add up bills, handle money and memorise prices. Bar staff often have to deal with orders for a large number of items, each with its associated price, and by custom they are expected to add up the cost mentally.

Work with computers

144 Relatively few school leavers are likely to work directly with a computer. Their work will usually be at clerical or operator level dealing with the input and output of data, though some leavers with A-level qualifications obtain posts as junior programmers. The preparation of data for input to a computer entails the strict discipline of presenting data accurately in the required format; the handling of computer output often involves extracting data from tables which contain more information or more figures than are needed at that moment. These tasks demand little in the way of mathematical expertise apart from the need to feel 'at home' with the handling of numerical information. In some cases it is also necessary to be able to carry out straightforward artithmetical calculations which may involve the use of decimals and percentages.

Nursing

145 Except in a very few specialised cases, it is not possible to start training as a nurse until the age of 18 but many of those who start their training at this age will have left school at 16. The minimum entry qualifications for training as a State Registered Nurse (SRN) are stated in terms of passes at O-level or CSE grade 1, mathematics not being a compulsory subject. An alternative method of entry is by passing the educational test of the General Nursing Council; this examination contains a section on mathematics. To become a State Enrolled Nurse (SEN) the minimum requirement is a good general education. Selection is on the basis of interview and sometimes also of an educational test. However, individual hospitals are free to set qualifications for entry to training as SRN or SEN which are higher than these minima and in some cases a qualification in mathematics is specified. The fact that there can be a two-year gap between leaving school and starting to train as a nurse can mean that, in some instances, computational skills have deteriorated through lack of use.

146 Almost all the mathematics which is required in nursing is concerned with measurement and recording, often in graphical form. This can include the measurement and recording of temperature, pulse rate, respiration, fluid input and output, gain or loss of weight. Measurements which have been recorded need to be interpreted. Measurement is now almost entirely in metric (SI) units and nurses need to be familiar with prefixes such as mega-, kilo-, milli- and micro- and to understand that in the SI system gradation of units is in multiples of a thousand. An understanding of ratio, proportion and percentage is essential.

147 Various formulae exist which have to be used and manipulated with ease and accuracy. These include, for example, the formula which a nurse uses to work out the amount of food to be given to a baby and the formula which is used to work out the dose to be given if a drug which has been prescribed is not available in the specified strength. Because nurses frequently work under

pressure, an ability to carry out calculations of this kind with speed, accuracy and confidence is essential.

148   However, although accuracy is important, so too is the ability to estimate and approximate which helps to ensure accuracy, especially if a calculator has been used. Errors of the order of a multiple of 10 can occur easily when working constantly with metric units and it is vital to detect them before harm results. Skills of visual estimation are also required of the kind which will lead to a realisation that a drip-feed bottle is emptying too quickly or that a syringe of unusual size appears to be required.

4 The mathematical needs of further and higher education

Further education

149 A wide range of both full-time and part-time courses is available within further education establishments for those who leave school at ages from 16 to 18. Many of the vocational courses which have a mathematical component operate under the regulations of the Business Education Council (BEC), the Technician Education Council (TEC), the City and Guilds of London Institute (CGLI) and the Royal Society of Arts (RSA). CGLI and RSA have been in existence for many years; BEC was set up in 1974 and TEC in 1973. The role of BEC and TEC is to plan, administer and keep under review the establishment of a national system of non-degree courses. BEC is concerned with those whose occupations fall, or will fall, within the broad areas of business and public administration. TEC caters for those in, or about to enter, all levels of technician occupation in industry and elsewhere. BEC and TEC courses have been designed to subsume the National Certificate and Diploma Courses (ONC, OND, HNC, HND).

Types of course

TEC courses
150 Courses for technicians at all levels in industry and elsewhere are provided by means of the TEC programmes. These are arranged within three sectors—engineering, construction and science-based occupations. The 16 year old school leaver will follow the course for the TEC Certificate or Diploma. The Diploma is an extension of the Certificate; in general it provides broader coverage at the same level and not study in greater depth. Both programmes may be taken by either full-time or part-time study. The certificate and diploma courses are made up of units of study offered at different levels. The Level I mathematics unit is common to programmes in all three sectors and is designed to be appropriate for those who have a minimum of CSE grade 3 or O-level grade E in relevant subjects. Level II units follow on from Level I but in some cases, including mathematics, may be entered directly by those who have CSE grade 1 or O-level grade C or above.

151 The mathematical content of the TEC Level I unit and the associated technological units usually extends beyond the mathematics which technician apprentices will require during their first year at work. Some companies insist that technician apprentices who are qualified to enter the Level II course in mathematics must nevertheless take the Level I unit in order to consolidate their understanding of topics which are specifically relevant to their work. In some cases this insistence stems from complaints by company training officers about lack of competence among some O-level entrants in certain areas of mathematics which are considered to be of particular importance. As we

explained in paragraph 127, some firms make no distinction between craft and technician apprentices in the first year of training and CSE or O-level qualifications will usually determine whether an apprentice follows a TEC or CGLI course. However, it does not follow that all who start a TEC programme will eventually be classified as technicians; similarly, some of those who start a CGLI course may advance to become technicians.

152 TEC has published a mathematical 'Bank of Objectives' as a resource to be used in the preparation of mathematics units; it has also issued a set of 'standard units' based on the materials in the Bank. Colleges can make use of these standard units if they wish to do so, but are at liberty, subject to validation, to modify these units or to devise their own units to meet local needs. In the latter case they are able, in consultation with local firms, to make use of the Bank of Objectives. However, in practice many colleges include the standard units within the certificate and diploma courses which they offer.

153 The TEC units and Bank of Objectives are kept under review in order to attempt to match school syllabuses and to meet industrial requirements. Both units and Bank were revised in 1980 after extensive discussion with colleges and industry and in the light of comments from mathematics panels of GCE and CSE boards and from the mathematics committee of Schools Council. Recent changes in the standard units for Level I and Level II include the removal of slide rule and logarithm tables for purposes of calculation and greater emphasis on the correct use of calculators.

BEC courses
154 BEC courses are designed to provide a foundation of vocational education for a range of related careers and to meet the needs of one or more of four Boards—Business Studies, Financial Sector Studies, Distribution Studies, Public Administration and Public Sector Studies.

155 In written evidence to us, BEC stated:

> The Council has been saddened and concerned to receive consistent reports of the inability of secondary school leavers to cope with basic arithmetical tasks requiring manipulative skills or to show that they have a 'feel for the order of magnitude of a quantity', or to apply basic quantitative concepts and techniques to business problems. Accordingly the Council has included compulsory quantitative studies in all of the courses leading to its awards.

All BEC courses incorporate four 'central themes' of which one is 'a logical and numerate approach to business problems'. The courses are designed on a modular basis and 'cross-modular assignments' form part of each course.

156 BEC General Level and National Level courses are open to 16 year olds. The lower of these is the General Level course which is designed on a 'fresh start' policy and requires no formal examination qualification on entry. However, many entrants to this course possess examination qualifications at a level which is below the minimum of four O-level grade C or equivalent required for entry to the National Level course.

157   The General Level course contains a compulsory module entitled *Business calculations* which is examined separately. The emphasis in this module is on accurate and efficient performance of routine business calculations, ability to interpret numerical data and information, and ability to make use of numerical skills in the solution of business problems. The syllabus attempts to reflect the requirements of junior employees in clerical-type jobs and includes the operations of addition, subtraction, multiplication and division applied to whole numbers, decimals and fractions; metric and imperial units used in business; the purpose and use of approximations; the nature and function of percentages; the use of the average in business; the use of appropriate visual presentation of data. The learning objectives include 'appropriate and effective use of calculators'.

158   The entry requirement of four O-levels or equivalent for the BEC National Level course is not subject specific and so the mathematical attainment of those who enter this course from school varies widely. Those who proceed to the course as a result of having achieved credit standard at General Level will already have taken the *Business calculations* module. The National Level course includes a compulsory module entitled *Numeracy and accounting*. The 'numeracy' part of this module includes some, but not all, of the *Business calculations* module at General Level; notable omissions are metric and imperial units and the purpose and use of approximation. It also includes simple algebraic operations; construction and interpretation of graphs, representation of data in tabular and diagrammatic form; calculation and interpretation of averages; calculation of weighted averages; method used to construct selected index numbers, for example retail price index.

CGLI and RSA courses

159   Courses at craft level for those employed in technical jobs are offered by CGLI. The mathematics within these courses is not usually treated as a separate topic, nor is it usually examined separately. Instead, it is regarded essentially as a tool for vocational studies and is commonly taught as part of these studies as and when necessary.

160   In order to provide for the needs of less-academic students, CGLI has recently introduced a number of Foundation Courses, each focussed in a broad vocational area. The main aims of these courses are "to improve basic educational skills such as literacy and numeracy; to ease the transition from full-time education into the world of work; and to provide students with the basis on which they can make a more informed choice of career". These courses are full-time and may be taken at school, at FE college or as 'link-courses', in which the course is partly at school and partly at college. Each course contains a numeracy component which is developed in a context relevant to the vocational focus of the course. During the time we have been working, CGLI has introduced a course entitled *Numeracy*, which may be taken at school or at college.

161   In 1980 RSA introduced a Vocational Preparation (Clerical) course which is aimed at those initially seeking employment at operative or

equivalent levels. Successful completion of the course secures a profile-certificate which attests, as part of *Communication*, competence in various areas of arithmetic. Among the single-subject examinations offered by RSA are arithmetic, mathematics and machine calculating. These latter examinations may be taken from school or college.

The match between school, employment and further education

162   Ideally a school leaver should experience continuity of learning in mathematics after moving from school to further education and employment. However, this is by no means a simple process. Further education courses are designed for broad categories of employees and entry requirements are framed in such a way as to allow admission to young people of as wide a range of educational attainment as possible. Furthermore, the variety which exists in the content of O-level and CSE syllabuses and the fact that attainment of a given grade does not indicate competence in any specific part of the syllabus make it difficult to identify with any degree of certainty topics with which most of the entrants to a particular course will be familiar. It is probably for this reason that we have been told that many courses, especially those which do not require an O-level or equivalent qualification in mathematics, often start at a very elementary level but then move so quickly that weaker students find great difficulty in keeping up.

163   We have already drawn attention in paragraph 130 to the fact that the mathematical demands of FE courses are likely to be considerably more than the demands of the job itself. One reason can be that in some cases it is necessary to go beyond immediate requirements in order to develop confidence and familiarity with essential topics. It is also the case that many courses are intended to provide not only the specific skills which are needed in the early years of employment but also a base for forty or more years of working life. Nevertheless, mathematical skills which are not used regularly can very easily atrophy, especially if they have proved difficult to comprehend, and so may not prove to be available when they are needed. It seems also to be the case that promotion can often lead to the use of less mathematics rather than more, because time is spent on supervisory and other duties.

164   We noted in paragraph 130 that craft and technician apprentices are sometimes required to follow the same initial training course. In some cases, apprentices who will eventually be designated as craftsmen are required in FE colleges to undertake the technician courses which are academically more demanding and for which they may not be adequately prepared. **We believe that it is with this type of entrant that the mismatch between the mathematics content of FE courses and the future demands of the job gives most cause for concern**. A comparable problem can arise in the case of entrants to BEC National Level courses whose four or more O-levels do not include mathematics; however, we have no evidence to indicate that this is so great a problem. The diversity of school syllabuses can also lead to problems of mismatch between college courses and courses which have been followed at school.

165 Although the range of mathematical ability can be particularly wide among those on craft courses, the fact that mathematics is not usually examined separately on CGLI courses makes it possible for students to avoid the more mathematical questions in examinations and still obtain good grades. This would seem to underline the fact that many craftsmen need to use only a very limited range of mathematical skills.

166 The Bath and Nottingham studies found that attitudes towards mathematics among students at FE colleges were very often more favourable than had been the case when they were at school. In some instances this seemed to arise from the fact that the applications of mathematics were more immediately apparent. Even when this was not the case there were some who persevered because they felt that the mathematics they were having to learn was certain to be needed at some stage or it would not have been included in the college course.

The mathematical requirements of higher education
Non-university sector

167 Within the non-university sector of higher education in England and Wales there are thirty Polytechnics and more than sixty Colleges of Higher Education, which offer a wide range of courses at degree level. Many of these courses are validated by the Council for National Academic Awards but some universities validate Bachelor of Education and other degrees offered at colleges within their locality. Polytechnics and colleges also offer a considerable range of full-time and sandwich courses at sub-degree level. The majority of these have a specific vocational slant, which is further reinforced by work experience when the course includes 'sandwich' placement in an appropriate firm. Some courses are aimed directly at the membership grades of various professional bodies (see also paragraphs 184 to 187).

168 Entry requirements for both degree and sub-degree courses are usually stated in terms of success at O- and A-level in appropriate subjects, which may include mathematics, or the successful completion of other relevant courses such as the BEC National Certificate or Diploma or an appropriate TEC Certificate or Diploma.

169 There is at present no single body which is concerned with the administration of, or entry to, the complete range of courses in the non-university sector of higher education. We have sought information about students in this sector but the information which we have been able to obtain is not classified and analysed in a way which makes it possible to identify the mathematical qualifications of students on either degree or sub-degree courses.

University sector

170 However, within the university sector of higher education, detailed information about the A-level qualifications of undergraduates entering degree courses at universities in the United Kingdom is collected by the Universities Statistical Record (USR). It has therefore been possible to establish the number of undergraduates with an A-level qualification in mathematics who have entered universities in England and Wales in recent years and also the degree courses which they have chosen. In the following paragraphs we

discuss the general information which these figures provide before considering certain degree courses for which a specific mathematical entry qualification is normally required. The figures which we quote refer to undergraduates at universities in England and Wales with home fee-paying status;
students from overseas are not included. We estimate that the statistics we
give for 1979 refer to about three-quarters of all those with home fee-paying
status who started degree courses at institutions in England and Wales.

171 In 1979 almost 92 per cent of all new entrants with home fee-paying
status to first degree or first degree and diploma courses at universities in
England and Wales had entry qualifications based on A-levels. About $2\frac{1}{2}$ per
cent, of whom just over half were entrants to courses in engineering and
technology, had qualifications based on National Certificates or Diplomas;
the remainder had a variety of other qualifications, including qualifications
gained at other universities or in other countries. For mathematical studies
and most science subjects some 95 per cent of entrants had an A-level qualification; for engineering and technology the figure was about 85 per cent,
reflecting the greater numbers who enter these courses on the basis of
National Certificates or Diplomas.

172 These proportions have remained substantially the same since 1973, the
first year for which we quote figures. Between 1973 and 1979 the total number
of undergraduates increased by just over 28 per cent, though there were considerable variations from subject to subject. The number reading mathematical studies increased by about 19 per cent, physical sciences by about 17 per
cent (but physics itself by about 28 per cent) and engineering and technology
by about 34 per cent.

Undergraduates with A-level mathematics
173 Mathematics is the only subject other than music which it is possible to
offer as either one or two subjects at A-level. This is commonly referred to as
taking 'single-subject' or 'double-subject' mathematics and we shall use these
terms. For the present it is sufficient to be aware that both single-subject and
double-subject mathematics are available at A-level; we discuss the differences which exist in the structure of the double-subject examinations and in the
syllabuses for both single- and double-subject examinations in Chapter 11.
Those who take the double subject cover a more extended syllabus than those
who take the single subject. It does not follow that they are necessarily more
able mathematically than those who take the single subject; however, the fact
that they will almost certainly spend more time on mathematics during the
A-level years enables them not only to cover more ground but also to develop
increased confidence and competence in the content of the single-subject
syllabus. For many years almost all of those who took double-subject mathematics combined it with A-level physics; some also took A-level chemistry to
give a total of four subjects in all. Many still take these traditional combinations but in recent years it has become increasingly common to combine
both single- and double-subject mathematics with a wide variety of other subjects.

174 The numbers of men and women entering universities in England and
Wales from 1973 to 1979 who had an A-level qualification in mathematics*
are shown approximately in Figure 2 (see also Appendix 1, Table 27). We may
note that throughout these years the percentage of men with an A-level qualification in mathematics has been about twice that of women.

175 We now examine the degree subjects studied by entrants to degree
courses who had one or more A-levels in mathematics. The information for
1973 and 1979 is given in Table A (see also Appendix 1, Tables 28 and 29).

Table A *Distribution between subject groups of entrants to degree courses at
universities in England and Wales who had one or more A-levels in
mathematics*

	No of entrants with A-level mathematics		% of all entrants with A-level mathematics	
	1973	1979	1973	1979
Subject group:[1]				
Engineering and technology	5122	7045	*27.1*	*28.0*
Physical sciences	3717	4185	*19.6*	*16.6*
Mathematical studies	2643	3107	*14.0*	*12.4*
Medical and dental	1285	2095	*6.8*	*8.3*
Biological sciences	1097	1656	*5.8*	*6.6*
Other sciences	1506	1646	*8.0*	*6.5*
Business studies	863	1621	*4.6*	*6.5*
Geography	372	454	*2.0*	*1.8*
Other subjects	2315	3338	*12.2*	*13.3*
All subject groups	18920	25147	*100*	*100*

Source: Universities Statistical Record

[1] Degree courses included in each subject group are given in Appendix 1, paragraph A21.

176 We present this information in a different way in Figure 3 on page 50
which draws attention to the proportion of entrants to each group of courses
who had an A-level qualification in mathematics. The most significant
changes between 1973 and 1979 have been in respect of those starting medical
and dental studies, in which the proportion has increased from 33 per cent in
1973 to 44 per cent in 1979, and those starting courses in business and management studies, economics and accountancy, in which there has been an increase
from 42 per cent to 54 per cent.

Undergraduates with double-subject mathematics

177 There has been a marked change in the proportion of entrants to universities in England and Wales with a double-subject qualification in mathematics*. In 1973 almost 32 per cent of those entering with an A-level qualification in mathematics had a double-subject qualification; in 1979 the proportion had fallen to little more than 21 per cent. In absolute terms, although
there were over 6000 more entrants with A-level mathematics in 1979 than in
1973, the number of entrants with a double- subject qualification had dropped by about 650. The number and percentage change over this period is
shown in Table B on page 51 (see also Appendix 1, Table 28).

Figure 2 *Numbers of men and women entering universities in England and Wales from 1973 to 1979 who had an A-level qualification in mathematics*

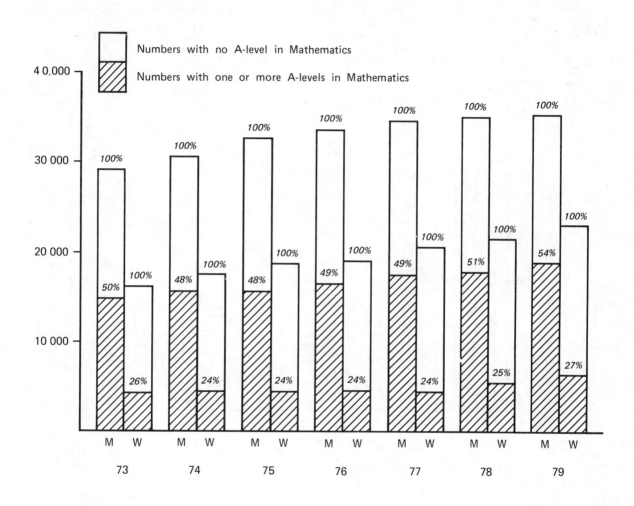

Source: Universities Statistical Record

Figure 3 *Proportion of entrants to each group of courses who had one or more A-levels in mathematics: Universities in England and Wales 1973 and 1979*

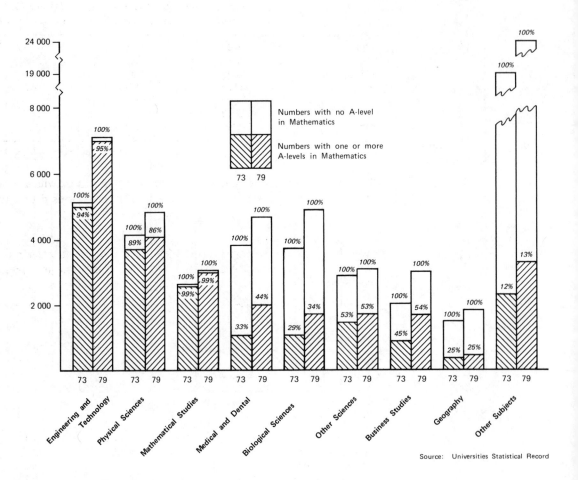

Source: Universities Statistical Record

Table B *Levels of mathematical qualification of entrants to degree courses at universities in England and Wales*

	No of entrants with one or more A-levels in mathematics	No of entrants with double-subject A-level in mathematics	No with double-subject as % of all with any A-levels in mathematics
1973	18920	6006	*31.7*
1974	18783	5263	*28.0*
1975	20089	5446	*27.1*
1976	20920	5501	*26.3*
1977	22060	4892	*22.2*
1978	23222	5056	*21.8*
1979	25147	5341	*21.2*

Source: Universities Statistical Record

178 There have also been changes in the way in which those with double-subject mathematics have been distributed among the various degree courses. This is shown in Table C (also Appendix 1, Tables 28 and 29).

Table C *Distribution between subject groups of entrants to degree courses at universities in England and Wales who had double-subject A-levels in mathematics.*

	No of entrants with double-subject A-level mathematics		% of all entrants with double-subject A-level mathematics	
	1973	1979	1973	1979
Subject group:[1]				
Engineering and technology	1617	1714	*26.9*	*32.1*
Physical sciences	1048	968	*17.5*	*18.1*
Mathematical studies	2103	1728	*35.0*	*32.4*
Medical and dental	139	161	*2.3*	*3.0*
Biological sciences	61	47	*1.0*	*0.9*
Other sciences	422	281	*7.0*	*5.3*
Business studies	185	177	*3.1*	*3.3*
Geography	34	19	*0.6*	*0.4*
Other subjects	397	246	*6.6*	*4.6*
All subject groups	6006	5341	*100*	*100*

Source: Universities Statistical Record

[1] Degree courses included in each subject group are given in Appendix 1, paragraph A21.

179 About 80 per cent of entrants with a double-subject qualification in mathematics at A-level read engineering and technology, physical sciences or mathematical studies. Figure 4 illustrates the way in which entrants with a double-subject qualification were distributed between these three areas of study in 1973 and 1979. (See also Appendix 1, Table 30).

Figure 4 *Numbers of university entrants in England and Wales admitted on the basis of A-levels. Subject choices and numbers with double-subject A-level qualifications in mathematics: 1973 and 1979*

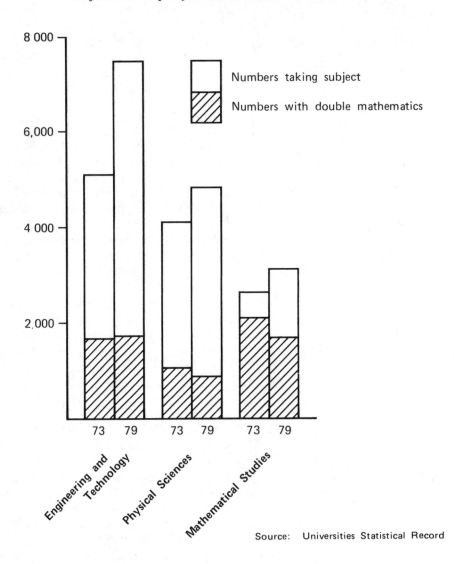

Source: Universities Statistical Record

Degree courses in mathematical studies

*Mathematical studies includes degree courses in mathematics, statistics, computer science, combinations of these and also a variety of courses which combine mathematics with subjects other than these. Analysis of information provided by USR about first degrees awarded in 1979 within the field of mathematical studies shows that about 65 per cent were in mathematics only, about 25 per cent in computer science, either wholly or in part, and about 8 per cent in statistics.

180   We wish to draw attention to the drop in the proportion of those reading mathematical studies* at universities in England and Wales who have a double-subject qualification and to its implications. In 1973 the proportion was almost 80 per cent; in 1979 this had dropped to 55 per cent. Although there are some universities at which it is still the case that a very high proportion of those reading degrees in mathematics have taken the double subject at A-level, our own enquiries have established that there are others at which substantially less than half of those reading mathematics have a double-subject qualification. We believe that this information may come as a surprise to many people in both universities and schools. It is not within our terms of reference to comment on its implications for those who teach mathematical studies in universities; the implications for those who teach in schools are very great.

181   It is very commonly supposed that it is almost essential to have taken double-subject mathematics at A-level in order to read mathematics successfully at university. However, **it is very important that those who teach mathematics in sixth forms and those who advise pupils about their choice of degree course should realise that there are now universities in which more than half of those reading mathematics are doing so from a basis of single-subject A-level. It follows that they should not dissuade pupils who have taken only the single subject at A-level from applying to read degree courses in mathematics.** We see no likelihood that the demand for mathematics graduates will decrease—indeed, we believe that the demand will continue to grow—and those whose interests and abilities lie in this field need every encouragement to study mathematics at degree level. As we pointed out in paragraph 173, it does not follow that those who have taken only the single subject are necessarily less able at mathematics than those who have taken the double subject and so less fitted to embark on a mathematics degree course. There seems no doubt that at most universities they will be increasingly likely to find themselves in the company of others who are similarly qualified.

Degree courses in engineering and technology

182   A knowledge of mathematics is essential for the study of engineering and of most other technological subjects. We drew attention in paragraph 172 to the fact that the number of entrants to courses in engineering and technology increased by 34 per cent between 1973 and 1979 whereas the overall university entry increased by 28 per cent. This increase has been much greater than the increase in total entry to all other mathematics and science courses, which has risen by only 20 per cent. Furthermore, despite an overall drop during this period in the total number of entrants to universities in England and Wales with a double-subject qualification in mathematics, there has been an increase in the proportion of these entrants who have chosen to read engineering and technology, and also a small absolute increase in their numbers. Tables A and C show that degree courses in engineering and technology are attracting an increasing proportion of university entrants with an A-level

qualification in mathematics and, in particular, of those with double-subject mathematics. The proportion of entrants with double-subject mathematics is, however, decreasing both within engineering and technology courses and overall.

The mathematical requirements of professional bodies

183 Although not directly within our terms of reference, we have given some attention to the mathematical requirements of professional bodies. Many of those engaged in professional activities seek membership of the appropriate professional institution or association. In some cases membership of such a body is a necessary qualification for professional advancement; in other cases membership, although not essential for career purposes, provides opportunity to keep abreast of current developments by reading publications, attending meetings and taking part in the work of committees. Most institutions conduct their own examinations, commonly in two or three parts, for admission to membership, which is usually offered at more than one grade. The possession of an appropriate academic qualification often secures exemption from some or all of these examinations but admission to higher grades of membership normally requires evidence of relevant professional experience.

184 A number of the professional bodies who have written to us have stated the mathematical requirements for direct entry to their various grades and have also supplied details of their own examinations. When entry is at graduate level, it is usually assumed that any necessary mathematics will have been covered either at school or during the degree course and no further mathematical requirement is stipulated. However, one exception to this is the Institute of Actuaries whose final examinations require a considerable extension of mathematical and statistical knowledge and its application. When entry to a professional body is at lower levels, any mathematical requirement is normally stated in terms of success at A- or O-level; any further mathematics which is required is then included within subsequent professional study.

185 Almost all the professional bodies who submitted evidence stressed the importance of being able to apply computational skills confidently in a variety of ways. These include accuracy and speed in mental calculation and ability to check the reasonableness of answers; in some cases extended and complex calculations are necessary. Specific calculations identified by bodies whose members are concerned with commerce include interest, discount and value-added tax, cash flow, costing and pricing, and budgetary control; it is frequently necessary to be able to deal with both metric and imperial units. There were also many references to the need to be able to interpret data with understanding.

186 Most institutions take for granted the mathematical foundation provided by an entrant's previous study. Any mathematics included within professional examinations is usually limited either to topics of a specialist nature which are unlikely to have been studied before entry to the profession or to applications of mathematics in unfamiliar contexts. This is especially

true of professional bodies whose members are concerned with business and commerce. The examinations of these bodies frequently include applications of statistics, and to a lesser extent techniques of operational research, which are used within the particular profession. The collection, classification, presentation and analysis of data, use of probability distributions, hypothesis testing, correlation and regression analysis, survey methods and sampling techniques all occur frequently within the syllabuses of professional examinations. This emphasis on statistics no doubt reflects the fact that at the present time few school leavers will have studied the subject to any depth.

Part 2

5 Mathematics in schools

187 In the second part of our report we discuss the teaching and learning of mathematics in schools as well as methods which are used to assess attainment. Before turning to particular aspects such as mathematics in the primary and secondary years we consider some matters which are fundamental to the teaching of mathematics to pupils of all ages, and also certain matters which arise as a consequence of the discussion in earlier chapters, of the submissions which we have received and of our own experience. In order to provide a background we start by drawing attention to the levels of attainment in mathematics which are to be expected of school leavers, so that readers may bear in mind the proportions of the school population to which the different parts of our discussion relate; we also consider the attitudes towards mathematics which pupils develop during their schooldays and the mathematical attainment of girls.

*A brief summary of the Review may be purchased from the Shell Centre for Mathematical Education, University of Nottingham; see also paragraph 756.

188 In this part of our report we draw on *A review of research in mathematical education* which summarises the results of the study carried out for our Committee under the direction of Dr A Bell of the University of Nottingham and Dr A Bishop of the University of Cambridge. For the sake of brevity, we shall henceforward refer to it as the *Review of research.**

Attainment in mathematics

189 We believe that there is widespread misunderstanding among the public at large as to the levels of attainment in mathematics which are to be expected among school leavers. At the present time about a quarter of the pupils in each year group achieve O-level grade A, B or C or CSE grade 1; about a further two-fifths achieve CSE grade 2, 3, 4 or 5: the remainder, amounting to almost one-third of the year group, leave school without any mathematical qualification in O-level or CSE. These figures are not surprising; they reflect the proportions of the school population for whom O-level and CSE examinations are intended and have been designed. At a higher level, between 5 and 6 per cent of the pupils in each year group achieve an A-level qualification in mathematics; about 1 pupil in 200 reads a degree course in mathematical studies.

190 Figure 5 illustrates in approximate diagrammatic form the 'mathematical attainment profile' of those in England and Wales who left school in 1979, and of those of school age who completed A-level courses in FE or tertiary colleges in that year. It is based on figures collected in the annual 10 per cent survey of school leavers (see Appendix 1, paragraph A3) and on an

Figure 5 *'Mathematical attainment profile' for leavers in 1979*

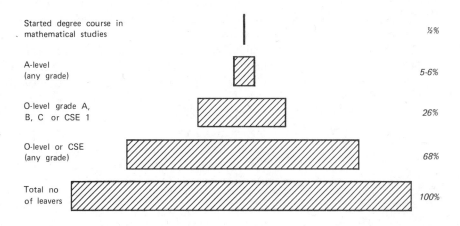

estimate of the numbers completing A-level courses in mathematics in FE and tertiary colleges, since information available about the examination performance of students in these colleges does not identify separately those who are of school age.

191 The number of pupils who have been studying mathematics at A-level in schools and sixth form colleges in England has increased steadily in recent years both in absolute terms, as a percentage of all pupils, and as a percentage of all pupils taking A-level courses. In the school year 1973–74, some 43 per cent of boys and 17 per cent of girls taking A-level courses in the first year of the sixth form were studying A-level mathematics. In the school year 1979–80, these figures had risen to almost 51 per cent of boys (approximately 41 000 and to 23 per cent of girls (approximately 17 000).

192 It is not possible to obtain comparable statistics for those of sixth-form age who are taking A-level courses in FE and tertiary colleges, but we have no reason to believe that the picture would be significantly different.

Comparison of examination results in English and mathematics

193 In order to provide more detailed information about mathematical performance in CSE, O-level and A-level, the DES has at our request analysed in a variety of ways the information about results obtained in these examinations which was supplied in 1977, 1978 and 1979 by schools in England and Wales as part of the annual 10 per cent survey of school leavers in both maintained and independent sectors. This has provided more detailed information about examination performance in mathematics than has hitherto been available. Because the survey relates to school leavers in a given year who may be aged 16, 17, 18 or, very occasionally, 19, the information does not relate to a complete year group. Nevertheless, because the patterns for 1977, 1978 and 1979 are very similar, we believe that the picture they provide is unlikely to differ significantly from that which would emerge if it were possible to obtain information relating to a complete year group.

194 Some of those who have written to us have drawn attention to figures published each year in DES *Statistics of education* Vol 2 which show that the proportion of pupils who achieve O-level grade A, B or C or CSE grade 1 in English at some stage during their school career is much higher than that of those who achieve these grades in mathematics. In some submissions it has been suggested that the standard required in mathematics examinations is therefore too high. In consequence we have also obtained information relating to CSE and O-level results in English for these three years.

195 Figure 6 illustrates approximately the O-level and CSE performances in English and mathematics or arithmetic of those leaving school in these three years. The figures on which it is based are set out in detail in Appendix 1, Tables 11 and 12; paragraphs A7 and A9 of this Appendix describe the procedure which has been used to ensure that those who have taken both O-level and CSE in the same subject or, for instance, both mathematics and arithmetic are only included once. The letters in the columns of Figure 6 refer to O-level grades A to E, the numbers to CSE grades 1 to 5. It is reasonable to

Figure 6 *Proportions of pupils awarded O-Level grades A to E and CSE grades 1 to 5 in English and Mathematics/Arithmetic*

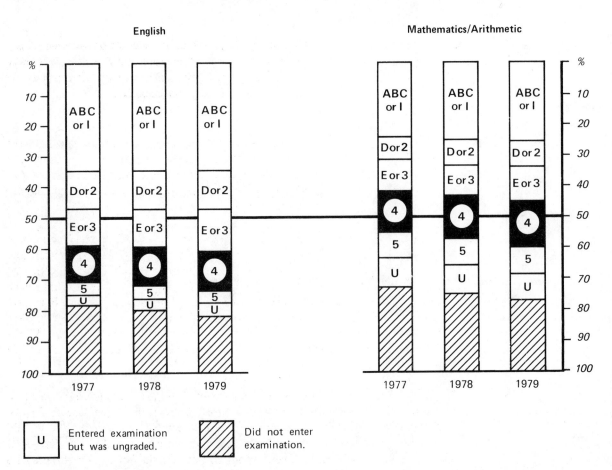

Aspects of secondary education in England. A survey by HM Inspectors of Schools. HMSO 1979.

assume that these figures provide a valid comparison between English and Mathematics for the school population as a whole because, as the report of the National Secondary Survey* shows in respect of the maintained sector, virtually all pupils study English and mathematics up to the age of 16 and enter for O-level or CSE in these subjects if they have the ability to do so. Among those who leave school at Easter in any year there will presumably be a small number who could have achieved a graded result at O-level or CSE if they had made the attempt, but we believe this number to be so small that it does not affect the overall picture significantly.

196 These figures show that from 1977 to 1979 there was a small increase in the proportion of school leavers obtaining a graded result in mathematics of some kind. CSE regulations state that in any subject "a 16 year old pupil of average ability who has applied himself to a course of study regarded by teachers of the subject as appropriate to his age, ability and aptitude may reasonably expect to secure grade 4". It is clear that the position of grade 4 in mathematics is as nearly in accordance with this definition as could reasonably be expected. The proportion of O-level and CSE grade 1 passes has remained constant at about a quarter, and also reflects the proportion of the school population which is expected to obtain these grades. In this sense it cannot be maintained that the standard required in mathematics examinations at this level is too high.

197 **It is, however, the case that the positions of the grade boundaries in English are very different from those in mathematics and it is therefore to be expected that there will be many pupils who will achieve significantly better grades in English than they will achieve in mathematics.**

Attitudes towards mathematics

198 As we stated in Chapter 2, the research study into the mathematical needs of adult life revealed the extent to which the need to make use of mathematics could induce feelings of anxiety and helplessness in some people. It also revealed that many people had far from happy recollections of their study of mathematics at school.

*See paragraph 59

199 In the course of their work, those involved in the Bath and Nottingham studies* gathered reactions from a large number of young employees, and from some who were older, to the mathematics teaching which they had received at school. The views expressed are, of course, likely to have been influenced by subsequent experience in employment, particularly in respect of what was seen as the usefulness or otherwise of particular topics. Nevertheless, the remarks do record the attitudes towards mathematics of these young people and their experience at school as it has been remembered at a later date.

200 The report of the Bath study states that "we met many young people who had liked mathematics at school. Most of these, but not all, seemed to be among the mathematically more able as measured by school results. We did come across some who got low grades, or were ungraded or not entered at CSE level, who said they enjoyed mathematics; and one or two of these who

thought they were quite good at it, which we found encouraging. There were many who were non-committal. "Okay" was a frequent response. Mathematics was there, you had to do it and that was that". However, the studies found that there were also many, mainly but not exclusively among the less able, who had disliked mathematics and who had seen no point in it. Their criticism was mainly of two kinds, one concerning the content of the mathematics course and the other the way in which mathematics had been taught.

201 Formal algebra seems to have been the topic within mathematics which attracted most comment. Those engaged in the Bath study were "left with a very strong impression that algebra is a source of considerable confusion and negative attitudes among pupils". In some cases this was because the work had been found difficult to understand; in other cases it was felt that exercises in algebraic manipulation and topics such as sets and matrices had had little point. It is, however, of interest to note that some of those who found algebra difficult at school were finding it easier at college; "formulae make sense now". Many other topics including fractions, percentages, graphs, trigonometry, Pythagoras' theorem ("the name being remembered but not much else about it") were referred to unfavourably by some, but comments of this kind seemed often to be related to the usefulness or otherwise of these topics in the job with which the employee was concerned. For example, some clerical workers who had not understood trigonometry, and did not make use of it in their work, felt in retrospect that they should have done more about percentages.

202 Adverse comments on the teaching of mathematics often concentrated on an alleged inability on the part of some teachers to explain clearly, on a tendency to ignore some of those in the class, on an unwillingness to answer questions and on moving through the course too quickly. There was criticism also of teachers who had not required their pupils to do sufficient work and of teachers who had been unable to state the purpose of the work which was being done—"do it to pass your exams". Many of those interviewed said that they had been given little or no practice in mental calculation at secondary school, especially after the second year. The report of the Bath study comments that "on a number of occasions the feeling came through to us very strongly that the sort of teacher appreciated by many pupils is one who can control the class, expects pupils to work hard, explains things briefly and clearly, gives plenty of practice and is prepared to help pupils individually". Although this description does not, as we explain in paragraph 243 of this chapter, include all the elements which should be present in good mathematics teaching, we nevertheless believe that the qualities to which it draws attention are essential in those who teach mathematics, if pupils are to develop good attitudes. We believe that a somewhat different criterion, stated by one young employee—"any teacher who can make maths interesting must be good"—also provides food for thought.

203 The comments of young employees showed that very often, but not always, inability to do mathematics and not liking it seemed to go together. Equally, success more often than not seemed to lead to a favourable attitude;

"you enjoy it when you can do it". Nevertheless, the belief that mathematics was useful was shared by many of those who disliked mathematics as well as by those who liked it. Most of those interviewed, whether young or old, saw mathematics in vocational utilitarian terms. Very few saw it as serving any wider purpose, especially of what might be termed a cultural or general educational kind.

204 Very few who had taken mathematics to A-level or beyond were included among those who were interviewed, nor were there very many who had taken any subjects at A-level. The attitudes of such people might well be very different, but those who take A-level form only a small minority of the school population.

205 It is to be expected that most teachers will attach considerable importance to the development of good attitudes among the pupils whom they teach and so it is interesting to compare the comments made by young employees with the findings of research studies. The *Review of research* discusses the considerable amount of research on pupils' attitudes which has been carried out in recent years, including that which is being undertaken in England and Wales as part of the work of the Assessment of Performance Unit. It states that "it is not easy to pick out points which summarise all the research on attitudes to mathematics. Strongly polarised attitudes can be established, even amongst primary school children, and about 11 seems to be a critical age for this establishment. Attitudes are derived from teachers' attitudes (though this affects more intelligent pupils rather than the less able) and to an extent from parents' attitudes (though the correlation is fairly low). Attitude to mathematics is correlated with attitude to school as a whole (which is fairly consistent across subjects) and with the peer-group's attitude (a group-attitude tends to become established). These things do not seem to be related to type or size of school or to subject content. Throughout school, a decline in attitudes to mathematics appears to go on, but this is part of a decline in attitudes to all school subjects and may be merely part of an increasingly critical approach to many aspects of life".

206 The *Review of research* points out that research studies also show that there is a strong tendency among pupils of all ages to believe mathematics to be useful but not necessarily interesting or enjoyable. "There appears to be an identifiable (although small) correlation between attitude and achievement: it is not clear, however, in what way attitude and achievement affect one another. This does not necessarily contradict the teacher's perception that more interesting and enjoyable work will lead to greater attainment. For one thing, the research does not deal with changes in achievement which might result within particular individuals or classes from improvements directed towards attitude; it rather shows that, broadly speaking, the set of people who like mathematics has only a relatively small overlap with the set of those who are good at it. However, research certainly suggests caution against over-optimism in assuming a very direct relation between attitude and achievement." **We have already noted the interest displayed by the Assessment of Performance Unit in respect of attitudes toward mathematics. It is clear that there is need for continuing investigation in this field.**

Parents and schools

207 Parents can exercise, even if unknowingly, a considerable influence on their children's attitudes towards mathematics. Encouragement to make use of mathematics during normal family activities, for example to weigh and measure, to use money for shopping and to play games which involve the use of dice or the keeping of scores, can assist children to develop familiarity with numbers and confidence in making use of them. However, in some cases parents can expect too little; "don't worry, dear, I could never understand mathematics at school either". In other cases parents can expect too much of their children and so, as we have noted in paragraph 24, exert pressure which can lead to failure and consequent dislike of mathematics. It can happen, too, that parents fail to understand the purpose of the mathematics which their children are doing and so make critical remarks which can also encourage the development of poor attitudes towards mathematics in their children. **We believe that it is therefore important that schools should make active efforts to enlist the help of parents by explaining the approaches to mathematics which they are using and the purposes of mathematical activities which parents themselves may not have undertaken while at school.** Schools should also encourage parents to discuss with teachers their children's progress in mathematics, so that both parents and schools may work together to assist the mathematical development of pupils.

The mathematical performance of girls

208 In recent years increasing attention has been paid to the fact that, as measured by the results of public examinations, the overall level of performance of girls in mathematics is significantly lower than that of boys. In this section we discuss this matter in general terms; a more detailed discussion is contained in Appendix 2, which includes relevant statistics and also reference to some of the research studies into the mathematical performance of girls which have been carried out in recent years both in this country and in other parts of the world.

209 Concern about differences in the performance of boys and girls is not confined to the United Kingdom. This was demonstrated clearly at the Fourth International Congress on Mathematical Education, held at the University of California at Berkeley in August 1980, at which papers about the mathematical performance of girls were presented by participants from many countries. We include in this section some of the ideas which were put forward in these papers and in the discussions which followed them.

210. It is not easy to establish why girls should perform less well than boys at mathematics and many possible reasons have been suggested. Some of these relate to biological factors, some to child-rearing and social factors, some to factors within schools and some to career expectation. It has been suggested that among the child-rearing practices which may influence mathematical attainment is the fact that, whereas girls are usually given dolls and 'domestic' toys, boys are given significantly more scientific and constructional toys, which encourage the development of spatial concepts and problem solving activities. It has also been suggested that boys are encouraged to be more independent than girls, which may again encourage experiment and problem solving; and that adults respond to boys as if they find them more interesting

and more attention-provoking than girls. Among the social factors are the perception of mathematics as a 'male' activity (though more recent studies suggest that the sex-typing of mathematics is decreasing), and peer-group pressures which cause girls to fear that high attainment in mathematics will inhibit the development of their relationships with boys. The Committee for Girls and Mathematics has told us that "schools, the careers service and industry appear to have shown little initiative in encouraging or attracting girls with ability in mathematics into some of the fields where there are shortfalls of good applicants; . . . . . whereas a boy is usually counselled to work towards a qualification in mathematics as a career essential, this is not always the case with girls".

211 Factors related to school may also help to reinforce the impression that mathematics is a male domain. Men teachers in primary schools often take the older classes; at secondary level more men than women teach mathematics. The applications of mathematics which are found in many textbooks and examination questions reflect activities associated with men more often than they reflect activities associated with women. In primary schools it can be the case that boys engage in craft activities of a three-dimensional kind while girls do needlework, though we believe that differences of this kind exist in fewer schools now than was the case a few years ago. It also seems likely that there is, even if unconsciously, an expectation among teachers that girls will perform less well at mathematics than boys. Research studies, to which we refer in Appendix 2, suggest that during mathematics lessons teachers in secondary schools interact more with boys than they do with girls, give more serious consideration to boys' ideas than to those of girls and give boys more opportunity than girls to respond to higher cognitive level questions. It has also been found that high achieving girls may receive considerably less attention in mathematics lessons than do high achieving boys. In such circumstances, girls are likely to receive the message that they are not expected to perform as well as boys and to react accordingly.

212 We believe that the question of classroom interaction and expectation is of considerable importance. It was suggested at the Fourth International Congress that girls may have a greater need than boys to develop understanding through discussion and that many girls consider that mathematics teachers do not listen sufficiently to what girls say in response to questions. We shall argue later in this chapter that discussion should be an essential part of all mathematics teaching; it seems likely that in classrooms where it is not practised sufficiently girls are liable to be especially disadvantaged. Even in classrooms in which there is plenty of discussion teachers may need to take steps to ensure that boys do not dominate the discussion and that girls are given opportunity to play their full part. We may note at this point that, whereas girls perform less well than boys at mathematics, they perform better overall than boys at English. We believe that this gives further support to the suggestion that girls need to undertake extended discussion in order to clarify their ideas and understanding and that plenty of verbal interaction is essential if they are to learn mathematics satisfactorily.

213 Research studies suggest that, whereas boys more often attribute their successes in mathematics to ability and their failures to lack of effort or bad luck, girls more often attribute their success to hard work or good luck and their failures to lack of ability. If this is the case, it underlines the need to do all that is possible to encourage girls to develop confidence in their mathematical powers. It is sometimes suggested that girls succeed better at mathematics when they are taught in single-sex groups but we are not aware of any studies which establish this fact. Although it is possible to identify some girls' schools in which levels of mathematical attainment are high, it is often the case that there are other factors, such as the fact that the school is selective, which may provide the explanation. We are aware of a small number of instances in which a secondary school has arranged to teach mathematics to girls and boys in separate classes but the scale on which grouping of this kind has been carried out is as yet too small for it to be possible to draw conclusions.

*HMI Series: Matters for discussion 13. *Girls and science*. HMSO 1980.

214 The HMI booklet *Girls and science** discusses the attitudes of girls to science and various matters relating to the approaches to teaching which are helpful in encouraging more girls to take an interest in science. **We believe that much of this discussion is also relevant to the teaching of mathematics to girls and we commend the booklet, which also contains an appendix on 'girls and engineering', for study by those who teach mathematics.**

215 In Appendix 2 we suggest a number of strategies which we believe may contribute to improvement in the mathematical performance of girls. We wish here to draw attention to two of these. **The first is the need for appropriate careers guidance to be available to girls at an early stage so that their attention can be drawn to the fact that lack of an appropriate mathematical qualification can exclude entry to many fields of employment. The second is the need for teachers to be aware of the differences in mathematical performance which at present exist between boys and girls.** If teachers hold this fact in mind and make a conscious effort to ensure that they do not make use of teaching methods which are liable to put girls at a disadvantage, we believe that an improvement in the overall performance of girls could be achieved.

216 Some programmes have recently been set up in the United States of America which are designed to improve the motivation and attitudes of girls towards mathematics. One of these, the *Math/Science Network*, operates mainly in California and works to encourage greater participation by girls in "math-based fields of work and study" by holding conferences for teachers, providing careers information and developing appropriate activities in schools. A second, *Women and Mathematics*, is sponsored by the Mathematical Association of America and operates more widely. It attempts to change attitudes by arranging for women who are using mathematics in interesting careers to speak at high-school conferences. The group also organises conferences for careers advisers. We have been told that these initiatives are meeting with some success. It may be that a similar initiative would be helpful in the United Kingdom.

The teaching of mathematics through the medium of Welsh

217 The number of children who are taught mathematics through the medium of Welsh has increased considerably in recent years. Some of these children come from homes in which Welsh is the family language but increasing numbers are entering Welsh-medium primary schools from homes in which the family language is English. In our meetings with teachers in Welsh-medium schools it has been made clear to us that the problems of teaching mathematics through the medium of Welsh to children who are Welsh speaking are fundamentally no different from the problems of teaching mathematics in English to children who are English speaking.

218 There is, however, one major difficulty which exists for those who teach mathematics through the medium of Welsh. This is the great shortage of mathematics textbooks which are written in Welsh. There is at present no complete primary mathematics scheme in Welsh. Such materials as are available are confined mainly to infant level; other materials are fragmentary and deal only with parts of the mathematics curriculum. This means that teachers have to prepare most of their own classroom materials. This not only takes a great deal of time but also results in a standard of presentation which is less good than that of a published text. For this reason there are some classrooms in which, although the teaching and discussion are in Welsh, children work from books which are written in English.

219 Work is in progress on the production of a secondary mathematics course written in Welsh. Books covering the first three years of the course have now been published and material for the fourth year is to appear shortly. However, even when the whole course has been published, it will be the only one which is available and will have to serve the needs of all pupils, whatever their level of attainment. This cannot be considered to be a satisfactory situation.

220 **If it is government policy to support the provision of Welsh-medium schools, it is essential that a suitable supply of teaching materials is made available.** It has been suggested to us that, because these materials are needed urgently, a first step should be to commission the translation into Welsh of at least two mathematics courses which have already been published in English. This is a proposal which we support. **We believe that it will be necessary for special funding of some kind to be made available for this purpose because of the inevitably small market for teaching materials written in Welsh and the need for some kind of subsidy if they are to be produced for sale at a price which is not unreasonably high.**

The teaching of mathematics to those for whom English is not the first language.

221 In recent years there has been a considerable increase in the number of children in schools for whom English is not the first language. Even when, as is increasingly the case, these children, who are mainly from European, African and Asian family backgrounds, have been born in Britain, they very often start their schooling as only partial English speakers; sometimes they are not able to speak English at all. In the early stages, therefore, it is necessary to ensure that the language demands of discussion and of written materials such

as work cards take account of the vocabulary and language structures available to these children. It is, however, important that they should take part in oral work both to assist the development of their general language skills and also to enable them to become familiar with the language which is used in mathematics.

222   Especial care is likely to be needed with young children in the early stages of naming numbers and counting. Almost all European languages show irregularities in the naming of some or all of the numbers between 10 and 20. In English, for example, numbers from 13 to 19 are spoken 'back to front' compared with numbers from 20 onwards, so that we say 'twenty-*four*' but '*four*teen'; the words 'eleven' and 'twelve' are even more irregular. In the major Asian languages, each number up to 40 has its own name. Those who teach children whose first language is not English therefore need to take steps to find out about the number system which is used in the countries from which the children's families originate so they can be aware of the kind of difficulties which may arise.

223   With older pupils it is necessary to ensure that difficulty in speaking and understanding English does not lead to placement in a mathematics teaching group whose level is too low. The notation which is used in mathematics, and also the way in which numerals and other mathematical symbols are printed, is the same in very many countries, including some in which not only the language but also the written script is different. For this reason these pupils may well be able to work at a higher level in mathematics than is, for the time being, possible for them in some other subjects in which lack of fluency in English leads to greater difficulties.

224   It is possible to make positive use of mathematical ideas drawn from other cultures, especially when discussing shape and space. For example, many of the Rangoli patterns which are used by Hindu and Sikh families to decorate their homes on important occasions have a geometrical basis in which symmetry plays a major part. Practice in drawing patterns of this kind can help to develop geometrical concepts. Again, the intricate patterns which decorate many Islamic buildings are formed by fitting together various geometrical shapes. Patterns of this kind can be examined and discussed and children can then create patterns of their own. As children grow older, it is possible to discuss the ways in which the numerals which we now use have developed from those which were originally used in eastern countries, and the contributions to the development of mathematics which have come from different countries and different cultures.

The teaching and learning of mathematics

225   In the preceding chapters we have shown that, in broad terms, it is possible to sum up much of the mathematical requirement for adult life as 'a feeling for number' and much of the mathematical need for employment as 'a feeling for measurement'. Underlying both of these, and essential to their

development, is the need to establish confidence in the use of mathematics while at school.

226   However, we do not believe that mathematical activity in schools is to be judged worthwhile only in so far as it has clear practical usefulness. The widespread appeal of mathematical puzzles and problems to which we have already referred shows that the capacity for appreciating mathematics for its own sake is present in many people. It follows that mathematics should be presented as a subject both to use and to enjoy.

227   The study of shape and space, of graphical methods of presenting information and of the properties of number such as those of even, odd, prime and square numbers, provides opportunity to develop the powers of 'abstraction' and 'generalisation' and their expression in algebraic form on which higher level mathematics depends. Although many pupils will not attain to work at this level, nevertheless we believe that all should have opportunity to gain some insight, however slight, into the generalised nature of mathematics and the logical processes on which it depends. At all ages, pupils should be encouraged to look for 'pattern' in the results they obtain and to explain this in words even though they may not be able to express in algebraic terms what they have observed. Whatever the level of attainment of pupils, carefully planned use of mathematical puzzles and 'games' can clarify the ideas in a syllabus and assist the development of logical thinking; "if I move this piece to that position, then I shall be able . . .", "because those two numbers are even, this one cannot be odd". The determination to command a computer to do one's bidding can be a further potent force in encouraging pupils to think logically and mathematically. Activities of all these kinds require pupils to think about numbers and the processes of mathematics in ways which are different from those encountered within the more usual applications of mathematics and so enable confidence and understanding to increase.

228   **Mathematics is a difficult subject both to teach and to learn.** One of the reasons why this is so is that mathematics is a hierarchical subject. This does not mean that there is an absolute order in which it is necessary to study the subject but that ability to proceed to new work is very often dependent on a sufficient understanding of one or more pieces of work which have gone before. Whether or not it is true, as is sometimes suggested, that each person has a 'mathematical ceiling' (and so far as we are aware no research has been undertaken to establish whether or not this is the case), it is certainly true that children, and adults, learn mathematics at greatly differing speeds. A concept which some may comprehend in a single lesson may require days or even weeks of work by others, and be inaccessible, at least for the time being, to those who lack understanding of the concepts on which it depends. This means that there are very great differences in attainment between children of the same age. A small number reach a standard which enables them to study mathematics at degree level but many others have time to advance only a very short distance along the mathematical road during their years at school. Because of the hierarchical nature of mathematics these pupils do not reach a

position from which they are able to tackle the more abstract branches of the subject with understanding or hope of success, though some can and do continue their advance after they have left school.

229 Mathematics is also a subject which requires hard work and much practice, whatever one's level of attainment may be. It can be straightforward to understand the solution of a problem which someone else has worked out; it is usually very much more difficult to discover a solution by oneself. Indeed, it is 'getting started' which is often the most difficult part of solving a mathematical problem and it is easy to underestimate the qualities both of determination and of imagination which can be required.

230 One of the reasons why it is difficult to teach mathematics is the fact that attainment and rate of learning vary so greatly from pupil to pupil. If the pace of the teaching is too fast, understanding is not able to develop; on the other hand, if the pace is too slow pupils can become bored and disenchanted. The amount of ground which it is appropriate to cover in any one period of work on the same topic also varies with the attainment of the pupils. Those whose attainment is high are often able to advance a considerable distance at one time but those whose attainment is low need to advance by smaller stages and to return to the topic more frequently. The achievement of a correct balance in these matters requires skilled professional judgement and presents problems to the teacher which should not be underestimated. **Whatever their level of attainment, pupils should not be allowed to experience repeated failure.** If this shows signs of occurring, it is an indication that the advance has continued too far and that a change of topic is needed.

Understanding

*cf R R Skemp. *The psychology of learning mathematics.* Penguin 1971.

231 In recent years there has been considerable discussion of the nature of mathematical understanding. There is general agreement that understanding in mathematics implies an ability to recognise and to make use of a mathematical concept in a variety of settings, including some which are not immediately familiar. A distinction* is sometimes made between 'relational understanding' (in brief, both knowing what to do in particular cases and the relating of these procedures to more general mathematical knowledge) and 'instrumental understanding' (the rote memorising of rules for particular classes of examples without knowing why they work). However, mathematical understanding is not 'all or nothing'. It develops as knowledge of mathematics develops and needs to exist at a level which is sufficient for the work which is being done at the time. Thus, the level of understanding which is required for the study of mathematics in higher education is very different from that required by pupils at school. Indeed it is a common, and sometimes somewhat disconcerting, experience to those embarking on degree courses in mathematics to find that their understanding of topics which they have tackled with apparent success at school is questioned and shown to be insufficient. For this reason the distinction between relational and instrumental understanding can never be clear cut, and in this sense it over-simplifies a

complex situation, but the distinction can be a helpful starting point for discussion of the nature of understanding.

232 Because understanding is an internal state of mind which has to be achieved individually by each pupil, it cannot be observed directly by the teacher. The fact that a pupil is able to solve a particular problem correctly does not necessarily indicate that understanding of the relevant concepts is present. A much better indication of the depth of understanding which exists can be obtained in the course of discussion, by means of appropriate practical work or through more general problem-solving activities. As understanding develops the teacher may need from time to time to challenge more deeply the understanding which already exists so as to make a pupil aware of a need to think more deeply and more critically.

233 Because, too, understanding develops gradually over a period of time as a result of successful experience, especially of a problem-solving kind, it is important that teachers should be aware of the fact that it can be counter-productive to continue work on the same mathematical topic for too long a period. Very often work on some other topic can provide insight from a different direction which will assist in the process of consolidating understanding and enabling ideas to mature.

Memory

234 The *Review of Research* discusses at length the question of memory and distinguishes between 'short term' or 'working' memory and 'long term' memory. Material held in short term memory fades in a matter of seconds unless it is deliberately held in consciousness; the amount of material which can be retained in this way is also limited. Once an item has been stored in long term memory it tends to be forgotten very slowly or not at all, though it may not necessarily be easy to retrieve.

235 Short term memory plays an important role in all tasks in which several attributes or items of information have to be considered simultaneously, for example in mental calculations, problem solving, the understanding of complex concepts and the construction or following of an explanation or argument; in other words, in most learning tasks. In order to carry out these tasks it is necessary to draw on information stored in long term memory. Research evidence makes it clear that information is stored better in long term memory if it is assimilated in such a way that it becomes part of a network of associated and related items which support one another. An everyday example of this is provided by the fact that some children whose ability to remember number facts appears to be weak are often able, because of their interest in and knowledge of sport, to remember without difficulty the scores in football or cricket matches which have been played weeks or even months earlier.

236 It follows that a child's long term memory tends to improve as he grows older and develops a network of items which contains more inter-connections into which new information can be fitted. Information which has been remembered in this way is also easier to retrieve because association between the purpose for which it is required and the information which is stored in the

memory provides a cue which triggers recall. Although it is possible to store in long term memory material which has not been integrated in this way, it is likely to be less well retained and, perhaps more importantly, to be more difficult to retrieve because fewer associations have been formed to act as retrieval cues. Furthermore, if relationship with other material has not been established, it is more difficult to use powers of reasoning to make complete material which has been remembered only in part.

237 For long term memory to be effective, 'rehearsal' is necessary; that is, recall of the material and the strengthening of its relationship with other facts which are already known. The more that such rehearsal can increase links with the existing network, the more effective it is likely to be. Thus retention of number facts is likely to be improved if practice in recall is associated with some kind of explanation or checking procedure; for example, that the sum of two odd numbers must be even or that any number in the '5 times table' ends in either 0 or 5.

Rote learning

238 We have received several submissions which have urged that more emphasis should be placed on 'rote learning'. The Oxford English Dictionary defines 'by rote' as "in a mechanical manner, by routine; especially by the mere exercise of memory without proper understanding of, or reflection upon, the matter in question; also, with precision, or by heart''. **There are certainly some things in mathematics which need to be learned by heart but we do not believe that it should ever be necessary in the teaching of mathematics to commit things to memory without at the same time seeking to develop a proper understanding of the mathematics to which they relate.** As our discussion of memory shows, such an approach is unlikely to meet with long term success.

239 However, the need to teach in a way which will help to develop long term memory and understanding need in no way be in opposition to, or at the expense of, the development of skills in computation and algebraic manipulation. It is important that children should practise routine manipulations until they can be done with an appropriate degree of fluency; this applies all the way from routines such as addition and subtraction to those required for A-level mathematics and beyond. Well-mastered routines are necessary in order to free conscious attention as much as possible so that it can focus on aspects of a task which are novel or problematic. Here again, we need to distinguish between 'fluent' performance and 'mechanical' performance. Fluent performance is based on understanding of the routine which is being carried out; mechanical performance is performance by rote in which the necessary understanding is not present. Although mechanical performance may be successful in the short term, any routine which is carried out in this way is much less likely either to be capable of use in other situations or to be retained in long term memory.

Teaching methods

240 The *Review of research* points out that in the teaching of mathematics it is possible to distinguish between three elements—facts and skills, conceptual structures, and general strategies and appreciation.

Facts are items of information which are essentially unconnected or arbitrary. They include notational conventions—for example that 34 means three tens plus four and not four tens plus three—conversion factors such as that '2.54 centimetres equals 1 inch' and the names allotted to particular concepts, for example trigonometrical ratios. The so-called 'number facts', for example $4 + 6 = 10$, do not fit into this category since they are not unconnected or arbitrary but follow logically from an understanding of the number system. *Skills* include not only the use of the number facts and the standard computational procedures of arithmetic and algebra, but also of any well established procedures which it is possible to carry out by the use of a routine. They need not only to be understood and embedded in the conceptual structure but also to be brought up to the level of immediate recall or fluency of performance by regular practice.

Conceptual structures are richly inter-connected bodies of knowledge, including the routines required for the exercise of skills. It is these which make up the substance of mathematical knowledge stored in the long term memory. They underpin the performance of skills and their presence is shown by the ability to remedy a memory failure or to adapt a procedure to a new situation.

General strategies are procedures which guide the choice of which skills to use or what knowledge to draw upon at each stage in the course of solving a problem or carrying out an investigation. They enable a problem to be approached with confidence and with the expectation that a solution will be possible. With them is associated *appreciation* which involves awareness of the nature of mathematics and attitudes towards it.

241 Research shows that these three elements—facts and skills, conceptual structures, general strategies and appreciation—involve distinct aspects of teaching and require separate attention. It follows that effective mathematics teaching must pay attention to all three.

Classroom practice

242 We wish now to discuss the implications of the previous sections for work in the classroom. **We are aware that there are some teachers who would wish us to indicate a definitive style for the teaching of mathematics, but we do not believe that this is either desirable or possible.** Approaches to the teaching of a particular piece of mathematics need to be related to the topic itself and to the abilities and experience of both teachers and pupils. Because of differences of personality and circumstance, methods which may be extremely successful with one teacher and one group of pupils will not necessarily be suitable for use by another teacher or with a different group of pupils. Nevertheless, we believe that there are certain elements which need to be present in successful mathematics teaching to pupils of all ages.

243 **Mathematics teaching at all levels should include opportunities for**
- exposition by the teacher;
- discussion between teacher and pupils and between pupils themselves;
- appropriate practical work;
- consolidation and practice of fundamental skills and routines;
- problem solving, including the application of mathematics to everyday situations;
- investigational work.

In setting out this list we are aware that we are not saying anything which has not already been said many times and over many years. The list which we have given has appeared, by implication if not explicitly, in official reports, DES publications, HMI discussion papers and the journals and publications of the professional mathematical associations. Yet we are aware that although there are some classrooms in which the teaching includes, as a matter of course, all the elements which we have listed, there are still many in which the mathematics teaching does not include even a majority of these elements.

244 We believe that one of the reasons for this may be that a brief statement such as "mathematics teaching should include opportunities for investigational work" does not explain sufficiently what is intended. We wish, therefore, to consider more fully each of the elements which we have listed.

Exposition
245 Exposition by the teacher has always been a fundamental ingredient of work in the classroom and we believe that this continues to be the case. We wish, though, to stress one aspect of it which seems often to be insufficently appreciated. Questions and answers should constitute a dialogue. There is a need to take account of, and to respond to, the answers which pupils give to questions asked by the teacher as the exposition develops. Even if an answer is incorrect, or is not the one which the teacher was expecting or hoping to receive, it should not be ignored; exploration of a pupil's incorrect or unexpected response can lead to worthwhile discussion and increased awareness for both teacher and pupil of specific misunderstandings or misinterpretations.

Discussion
246 By the term 'discussion' we mean more than the short questions and answers which arise during exposition by the teacher. In the National Primary Survey report* we read "In some cases, particularly in the older classes, more attention could usefully have been given to more precise and unambiguous use of ordinary language to describe the properties of number, size, shape or position". The National Secondary Survey report† noted that "the potential of mathematics for developing precision and sensitivity in the use of language was underused". The ability to 'say what you mean and mean what you say' should be one of the outcomes of good mathematics teaching. This ability develops as a result of opportunities to talk about mathematics, to explain and discuss results which have been obtained, and to test hypotheses. Moreover, the many different topics which exist within mathematics at both primary and secondary level should be presented and developed in such a way that they are seen to be inter-related. Pupils need the explicit help, which can only be given by extended discussion, to establish these relationships; even pupils whose mathematical attainment is high do not easily do this for themselves.

Practical work
247 Practical work is fundamental to the development of mathematics at the primary stage; we discuss this in detail in the following chapter. It is too often

*Primary education in England. A survey by HM Inspectors of Schools. HMSO 1978.

†*Aspects of secondary education in England.* A survey by HM Inspectors of Schools. HMSO 1979.

assumed that the need for practical activity ceases at the secondary stage but this is not the case. Nor is it the case that practical activity is needed only by pupils whose attainment is low; pupils of all levels of attainment can benefit from the opportunity for appropriate practical experience. The type of activity, the amount of time which is spent on it and the amount of repetition which is required will, of course, vary according to the needs and attainment of pupils. The results of the practical testing carried out by the Assessment of Performance Unit and described in the reports of both primary and secondary tests* illustrate clearly the need to provide opportunities for practical experience and experiment for pupils of all ages.

*Assessment of Performance Unit. *Mathematical development: Primary survey report No. 1* and *No. 2*. HMSO 1980 and 1981; and *Secondary survey report No. 1* HMSO 1980.

Practice

248 All pupils need opportunities to practise skills and routines which have been acquired recently, and to consolidate those which they already possess, so that these may be available for use in problem solving and investigational work. The amount of practice which is required varies from pupil to pupil, as does the level of fluency which is appropriate at any given stage. However, as we have pointed out already, practice of fundamental skills is not by itself sufficient to develop the ability to solve problems or to investigate—these are matters which need separate attention.

Problem solving

249 The ability to solve problems is at the heart of mathematics. Mathematics is only 'useful' to the extent to which it can be applied to a particular situation and it is the ability to apply mathematics to a variety of situations to which we give the name 'problem solving'. However, the solution of a mathematical problem cannot begin until the problem has been translated into the appropriate mathematical terms. This first and essential step presents very great difficulties to many pupils—a fact which is often too little appreciated. At each stage of the mathematics course the teacher needs to help pupils to understand how to apply the concepts and skills which are being learned and how to make use of them to solve problems. These problems should relate both to the application of mathematics to everyday situations within the pupils' experience, and also to situations which are unfamiliar. For many pupils this will require a great deal of discussion and oral work before even very simple problems can be tackled in written form.

Investigational work

250 The idea of investigation is fundamental both to the study of mathematics itself and also to an understanding of the ways in which mathematics can be used to extend knowledge and to solve problems in very many fields. We suspect that there are many teachers who think of 'mathematical investigations' as being in some way similar to the 'projects' which in recent years have become common as a way of working in many areas of the curriculum; in other words, that a mathematical investigation is an extensive piece of work which will take quite a long time to complete and will probably be undertaken individually or as a member of a small group. But although this is one of the forms which mathematical investigation can take, it is by no means the only form nor need it be the most common. Investigations need be neither lengthy

nor difficult. At the most fundamental level, and perhaps most frequently, they should start in response to pupils' questions, perhaps during exposition by the teacher or as a result of a piece of work which is in progress or has just been completed. The essential condition for work of this kind is that the teacher must be willing to pursue the matter when a pupil asks "could we have done the same thing with three other numbers?" or "what would happen if ...?" Very often the question can be resolved by a few minutes of discussion either with the pupil or with a group of pupils; sometimes it may be appropriate to suggest that the pupil or a group of pupils, or even the whole class, should try to find the answer for themselves; sometimes it will be necessary to find time on another occasion to discuss the matter. The essential requirement is that pupils should be encouraged to think in this way and that the teacher takes the opportunities which are presented by the members of the class. There should be willingness on the part of the teacher to follow some false trails and not to say at the outset that the trail leads nowhere. Nor should an interesting line of thought be curtailed because "there is no time" or because " it is not in the syllabus".

251 Many investigations lead to a result which will be the same for all pupils. On the other hand, there are many investigations which will produce a variety of results and pupils need to appreciate this. For example, the answer to the question "In how many different ways can you carry out this calculation on your calculator; which way requires the least number of steps?" depends on the particular model of calculator which is used, and pupils who undertake an investigation of this kind will produce a variety of answers, all of which may be equally valid. Mathematical puzzles of various kinds also offer valuable opportunities for investigational work. Even practice in routine skills can sometimes, with benefit, be carried out in investigational form; for example, 'make up three subtraction sums which have 473 as their answer'. The successful completion of a task of this kind may well assist understanding of the fact that subtraction can be checked by means of addition.

252 It is necessary to realise that much of the value of an investigation can be lost unless the outcome of the investigation is discussed. Such discussion should include consideration not only of the method which has been used and the results which have been obtained but also of false trails which have been followed and mistakes which may have been made in the course of the investigation.

Some specific aspects

253 We turn now to certain more detailed aspects of the teaching and learning of mathematics which are relevant to work in both primary and secondary schools and which we consider to be of sufficient importance to consider further at this stage.

Mental calculation
254 **We have already referred several times to the need to be able to carry out straightforward calculations mentally.** 'Mental arithmetic' was once a regular part of the mathematics taught in both primary and secondary schools; very often it occurred as a separate heading in school reports. It is

clear that it now occupies a far less prominent position within most mathematics teaching; reports which have come to us confirm the comments of young employees, to which we referred in paragraph 202, that in some classrooms it is no longer practised at all. We believe that one reason for this change is the increasing use of individual learning programmes in which a pupil works for much of the time on his own using prepared materials, very often in the form of work cards or work sheets. This method of working reduces opportunity for discussion and oral work generally. Again, most primary classes and many secondary classes, especially in the earlier years, contain pupils of a very wide range of ability. It is difficult to find mental questions which are suitable for all the pupils in such classes and so, because of an increasing awareness of the effects which a sense of failure can produce, teachers have tended to avoid methods of working which are liable to draw attention to a child's lack of success. Many of our readers will no doubt recall the 'ten quick questions' at the beginning of a mathematics lesson. For the mathematically able they could be a source of enjoyment and challenge but for those whose mathematical ability was limited they were much more likely to lead to increasing loss of confidence, increasing antipathy to mathematics and sometimes even to feelings of humiliation which would long be remembered.

255 We believe that the decline of mental and oral work within mathematics classrooms represents a failure to recognise the central place which working 'done in the head' occupies throughout mathematics. Even when using traditional methods of recording calculations on paper, the written record is usually based on steps which are done mentally. For instance, the written calculation $27 + 65$, carried out by the method which is traditionally taught, requires the mental calculations $7 + 5 = 12$ and $1 + 2 + 6 = 9$; a relatively simple division sum can involve mental experiment with various multiples of the divisor before the correct one is chosen.

256 However, a more important reason for including the practice of mental calculation is the now well established fact that those who are mathematically effective in daily life seldom make use 'in their heads' of the standard written methods which are taught in the classroom, but either adapt them in a personal way or make use of methods which are highly idiosyncratic. It is, for example, common when carrying out calculations mentally to deal with the hundreds or the tens first and then the units afterwards; for instance, in the example we quoted in the previous paragraph, to use the sequence '$20 + 60 = 80, 5 + 7 = 12, 80 + 12 = 92$'; we would stress however that this is only one of several possible methods which can be used to carry out this calculation mentally. Again, when adding sums of money mentally, it is common to add the pounds first and then the pence, rather than deal with the pence first and then the pounds as is usually done when working with pencil and paper. Although many pupils come to realise by themselves that methods which may be convenient on paper are often not well suited to use 'in their heads', we believe that in the case of many other pupils it is necessary for the teacher to point this out explicitly and to discuss at length the variety of methods which it is possible to use. However, no attempt should be made to force a single

'proper method' of performing mental calculations; pupils should be encouraged to make use of whatever method suits them best. Teachers should also encourage pupils to reflect upon the methods which they develop for themselves so that facility in mental computation can be consolidated and extended.

Estimation

257 LEA guidelines and schemes of work in both primary and secondary schools almost always include a reference to the need for pupils to be able to estimate and sometimes state more specifically that pupils should be encouraged or required to write down an approximate answer before carrying out a calculation. **The earlier chapters of this report make it clear that ability to estimate is important not only in many kinds of employment but in the ordinary activities of adult life.**

258 However, from our own observations and from what we have been told by others, we believe that, even though it may be advocated in guidelines and schemes of work, estimation is not practised in very many classrooms. In our view this is not because of unwillingness on the part of teachers to encourage and develop this skill in their pupils but because of a failure to realise how much is implied by the words 'ability to estimate' and how long it takes to develop this ability.

259 There are several aspects of estimation. One is that of obtaining, before a calculation is carried out, a 'rough answer'; in other words, an answer in rounded terms which will enable a check to be made that the result of the calculation is of the correct 'order of magnitude'. Estimation of this kind is probably most commonly applied to the operations of multiplication and division; it concentrates on ensuring that the result of the calculation is not, for example, ten or a hundred times too large or too small. The method usually involves working 'to one significant figure', for example replacing 26×52 by 30×50 in order to obtain an approximate answer. However, for very many children this replacement is conceptually far more difficult than might be expected and the ability to work in this way takes a long time to develop. A much more elementary application is that the sum of two numbers, each less than 50, must be less than 100. We may, perhaps, describe this as 'awareness' rather than estimation but it is awareness of this kind which is needed if a 'feel' for number and the ability to estimate are to develop. In our view they are most likely to develop as a result of a great deal of discussion and 'thinking aloud' by both teacher and pupils, coupled with appropriate practical work on measurement.

260 A second aspect of estimation may be described as 'realising whether the answer is reasonable'. Many pupils find it difficult to develop this ability. In some instances it is concerned, as is the previous aspect, merely with computation but, as is frequently the case when using a calculator, after the calculation rather than before. Again, most mathematics teachers will recall instances in which, for example, a pupil has not only given as the area of a table top a measurement which more nearly corresponds to the area either of a

postcard or of a football pitch, but has also failed to realise the incongruity. This type of mistake shows a lack of appreciation of the size of units and of their relation to everyday objects. There is need for pupils to establish everyday equivalences—for instance, to know that in most living rooms the door is about 2 metres high—and for the teacher to discuss and ask questions; "would it go through the doorway?", "could you lift it?". It is to be hoped that, at a later stage, this kind of estimation will become part of 'common sense', but we believe that its development needs conscious attention.

261. A related aspect, that of estimating measurements of various kinds, is most evidently the aspect in which success is achieved by continuing usage and practical experience. Very specific skills of estimation, such as those of the decorator who is able to gauge by eye the quantity of paint or wallpaper he will require for a given task, have in the end to develop on the job. Nevertheless, throughout their time at school pupils should be encouraged to practise the estimation of lengths, areas, capacities and weights.

262 As we pointed out at the beginning of this section, work of the kind we have described in this section does not at present figure prominently in most school mathematics courses. There is therefore little experience available on which to draw or evidence of approaches which are known to have been successful. **We believe that this is an area in which further study is required.**

Approaches to calculation

263 We have already drawn attention to the fact that in adult life written calculations are often carried out in ways which are different from those traditionally taught in the classroom. Although there are occasions on which it can be both quick and convenient to carry out written calculation in the traditional way, 'back of an envelope' methods are often not only quicker but also more straightforward. This raises important questions concerning the approach to computation which should be adopted in the classroom and the way in which calculations should be recorded on paper.

264 At the primary stage, computation is commonly introduced through the use of counting or structural materials and methods of recording develop in the first instance from the use of such apparatus; we discuss this further in the following chapter. By the age of 11 many children are able to carry out with confidence the operations of addition, subtraction, multiplication and division on whole numbers and to record their working in the traditional way. However, many other children do not reach this stage by the age of 11, especially in respect of multiplication and division.

265 It is common in secondary schools to insist that all pupils make use of the traditional methods of recording; these are often taught as standard routines to be followed. Nevertheless, even those pupils for whom this does not present undue difficulty need to be encouraged to develop alternative methods as well. These will frequently be similar to methods which are used to carry out mental calculations (see paragraph 256) or will make use of 'short cuts' of

various kinds. Discussion of such alternative methods with a group or a class provides valuable opportunity for developing confidence and 'feel' when doing number work.

266 In the same way, pupils should be encouraged to approach other calculations, such as those involving percentages, 'on their merits' and not be expected always to use a single standard routine. For example, in order to calculate value added tax (VAT) at 15 per cent on a given sum of money, many people are likely to find it easier to write down 10 per cent of the sum, halve it and add the two results together than to multiply by $\frac{15}{100}$. This method takes advantage of the particular relationship between 15 and 10 and also demonstrates understanding of the operation which is being carried out; if the rate of VAT were 8 per cent, most would probably find it easier to carry out the calculation in a single step by multiplying by $\frac{8}{100}$.

267 Some low-attaining pupils have great difficulty in carrying out and recording computation using standard routines. However, after discussion and practical work with appropriate counting materials, they are often able to carry out calculations successfully by making use of methods of their own devising. They should be encouraged to do this and should not be dissuaded from using their own 'best method' provided that they can do this reliably. It can be counter-productive and inappropriate to drill these pupils in standard routines from the time they enter secondary school.

268 The availability of the electronic calculator is another factor which needs to be taken into account when considering approaches to calculation, especially in the case of those who have difficulty in using the standard routines. We discuss this further in Chapter 7.

Measurement

269 Measurement is fundamental to the teaching and learning of mathematics because it provides a natural 'way in' both to the development of number concepts and also to the application of mathematics over a very wide field. Practice in ordering lengths, capacities and weights enables a young child to develop understanding of concepts such as 'more than', 'less than', 'longer than', 'shorter than'. After this the child learns first of all to use non-standard units such as handspans and cupfuls and then standard units to measure continuous quantities such as length and capacity.

270 All such forms of measurement are inexact and exist only within limits which can be specified. These limits may be chosen in accordance with the use which is to be made of the measurement or may be imposed by the limitations of the measuring instrument which is being used. *Mathematics: 5−11*[*]states:

*HMI Series: Matters for discussion 9. *Mathematics 5−11*. A handbook of suggestions. HMSO 1979.

> Children should be led to master the concept of 'betweenness'. If the child's answer to the question 'What is the time?' is 'Between five and ten past two', then the answer is absolutely true.

We wish to stress that in our view this concept is fundamental and children should from the earliest stages be encouraged, where appropriate, to record measurement in this way. "The jug holds more than 5 cupfuls but less than

Approximity of measure

6''; "this pencil is more than 12 cm long but less than 13 cm long". This approach to measurement leads to the necessity for the sub-division of units. When this has been understood it is possible to go on to discuss degrees of accuracy and the idea of measurement within a given tolerance.

271 Measurement of area, capacity and angle involve geometrical concepts of shape and space; the recording of measurement gives opportunity for the introduction of a wide variety of graphical methods. At later stages study of measurement can be used to introduce concepts such as those of scale, rate, and ratio, and the use of compound measures of the kind used for speed, density and pressure.

272 At all stages the teaching of measurement should be based on extensive practical work, including the use of a variety of measuring instruments, the use of compasses and other drawing instruments, and the construction of geometrical shapes and models. All children at all levels need experience of this kind. Some children will acquire facility in measurement and in the use of drawing instruments without difficulty but others will require a long time and much practice to achieve a reasonable competence. This time needs to be provided over a period of years.

Metrication

273 The change from imperial to metric units of measurement has not proceeded as quickly as had at one time been expected and, although considerable progress has been made, it is clear that both metric and imperial units will continue in use in England and Wales for some years to come. We have discussed some of the reasons for this in paragraph 82.

274 The continuing existence of the two systems of units has led to some confusion in schools in recent years. The guidance issued in 1974 by the DES and Welsh Office* drew attention to the "growing familiarity within the schools, particularly the primary schools, with the use of metric quantities". It acknowledged that "the greatest difficulty for schools is not metrication itself, but the fact that, for some purposes, the imperial system appears likely to remain in everyday use for some time to come". It suggested that schools should teach children to carry out calculations in metric units but should also enable them to maintain a general familiarity with imperial units. It also encouraged a policy of "thinking metric".

*Department of Education and Science *Administrative Memorandum 9/74*. Welsh Office *Administrative Memorandum 4/74*.

275 We believe that this advice continues to be sound and that the metric measures of length, weight and capacity should be used in both primary and secondary schools. Pupils should also learn to estimate in metric units. However, it is clear that it remains necessary for school leavers to have some knowledge of imperial units. We therefore consider that pupils in secondary schools should become familiar with the more common imperial units such as feet, inches, pounds, ounces, pints and gallons and should be able to use them for purposes of direct measurement. This means that they should be able to

measure length in feet, inches and fractions of an inch ($\frac{1}{2},\frac{1}{4},\frac{1}{8},\frac{1}{16}$) to measure weight in pounds and ounces, and to measure capacity in gallons and pints. Pupils should become aware of the approximate equivalences between these imperial units and the appropriate metric units, but should not normally be required to calculate in imperial units.

'The basics'

276 The word 'basic' appears in very many of the submissions which we have received. Various expressions are used: 'basic skills', 'basic computational skills', 'basic mathematics', 'basic numeracy', 'the basics'. The contexts in which these expressions are used suggest that, while not exactly synonymous, they are effectively different ways of describing what is thought to be much the same thing. Where a definition is given, it is almost always in terms of purely arithmetical skills, with stress on the operations of addition, subtraction, multiplication and division treated in isolation from application to real situations. Many submissions assume that the meaning of whichever expression is used is self-evident and that there is no need to go into details.

277 Knowledge or skills which are 'basic' are presumably needed as a *basis* either for the mathematics required in employment or in adult life or for further study. We have considered these requirements in the preceding chapters. Although many of the requirements may be considered to be 'elementary' in terms of their position within the hierarchy of learning mathematics and the stage of schooling at which they are first introduced, it does not follow that they are necessarily either simple or straightforward for most pupils to learn and, more importantly, to apply.

278 There is evidence that the public focussing of attention on standards in schools which has occurred in recent years has created pressure in some quarters for a 'back to basics' movement. This has encouraged some primary teachers and some teachers of low-attaining pupils in secondary schools to restrict their teaching largely to the attainment of computational skills. Some of the submissions which we have received advocate a 'back to basics' approach of this kind. However, we hope that the argument of this chapter makes it clear that the ability to carry out a particular numerical operation and the ability to know when to make use of it are not the same; both are needed. The mathematics of employment and of everyday life is always mathematics in context and is based largely on measurements of many kinds made in many different situations. Arithmetical skills are therefore a tool for use in situations which require an understanding of other areas of mathematics, for example the geometry of shape and space and graphical representation of various kinds. **An excessive concentration on the purely mechanical skills of arithmetic for their own sake will not assist the development of understanding in these other areas. It follows that the results of a 'back to basics' approach (as we understand the words) are most unlikely to be those which its proponents wish to see, and we can in no way support or recommend an approach of this kind.**

Modern mathematics

279 In Britain the beginning of 'modern mathematics' is usually associated with three conferences held in Oxford in 1957, in Liverpool in 1959 and in Southampton in 1961. These were among a number of conferences, held in Europe and the United States of America towards the end of the 1950s, which discussed the teaching of mathematics in schools and resulted in the setting up of a variety of mathematics curriculum development projects. It is interesting to note that the conferences held in Britain were financed by industry and paid considerable attention to modern industrial applications of mathematics. A direct outcome of the conference held in Southampton was the setting up of the School Mathematics Project (SMP) whose syllabuses and teaching materials are prepared by groups of practising teachers and which was again funded by industry in its early stages.

280 The Director of SMP wrote in his report for 1962-63 that "a major aim of the syllabus is to make school mathematics more exciting and more enjoyable, and to impart a knowledge of the nature of mathematics and its uses in the modern world. In this way, it is hoped to encourage more pupils to pursue further the study of mathematics, to bridge the gulf which at the moment separates university from school mathematics—both in content and in outlook—and also to reflect the change brought about in the world by increased automation and the introduction of electronic computers". However, this statement related to an O-level course designed for pupils whose mathematical attainment was in the top quarter of their age group and was written at a time at which the provision of 'modern' courses for pupils of lower attainment had not been planned or expected.

281 The intention of those who set out to develop SMP and other modern courses was to introduce changes both in the content of the mathematics syllabus for these higher attaining pupils and also in the teaching methods and approaches which were used. Teaching approaches were designed to encourage investigation, to emphasise the applications of mathematics, and to draw attention to the unified nature of mathematics rather than to its traditional division at school level into arithmetic, algebra and geometry; and the classroom materials which were produced took for granted that the teachers who used them would possess sufficient mathematical insight and experience to enable them to work in these ways. At first most teachers who wished to introduce a modern syllabus were able to attend in-service training courses at which the aims of modern mathematics courses were explained and teaching approaches discussed. However, the unexpectedly rapid expansion of modern mathematics courses meant that it was not long before many teachers were required to teach these courses without the benefit of introductory training. A further, and also very rapid, development was the extension of modern mathematics courses to pupils whose attainment was lower and the introduction of modern syllabuses in CSE mathematics examinations. Not all teachers possessed a sufficient mathematical background to enable them to appreciate the intentions underlying the new courses they were teaching. In consequence the material which was included in modern courses was often not presented as part of a unified structure but as a collection of disconnected

topics whose relevance to the mathematics course as a whole did not become apparent to pupils.

282 However, in our view the introduction of certain topics which had not previously been included in most mathematics courses has had a beneficial effect. We may cite as examples the increased emphasis on graphical work, the introduction of work based on the geometrical ideas of symmetry, reflection and rotation, the use of co-ordinates and the study of elementary statistics. Work on these topics has lent itself to more practical approaches to the teaching of mathematics and has proved to be within the capability of very many pupils. The same has not, however, been true of certain algebraic topics, which have proved difficult for many pupils to understand and whose purpose and use have not been evident to them. It is these topics, notably the algebra of sets and matrices, which have attracted considerable public attention and criticism and which have come, in the eyes of many people, to exemplify modern mathematics even though they form only a relatively small part of many courses.

283 During the last few years, a number of 'modern' O-level and CSE syllabuses have been modified so as to exclude some of the more abstract algebraic topics. At the same time many 'traditional' courses have also been modified to include such topics as elementary statistics and a greater emphasis on graphical work; as a result the differences between 'modern' and 'traditional' mathematics have become much less marked. **In our own discussions we have not thought in terms of traditional or modern mathematics nor has the evidence which we have received suggested that it is any longer profitable to do so.** Our discussion of mathematics teaching in the primary and secondary years which follows makes no distinction between the two because, in our view, it is no longer appropriate to make such a distinction; we believe that very many people now share this view.

6 Mathematics in the primary years

284 In this chapter we discuss the mathematical work of children between the ages of 5 and 11, whether they are in infant, junior, infant and junior, first or middle schools. We do not discuss the mathematical development of children of pre-school age.

285 We have received much evidence which is supportive of the work of primary schools and believe that the great majority of primary teachers are aware of their responsibility to provide a sound mathematical foundation for the children in their care. Some teachers have told us that they would welcome guidance in this task and we hope that the matters we discuss in this chapter will provide a basis for study and discussion in staff rooms and elsewhere. We have also received some comments which are critical of mathematics teaching in primary schools. Some of these refer to matters which we discuss in this chapter; others reflect a failure on the part of some of those who have written to us to appreciate the detailed and careful approach to the teaching of mathematics which is necessary, especially in the early years, in order that children may develop confidence and understanding.

The primary mathematics curriculum

*Board of Education. *Handbook of suggestions for teachers.* HMSO 1937

286 The use of practical methods in the primary classroom is sometimes thought to be of relatively recent introduction but this is not the case; work of this kind has been advocated for very many years. For example, in the *Handbook of suggestions for teachers*,* published in 1937 by the Board of Education, we read:

> First, by way of introduction, should come practical and oral work designed to give meaning to, and create interest in, the new arithmetical conception—through deriving it from the child's own experience—and to give him confidence in dealing with it by first establishing in his mind correct notions of the numerical and quantitative relations involved in the operation.

†*Mathematics in primary schools.* Schools Council Curriculum Bulletin No 1. HMSO 1965.

In the 1960s the work of the Nuffield Mathematics Project and the publication of Schools Council Curriculum Bulletin No 1: *Mathematics in primary schools*† gave added impetus to the use of approaches to mathematics which were based on practical experience. As a result, there has been a general widening of the mathematics curriculum in most primary schools during the last twenty years to include both a greater understanding of number and also work on measurement, shape and space, graphical representation and the development of simple logical ideas. **We believe that this broadening of the curriculum has had a beneficial effect both in improving children's attitudes to mathematics and also in laying the foundations of better understanding.**

Before we discuss these areas of work in more detail, we wish to emphasise, as we shall continue to emphasise, that work of this kind needs to be carefully structured and followed up by the teacher. Much of its value will be lost unless the work which has been done, and the results which have been obtained, are discussed with the children so as to establish the necessary concepts and make links with other pieces of work which have been undertaken.

287 **The primary mathematics curriculum should enrich children's aesthetic and linguistic experience, provide them with the means of exploring their environment and develop their powers of logical thought, in addition to equipping them with the numerical skills which will be a powerful tool for later work and study.** The practical and intuitive experience which should be the result of a course of this kind provides an invaluable base for further work in the secondary years. However, we do not believe that mathematics in the primary years should be seen solely as a preparation for the next stage of education. The primary years ought also to be seen as worthwhile in themselves—a time during which doors are opened onto a wide range of experience.

288 We believe that the public criticism of recent years, to which we referred in paragraph 278, has caused some teachers in primary schools to wonder whether they have been right to adopt the broader approach which we have described. We hope that our discussion in Chapter 5 of the fundamentals of mathematics teaching and learning will have convinced them, and also the critics of primary mathematics, that emphasis on arithmetical skills does not of itself lead to ability to make use of these skills in practical situations. It is only within a broadly based curriculum that the ability to apply mathematics is enabled to develop.

Practical work

289 **Practical work is essential throughout the primary years if the mathematics curriculum is to be developed in the way which we have advocated in paragraph 287.** It is, though, necessary to realise at the outset that such work requires a considerable amount of time. However, provided that the practical work is properly structured with a wide variety of experience and clear stages of progression, and is followed up by the teacher by means of questions and discussion, this time is well spent. For most children practical work provides the most effective means by which understanding of mathematics can develop. It enables them to think out the mathematical ideas which are contained within the various activities they undertake at the same time as they are carrying out these activities; and so to progress within each topic from the handling of actual objects to a stage in which pictures or diagrams can be used to represent these objects and then to a final stage at which symbols are used which can be manipulated in abstract ways. (We give an example of this in paragraph 305.)

290 Children vary greatly in the amount of time which they take to move through these stages. It is as harmful to insist that one child should continue to use practical materials for a process which he understands and can carry out by using symbols as to insist that another should proceed to diagrammatic or

symbolic representation before he is able to carry out the process by using practical materials. It is therefore a mistake to suppose that there is any particular age at which children no longer need to use practical materials or that such materials are needed only by those whose attainment is low. It is not 'babyish' to work with practical materials while the need exists and we believe that many children would derive benefit from a much greater use of these materials in the later primary years than occurs in many classrooms.

Measurement

291 In paragraph 269 we emphasised the importance of practising measurement of all kinds and explained the approach to measurement which we believe to be necessary. The measurement of length, capacity, weight, area and time should be part of every child's experience; some children will extend their work to the measurement of angle and of speed, volume and density. Practice in measurement needs to be associated with practice in estimation so that children gain an appreciation of the size of units and of their relation to everyday objects. Measurement should be linked with number work and attention drawn, for example, to the fact that the mathematical structure of metres, decimetres and centimetres is identical to that of hundreds, tens and units. This enables a ruler or tape measure to be used as a portable 'number-line'; this can be helpful where numbers have to be added or subtracted.

Shape and space

292 All children should have experience of work with a variety of plane shapes and solids. Here again, progression is essential. After the early stages of drawing round, cutting out, folding and colouring a variety of shapes, children should form them into patterns and then experiment to discover which shapes will 'tessellate' (that is, fit together without leaving any spaces between adjoining shapes) and which will not. They can also explore the ideas of symmetry, rotation and reflection. Geometrical knowledge grows through investigations of this kind and the properties of geometrical shapes become apparent as different patterns are constructed. From time to time the teacher should draw on the knowledge and experience which the children have gained in order to discuss these properties explicitly. For some children the idea of proof can start to develop as they seek to discover, for example, why triangles and quadrilaterals of all kinds will tessellate but many other shapes will not. The construction of plane shapes and solids helps to develop skill and accuracy in the use of instruments for measuring and drawing and also the ability to visualise three-dimensional figures. Almost all children find pleasure in working with shapes, and work of this kind can encourage the development of positive attitudes towards mathematics in those who are finding difficulty with number work. The rich variety of practical work which is possible in the primary years provides a foundation on which the more formal geometry of the secondary years can be based.

Graphical work

293 Throughout the primary years attention should be paid to methods of presenting mathematical information in pictorial and graphical form, and also to interpreting information which is presented in this way. It can often be

the case that graphical work lacks variety and progression, so that older children are limited to drawing graphs which differ little from those which are to be found in infant classrooms. Children need experience of a wide variety of graphical work; the mere drawing of graphs should not be over-emphasised. It is essential to discuss and interpret the information which is displayed both in graphs which children have themselves drawn and also in graphs which they have not. Children should be encouraged to collect examples of graphs and charts from newspapers, magazines and books, to discuss in detail what they depict and to make deductions from them. Work of this kind often enables children to link their work in mathematics with their work in other areas of the curriculum.

294 As well as drawing graphs which provide factual information, some children should be able to construct graphs which display mathematical relationships such as those of the multiplication tables or of the growth of squares and cubes. Games such as 'Battleships' can be used to introduce the idea of co-ordinates to identify spaces and, later, single points; it then becomes possible to record graphically relationships such as 'pairs of numbers which add up to 10'. Many of these activities provide a basis from which later work in algebra can develop. When discussing graphs which have been constructed, attention can be drawn to the relationships which they display; for instance, "every time we added another 20 grams the length of the spring increased by 3 centimetres".

Logic
295 "To speak of logic in connection with young children may surprise some people, but no highly theoretical notions are involved. It is rather a matter of describing things accurately, noticing their resemblances and their differences, and saying how they are related to one another. In games and puzzles moves often have to be made according to rules, and finding the best moves involves logical thought".* In its most straightforward forms, the activity of sorting objects and of recording the results in diagrammatic form is practised in most infant classes and forms the basis on which the concept of number is built. As children become older, it can develop into the more sophisticated activity of sorting shapes which vary in colour, size and thickness according to their attributes such as, for example, 'large and blue', 'thin and square'. A wide variety of work with shapes of this kind can be undertaken in order to encourage precision of language and the development of logical thinking. Games such as noughts-and-crosses, dominoes, draughts etc can also encourage logical thinking; "if I go there ..., then he will ..., then I shall have to ...".

General activities
296 In addition to practical activities related to specific areas of the mathematics curriculum, of the kind which we have discussed in the preceding paragraphs, all children need experience of practical work which is directly related to the activities of everyday life, including shopping, travel, model making and the planning of school activities. Children cannot be expected to be able to make use of their mathematics in everyday situations unless they

*HMI Series: Matters for discussion 9. *Mathematics 5-11*. A handbook of suggestions. HMSO 1979.

have opportunity to experience these situations for themselves. For most children a very great deal of practical exploration and experience is needed before the underlying mathematical ideas become assimilated into their thinking. We emphasise again that discussion both with the teacher and with other pupils is a necessary part of this process. A few children pass through the various stages of mathematical development in rapid succession and need to advance to more challenging and more abstract work before they leave primary school. For the majority, however, the transition from the use of concrete materials to abstract thinking takes place slowly and gradually; and even those children to whom abstract thinking appears to come easily often need to undertake practical exploration at the beginning of a new topic.

297 There is another aspect of practical work which we have not yet discussed. This is the use of number apparatus as an aid to the understanding of the number system and of methods of computation. We deal with this in the following paragraphs.

Number and computation

298 The skills of mental and written computation are founded on basic concepts which need to be developed through measurement, shopping, the use of structural apparatus and many other activities. These concepts include the meaning of the operations of addition, subtraction, multiplication and division and the very important concept of place value (that is, for example, that the 2 stands for 2 units in the number 52, for 2 tens in the number 127 and for 2 hundreds in the number 263). Understanding of place value enables number facts stored in long term memory to be used as a means of carrying out calculations involving larger numbers; for example, the knowledge that $14 - 8 = 6$ can be used to work out $140 - 80$ or $54 - 8$. It is therefore essential that children should be helped to attain a secure and rapid recall of addition facts up to $10 + 10$ and the related subtraction facts, and of multiplication facts up to 10×10 and the related division facts. This knowledge, together with understanding of place value, provides a basis for calculation involving small or large numbers. The learning of the number facts to which we have referred needs to be based on understanding, but understanding does not necessarily result in remembering. A time comes, therefore, when most children need to make a conscious effort to commit these number facts to memory. We have, though, to be aware that there are some children who have not attained secure and rapid recall of addition and multiplication facts by the age of 11.

299 Understanding of place value needs to be developed not only by means of structural apparatus and the abacus but also by using as examples the structure of hundreds, tens and units which underlies both measurement (metres, decimetres and centimetres) and money (pounds, tenpences and pence). It should not be assumed that a child who understands the structure of hundreds, tens and units will necessarily be able with ease to make the generalisation to thousands and higher powers of 10. Many children need further practical experience with structural apparatus so that they can work out for themselves the meaning of large numbers and be able to carry out operations with them. Other steps in understanding of place value are the understanding

of tenths and hundredths, and of multiplication and division of both whole numbers and decimals by 10 or 100. These lead to more advanced computational skills such as long multiplication. They also provide a basis for developing the ability to approximate and to estimate the size of the answer which is to be expected as the result of carrying out a given calculation.

300 The skills of computation involving fractions are based on an understanding of the concept of equivalent fractions; that is, for example, that $\frac{5}{10}$ and $\frac{1}{2}$ have the same value. For many children understanding of this concept and of the notation for fractions are only just beginning to develop during the primary years. Furthermore, it is difficult to find everyday situations which require fractions to be added or multiplied and there seems to be little justification for teaching the routines for adding, subtracting, multiplying or dividing fractions to the majority of children during the primary years. Children should, however, become familiar in practical situations with terms such as 'one half of', 'one quarter of' and it is entirely appropriate, for example, to work out the number of children in 'half the class'. However, some children whose attainment is high are fully capable of understanding the equivalence of fractions and of applying this understanding to the solution of problems.

301 There is one aspect of computation which needs specific attention with most children towards the end of the primary years. This is the computation of time. Unlike other measures with which children become familiar, the relation between hours and minutes is based on 60 and not on 10 or 100. This means that children have to remember that routines which they normally use for addition and subtraction need to be modified if they are, for example, to be able to calculate the time taken for a journey which starts at 10.45am and ends at 1.30pm. They have also to be able to understand times expressed in terms of the 24-hour clock. It is not sufficient for computation of time to be practised in the abstract; it should be related to practical situations involving the planning of journeys and the use of timetables.

*Primary education in England. A survey by HM Inspectors of Schools. HMSO 1978.

302 Some of the submissions which we have received have suggested that many primary schools do not pay sufficient attention to developing computational skill in their pupils. We do not believe this to be the case, nor is the view supported by the report of the National Primary Survey* which says that "suitable calculations involving the four rules with whole numbers were practised in all classes at all ages". It also states that "in about a third of the classes, at all ages, children were spending too much time undertaking repetitive practice of processes which they had already mastered ... the efforts made to teach children to calculate are not rewarded by high scores in ... examples concerned with the handling of everyday situations. Learning to operate with numbers may need to be more closely linked with learning to use them in a variety of situations than is now common".

303 Work on calculation needs also to take account of the increasing availability of electronic calculators in many homes. We discuss this in the next chapter.

The beginnings of calculation

304 Young children should not be expected to move too quickly to written recording in mathematics. The forming of figures correctly on paper is a skill which needs to be learned. Until it has been mastered, attempts to carry out written calculation can inhibit the development of mathematical knowledge and understanding. In the early stages, therefore, mental and oral work should play a major part in learning. **It has been pointed out to us that, albeit with the best of intentions, some parents can exert undesirable pressure on teachers to introduce written recording of mathematics, and especially 'sums', at too early a stage, because they believe that the written record is a necessary sign of a child's progress.**

305 We believe that there are many who do not appreciate the number of stages through which a child must pass before even such an apparently simple 'sum' as 3 + 2 can be carried out with understanding. First of all the child must be able to recognise and form groups of three objects and two objects and to combine these to form a group of five. At the same time he will talk about what he is doing: "There are three toy cars—fetch two more—now there are five". When he carries out similar actions in different situations he needs to realise that essentially the same mathematics is contained in "There were three children in the room; two more came in; now there are five." and in "I had three crayons; Mary gave me two; now I have five". The next stage is to illustrate these in pictorial form; for instance, the crayons may be drawn as:

$$000 \quad 00 \quad 5$$

It is only when these experiences have all been assimilated and it has been realised that they all lead to the mathematical symbolisations

$$3 + 2 = 5 \quad \text{or} \quad \begin{array}{r} 3 \\ +2 \\ \hline 5 \end{array}$$

that this sum, and similar ones, can be properly understood in the abstract. Only then is it appropriate for the sum to be carried out without concrete materials. Children should then be encouraged to make up their own stories for some of the sums they are asked to do. Some children are able to pass through these different stages quite quickly but for others it can take a long time. A premature start on formal written arithmetic is likely to delay progress rather than hasten it.

Language

306 **Language plays an essential part in the formulation and expression of mathematical ideas.** In paragraph 246 we drew attention to the need to extend and refine the use of mathematical language in the classroom. Development

of this kind can only take place by means of continuing practice; from their earliest days at school children should be encouraged to discuss and explain the mathematics which they are doing. In the words of a submission we have received from the head of an infant school, "there is a need for more talking time . . . . . ideas and findings are passed on through language and developed through discussion, for it is this discussion after the activity that finally sees the point home".

307 Children vary greatly in the level of the language skills which they possess at the age of 5. Some are already familiar with words and expressions such as 'heavy', 'light', 'larger than', 'shortest' and with the concepts to which these relate, but many are not. All children need, as a first stage in their learning of mathematics, to develop their understanding of words and expressions of this kind by means of activities and discussion in the classroom, and this development of mathematical language should continue throughout the primary years.

308 It is important to be aware of the great variety of language which is used in connection with many of the mathematical operations which children will meet. For example, the instructions 'add 5 and 3', 'add 3 to 5', 'find the sum of 5 and 3', 'find the number which is 3 more than 5' all require the same mathematical operation to be carried out. These are only four of many ways in which it is possible to frame this instruction; the particular form of words which is used is most often the one which arises naturally in the context of the moment. Children need to be able to interpret these apparently different instructions, to use them in their own speech and thought, and eventually to be able to attach them all to the symbolic form 5 + 3. Unless children become familiar with the many different ways in which the same mathematical idea can be expressed and are able to recognise the same idea within different forms of words, they will not only have difficulty in dealing with computational examples of the kind which we have quoted but also in dealing with problems expressed in words.

309 Children whose grasp of language is not secure often try to overcome their difficulty by looking out for words such as 'more' or 'less' and using them as 'verbal cues' which they believe will indicate the operation they are required to carry out. However, this does not resolve the difficulty. For example, the two problems

> Janet has 5p and John has 3p *more* than Janet; how much money has John?

and

> Janet has 5p and John has 3p; how much *more* money has Janet than John?

each contain the word 'more' but the first requires addition and the second subtraction. In the problems we have quoted, the language has to provide a bridge between the real situation of comparing pocket money and the arithmetical operations which it is necessary to carry out in order to arrive at the answer. The somewhat stylised language which is often used in 'word problems' can make it difficult for children whose language and reading skills are weak to evoke the necessary mental image of the real situation and so choose the correct arithmetical operation. It is for this reason that they resort to 'verbal cues' and teachers need to be aware that this can happen.

310 Children need also to learn that certain words are used in mathematics in ways which are not the same as those in which they are used in everyday speech. We have been told of a visitor to a junior classroom who, in response to the question "what is the difference between 10 and 7?" was surprised to receive the answer "10 is even and 7 is odd" instead of the answer "3" which had been expected. 'Difference' is only one of many words with whose mathematical meaning children have to become familiar.

311 Reading skills in mathematics should be built up alongside other reading skills so that children can understand the explanations and instructions which occur in the mathematics books which they use. If the skills of reading mathematics are not developed, many children will evolve their own strategies for avoiding such reading. We have already referred to reliance on verbal cues. Another strategy is to avoid reading any explanatory passages which come at the beginning of a work card or textbook exercise and to start on the questions in the hope that they can be done without first studying the explanation or instruction which precedes them. Yet another is to seek help from a friend or the teacher. The policy of trying to avoid reading difficulties by preparing work cards in which the use of language is minimised or avoided altogether should not be adopted. Instead the necessary language skills should be developed through discussion and explanation and by encouraging children to talk and write about the investigations which they have undertaken. Children should also be encouraged to suggest their own problems and to express them in written form.

The use of books

312 Even although a child may without difficulty be able to read what is written in a mathematics textbook or on a work card, he may well find great difficulty in learning an unfamiliar piece of mathematics from the written word. This is likely to be the case however careful has been the choice of the language which is used. The ability to learn mathematics from the printed page is one which develops very slowly, so that even at the age of 16 there are few pupils who are able to learn satisfactorily from a textbook by themselves. At the primary stage new topics and concepts should always be introduced by appropriate oral and practical work and the necessary links with what has gone before established by discussion.

313 Nevertheless, textbooks provide valuable support for teachers in the day-by-day work of the classroom. They can provide a structure within which work in mathematics can develop and provide ideas for alternative approaches. They can be a source of exercises which have been carefully graded and are likely to provide revision exercises at suitable intervals. Accompanying teachers' manuals may suggest other kinds of work which should be undertaken alongside the exercises in the textbook and indicate ways in which the topic can be developed further for some pupils. However, it is always necessary to use any textbook with discrimination, and selections should be

made to suit the varying needs of different children. It may be better, too, to tackle some parts of the work in an order which is different from that in the book or to omit certain sections for some or all children. It should not be expected that any textbook, however good, can provide a complete course or meet the needs of all children; additional activities of various kinds need to be provided.

314 By the middle junior years some children are skilled readers and have become accustomed to acquiring information from books. Although the printed word is seldom a satisfactory means of introducing new mathematical concepts, the same limitation does not apply to the use of mathematical problems, puzzle and topic books, and books of this kind should be available in the classroom or school library. Their use can enable children to realise that mathematics is a living subject which is full of interest and of use outside the classroom, and can also contribute to the children's overall mathematical development. **Although some books of this kind are available, more are needed**; suitable topics would be the mathematics used in everyday life, the exploration of shape, communication by means of graphs and diagrams, the history and development of counting, calculation and measurement, and links between mathematics and science or art. More books of puzzles, problems and suggestions for investigations are also required.

Mental mathematics

315 We refer in this section to 'mental mathematics' rather than 'mental calculation' because we wish to include within our discussion both mental calculation and also the oral work which should play an important part in the teaching of primary mathematics. For the reasons to which we drew attention in paragraph 254 there has been a decrease in the use of mental mathematics in schools of all kinds in recent years and we believe that this trend should be reversed.

316 We have already explained that young children should not be allowed to move too quickly to written work in mathematics. It follows that, in the early stages, mental and oral work should form a major part of the mathematics which is done. As a child grows older, he needs to begin to develop the methods of mental calculation which he will use throughout his life; as we pointed out in paragraph 256 these will not necessarily be the same as the methods used on paper. Practice in the handling of money, the giving of change by 'counting on' in the way which is commonly used in shops, calculation of journey times and mental calculations involving measurement of various kinds should all start during the primary years. We may note that, although it is possible to practise written methods of computation as routines with little understanding of the underlying method, good mental methods have to be based on understanding of place value accompanied by recall of addition and multiplication facts; it follows that the practice of mental methods of computation will also assist in the understanding and development of written methods. Mental mathematics may also be used in the primary years to build up speed and confidence in the recall of basic number facts and to extend mathematical insights without the added complication of formal recording.

317 While one aspect of mental mathematics is work 'in the head', another is the promotion of mathematical discussion in the classroom. Exchanges between child and teacher, and between different children, should be encouraged. Even in a class in which an individual learning scheme is used for much of the time, there are some skills, puzzles and problems which are appropriate for every child no matter what stage of learning he may have reached and short class sessions can be arranged for work of this kind. If answers are recorded on paper, difficulties and weaknesses can be dealt with on an individual basis later so that the limited success of certain children is not drawn to the attention of the whole class as might be the case with oral answers. However, this does not preclude general discussion of certain problems; on the contrary some problems should be posed with general discussion in mind. Both children and their teachers learn from the different strategies and methods which other members of the class use and explain in answer to questions. It is valuable experience for children, and something which many children do not find easy at first, to explain the approach which has been used; even a wrong answer or a false start, if carefully handled by the teacher, can be illuminating when discussed. Different points of view offer considerable opportunities for exploring and increasing the depth of understanding of all members of the class. Sometimes, too, children can be asked to pose their own problems.

318 Mental mathematics can also be used as a means of introducing informally mathematical ideas which will later be developed in greater depth and perhaps in different ways by different children. Questions such as "if $2 \times \square + \Delta = 17$, what numbers can we write in the box and the triangle?" can be used to introduce algebraic ideas. The ability to visualise shapes in two and three dimensions can also be developed through mental work.

319 Even though an individual learning system may be in use the teacher will often assemble a small group to begin a new topic or to draw together common strands in work which is going on. On such occasions mental mathematics is easily and naturally introduced, both in the form of mental calculations and of questions which develop new ideas or bring together the work of the various children. Since such small groups will often consist of children of similar attainment, there is less chance of dispiriting failure and the size of the group makes it easier for the teacher to handle difficulties sensitively. With a small group of children the teacher can appropriately press them, according to their ability, to increase their speed of response and calculation. Repeated failure should be avoided; nevertheless children respond to questions which extend them and by careful questioning, either of the whole group or of individual children, confidence can be built up.

320 Whatever textbooks or work cards are used, the level of difficulty can never be matched exactly to every child's needs. Questioning and mental mathematics have their part to play in improving the match, in helping the weaker over difficulties and in increasing the challenge to those whose attainment is high. Mental mathematics also provides a means of developing the skills of estimation to which we referred in paragraphs 257 to 261.

Using mathematics to solve problems

321 All children need experience of applying the mathematics they are learning both to familiar everyday situations and also to the solution of problems which are not exact repetitions of exercises which have already been practised. When young children first come to school, much of their mathematics is 'doing'. They explore the mathematical situations which they encounter—perhaps sorting objects into different categories or fitting shapes together—and come to their own conclusions. At this stage their mathematical thinking may reach a high level of independence. As they grow older this independent thinking needs to continue; it should not give way to a method of learning which is based wholly on the assimilation of received mathematical knowledge and whose test of truth is "this is the way I was told to do it".

322 Mathematical explorations and investigations are of value even when they are not directed specifically to the learning of new concepts. Children should therefore be encouraged, for example, to work out the best way of arranging the seating for the audience at the school concert or to compare the cost of various packet sizes and brands of food for the classroom pets. The extent to which children are enabled to work in this way will depend a great deal on the teacher's own awareness of the ways in which mathematics can be used in the classroom and in everyday life.

323 The development of general strategies directed towards problem solving and investigations can start during the primary years. Children should therefore be given opportunity to become familiar with the processes which can be used in work of this kind. One of these is to *make a graphical or diagrammatic representation* of the situation which is being investigated; for example, if two dice are being thrown, the scores obtained can be recorded graphically. There may be a *pattern in the results* which are being obtained which can lead to the *making of a conjecture* to forecast later results; for example, 2 points on a circle can be joined by one line, 3 points can be joined in pairs by 3 lines, 4 points by 6 lines and so on. Efforts can then be made to *discover whether, and explain why, the conjecture is or is not correct*. It is sometimes appropriate to *set up an experiment*, for example to discover the length of a seconds pendulum, or to employ the strategy of *looking at a simpler related problem*; an example of this latter strategy is that the number of squares (of any size) on a full-sized chessboard may be too many to count, but a 2×2 and a 3×3 board are more manageable, and a pattern begins to emerge. It is necessary to *develop persistence in exploring a problem*, for example the number of different shapes which can be made from a given number of squares of the same size, and the ability to *record the possibilities* which have been tried. Finally, it is important to develop the *ability to work with others* in the discussion of possible approaches and to *be able to communicate progress* which has been made by means of words, diagrams and symbols.

324 Not a great deal is yet known about the ways in which these processes develop nor are suitable materials for teachers readily available. There is need for more study of children's spontaneous problem-solving activities and of the extent to which strategies and processes for problem solving can be taught.

Present knowledge suggests that, if children are not enabled to tackle problems which are at the right level for them to achieve success as the result of concentrated effort, their problem-solving abilities do not develop satisfactorily.

Links with other curricular areas

325   The experiences of young children do not come in separate packages with 'subject labels'; as children explore the world around them, mathematical experiences present themselves alongside others. The teacher needs therefore to seek opportunities for drawing mathematical experience out of a wide range of children's activities. Very many curricular areas give rise to mathematics. Measurement and symmetry arise frequently in art and craft; many patterns have a geometrical basis and designs may need enlarging or reducing. Environmental education makes use of measurement of many kinds and the study of maps introduces ideas of direction, scale and ratio. The patterns of the days of the week, of the calendar and of the recurring annual festivals all have a mathematical basis; for older children historical ideas require understanding of the passage of time, which can be illustrated on a 'time-line' which is analogous to the 'number-line' with which they will already be familiar. A great deal of measurement can arise in the course of simple cookery, including the calculation of cost; this may not always be straightforward if only part of a packet of ingredients has been used. Many athletic activities require measurement of distance and time. At the infant stage many stories and rhymes rely for their appeal on the pleasure of counting.

326   It would be easy to compile a much longer list of areas of the primary curriculum which provide opportunities for the use of mathematical skills; pressure of work in the classroom makes it much less easy for the teacher to make sure that advantage is taken of these opportunities when they arise. When planning the activities of the classroom, and especially any extended topic or project work, it is therefore necessary for the teacher to try to identify at the outset the mathematical possibilities which exist within the work which is planned. Not all of these will necessarily be realised but by planning in this way it becomes easier to make the most of whatever opportunities present themselves and perhaps also, by appropriate discussion, to draw attention to others.

327   We have not yet referred to one area of the curriculum which has clear and direct links with mathematics, that of science. Almost every investigation which is likely to be undertaken will require the use of one or more of the mathematical skills of classifying, counting, measuring, calculating, estimating, recording in tabular or graphical form, making hypotheses or generalising, and will provide opportunity for making use of mathematics in practical situations. Indeed, there is a great deal of overlap between practical mathematics and science in the primary years and many activities such as recording the growth of a plant or animal, measuring temperature and rainfall or investigating the chain-wheels of a bicycle could take place under

*Primary education in England. A
survey by HM Inspectors of
schools. HMSO 1978.
†Department of Education and
Science and Welsh Office. The
school curriculum. HMSO 1981.

either heading. The report of the National Primary Survey* draws attention
to the fact that too few of the schools visited had effective programmes for the
teaching of science. The Government paper *The school curriculum*† states
that it is intended to take further action in relation to science in schools. The
development of science teaching in primary schools will provide valuable
opportunities for developing the use of mathematics in practical ways; we
hope these will be exploited.

328   There is one matter to which attention will need to be paid. If, because of
lack of suitable expertise among other teachers in a school, the teaching of
science is undertaken on a specialist basis, it will be essential for those who
teach mathematics and those who teach science to work closely together so
that full advantage can be taken of the overlap between science and mathematics and links established between them.

329   **The overall aim must be to develop in children an attitude to mathematics
and an awareness of its power to communicate and explain which will result in
mathematics being used wherever it can illuminate or make more precise an
argument or enable the results of an investigation to be presented in a way
which will assist clarity and understanding.**

Attainment in mathematics
Children whose attainment is high

330   In this section we discuss the teaching of children who are within about
the top 10 per cent of their age group in terms of their attainment in mathematics. High attainment in mathematics is very often associated with high
attainment in other areas of the curriculum but it is important to be aware that
it can also exist in children whose performance in the rest of their work is no
more than average. The capacity for high attainment in mathematics can
sometimes become apparent at a very early age. A fascination with numbers
and a self-developed capability in the use of large numbers may be signs of
such capacity. However, some children do not display such ability until much
later. During the primary years a teacher may notice that a child learns and
grasps new ideas with great speed, that he shows energy and perseverance in
pursuing ideas, that he understands abstract concepts easily and is able to
make use of them in a variety of situations. All of these characteristics can be
signs of capacity for high attainment but it is possible for such capacity
to go unrecognised. This may be because poor linguistic or reading skills can
conceal capacity for mathematical attainment or, especially in the case of
gifted children (those whose attainment is within the top 2 to 3 per cent),
because of behaviour problems which are the result of boredom and frustration arising from work which is insufficiently demanding. It can also be the
case that gifted children seek to hide their powers so as not to appear different
from their fellows.

331 The Russian psychologist Krutetskii has listed some characteristics
which are often found during the primary years in children who are very

*V A Krutetskii. *The psychology of mathematical abilities in schoolchildren: survey of recent East European mathematical literature.* Trans J Teller; Ed J Kilpatrick and I Wirszup. University of Chicago Press 1976.

highly gifted mathematically.* These include ability to perceive and use mathematical information and grasp the inner structure of a problem; ability to think with clarity and economy when solving problems; ability to use symbols easily and flexibly and to reverse a mathematical process with ease; ability to remember generalised mathematical information, methods of problem solving and principles of approach. Although few children will display gifts of such a high order, this list indicates the directions in which high attainment in mathematics develops.

332 The problems which primary teachers can encounter in making suitable provision for high-attaining children are not always appreciated. It is not sufficient for such children to be left to work through a textbook or a set of work cards; nor should they be given repetitive practice of processes which they have already mastered. "The statement that able children can take care of themselves is misleading; it may be true that mathematically such children can take care of themselves better than the less able, but this does not mean that they should be entirely responsible for their own programming; they need guidance, encouragement and the right kind of opportunities and challenges to fulfil their promise."* High attaining children should combine more rapid progress through the mathematics syllabus with more demanding work related to topics which have already been encountered. In particular, they should be given opportunity to undertake activities and investigations which encourage the development of powers of generalisation and abstraction; older juniors may, for example, become able to express in terms of algebraic symbols the relationships which arise from graphical work or the investigation of number patterns. Geometrical work should also be encouraged. "Abler top juniors ... are capable of a considerable amount of geometry. Through the strength of their own intuition, they can cover the greater part of the geometry demanded by an O-level syllabus. They do, however, need suitable resource material including topic books and reference books, but not secondary school texts."†

*E Ogilvie. *Gifted children in primary schools.* The report of the Schools Council enquiry into the teaching of gifted children of primary age 1970/71. Macmillan Education, for the Schools Council 1973.

†HMI Series. Matters for discussion 9. *Mathematics 5-11.* A handbook of suggestions. HMSO 1979.

333 **There is undoubtedly a need to make specific provision for primary children whose mathematical attainment is high.** However, suitable resource material is not easily accessible for the use of teachers who are not mathematics specialists and more is needed. We believe that in some parts of the country arrangements have been made for a peripatetic teacher who is a mathematics specialist to teach groups of children in different schools from time to time. In some areas clubs which meet on Saturdays have been organised by LEAs, colleges or other agencies with considerable success; such clubs enable children to share their enthusiasm for mathematics with others who are likeminded. If these children's teachers are also able to take part, they are likely to be able to make use of some of the ideas which arise to enrich the curriculum of other children. Within a particular school it may be possible to re-group children for mathematics in order to provide more effectively for the higher attainers, and we believe that careful consideration should be given to this possibility. We return to this point in paragraph 350.

Children whose attainment
is low

334 Low attainment in mathematics can occur in children whose general ability is not low. Among the reasons for this can be inappropriate teaching, lack of confidence, lack of continuity, especially because of change of school, and frequent or prolonged illness; poor reading skills can also hinder progress in mathematics. In such cases it is essential to try to diagnose the reason for the 'mathematical blockage' and to remedy the lack of understanding which exists. Because the problem is likely to be individual to the child the diagnostic process should be very largely oral and practical so that by observation and discussion the teacher may establish which concepts are understood and which are not. The use of suitable diagnostic tests may also be helpful. **Failure can only be compounded if efforts are made to build further upon a foundation which does not exist**. Such efforts are likely only to result in confusion and lack of confidence because of continuing lack of success, and so lead to dislike of mathematics and further failure.

335 In the case of children whose low attainment in mathematics is associated with low general ability, the mathematics course needs to be specifically designed to build up a network of simple related ideas and their applications, so that the children can feel confident in their ability to make use in their daily lives of the mathematics which they know. Advance should be by very small stages with frequent opportunity for repetition and reinforcement. Use should be made of extensive and varied practical and oral work related to everyday situations such as measurement, shopping and the use of money. In this way efforts can be made to develop the necessary network of associations which is essential to long term memory.

336 Low-attaining children need extensive experience of counting objects of many kinds by grouping them in tens so that they become aware of the relative size of numbers. Work of this kind needs to be accompanied by the use of number apparatus to develop the idea of place value and an understanding of the different number operations. Familiarity with number and counting can also be developed by means of board and other games played with dice. Practice is also needed in counting money and using it for shopping. However, stress should not be placed on the development of number skills to the exclusion of other activities. Low-attaining children need to join with their fellows in experiencing the pleasure of simple work with shapes and the discovery of pattern. They should also undertake straightforward work of a graphical nature and learn to use a ruler and other simple geometrical instruments.

337 Children whose attainment is very low are often withdrawn from their normal class for part or all of the time to receive special attention from a remedial teacher. Such teachers usually concentrate largely on language work and are not always skilled in the teaching of mathematics or in the diagnosis of associated learning difficulties. Remedial teachers should nevertheless seek to develop the understanding and use of mathematical language alongside other language skills. It is of the utmost importance that the mathematical work of children in remedial classes should not consist of the practice of arithmetical skills in isolation but should be accompanied by discussion of the concepts on which these skills rest and of the ways in which they can be used in the chil-

dren's everyday lives. More time should be spent on oral and practical work than on written work.

338 The increasing availability of electronic calculators has made it all the more important that, in the teaching of low-attaining children, attention should be given to the development of concepts and applications. Once these are understood, it becomes possible to make use of a calculator to overcome lack of computational skill, but a calculator can be of no assistance until a child knows which arithmetical operation it is necessary to carry out.

Attainment at the age of 11

339 The mathematical course which a child will follow is normally set out in the mathematics syllabus of his school, which is often based on guidelines prepared by the LEA. In some cases syllabuses and guidelines indicate only a progression of topics while in other cases an attempt is made to suggest levels of attainment which are thought to be appropriate to a majority of children of a given age. The amount of guidance which is given as to the way in which different topics should be approached can also vary considerably. However, even when detailed guidance is given, it may not be heeded by a particular teacher. The report of the National Primary Survey* draws attention to the fact that "individual schools or teachers are making markedly individual decisions about what is to be taught based on their own perceptions and choices". We need therefore to be aware of the differences which are likely to exist between the mathematics which the syllabus intends should be taught, the mathematics which has actually been taught by a particular teacher and that part of what has been taught which a child has learned and understood. Furthermore, any test which is given to a child can measure performance only on that part of what he has learned to which the test questions relate; even then, he may not always be able to demonstrate his knowledge under test conditions.

Primary education in England. A survey by HM Inspectors of Schools. HMSO 1978.

* Assessment of Performance Unit. *Mathematical develpment. Primary survey reports No 1* and *No 2*. HMSO 1980 and 1981.

340 During recent years large-scale testing of children's performance in mathematics has been introduced in many countries. In addition to studying the reports of the Assessment of Performance Unit*, which provide information about children in England and Wales, and the results of some other tests carried out in England, we have been able to examine the results of large-scale tests carried out in Scotland and the United States of America. We have also studied the results of a survey of basic numeracy carried out in Australia. It is clear that the differences in attainment which exist between children of the same age in any one country are very much greater than the small variations which exist between the performance of an 'average' pupil from each country. It is also apparent that the general pattern of the development of number knowledge is very similar in all the countries whose tests we have studied. **We do not therefore believe that there are any grounds for thinking that the overall performance of children in England and Wales is markedly different from that of children in these other countries.**

341 Because different countries test pupils on a large scale at different ages, it has been possible to build up a composite picture, as demonstrated by the

results of written tests, of the performance in some numerical topics of an English-speaking child whose attainment is at about the fiftieth percentile and also of one whose attainment is much below average (at about the fifteenth percentile from the bottom). For the child whose attainment is at the fiftieth percentile, understanding of the topic of place value probably develops along these lines. At age 9 he can identify the largest or smallest of a set of numbers which range from units to thousands and he can write down the total number of postage stamps from a picture showing the stamps in blocks of 100, strips of 10 and single stamps. By age 11 he is clarifying his concept of place value in the range from thousands to tenths. He understands that the 2 in 12, 205 and 40.2 represents units, hundreds and tenths respectively but he cannot yet arrange the numbers 0.07, 0.02, 0.1 in ascending order of size. He can write down the number which is 1 less than 2010 and probably also the number which is 1 more than 6399. However, test results suggest that not until the age of 15 are at least half the children in a year group able to read a scale to two decimal places or to state that the 1 in the number 2.31 represents 1 hundredth. Although about half of the children in our schools will know more about place value than this 'average' child, the other half will know less and will be progressing through these stages more slowly. For example, the Concepts in Secondary Mathematics and Science study* found that pupils whose attainment was at the fifteenth percentile from the bottom had reached the third year of secondary school before they were able to give the number which is 1 more than 6399. On the other hand, there are a few 7 and 8 year olds who are able to answer this question and a few 9 year olds who are able to use two decimal places with understanding.

*K M Hart (Editor). *Children's understanding of mathematics: 11–16*. John Murray 1981.

342 It therefore seems that there is a 'seven year difference' in achieving an understanding of place value which is sufficient to write down the number which is 1 more than 6399. By this we mean that, whereas an 'average' child can perform this task at age 11 but not at age 10, there are some 14 year olds who cannot do it and some 7 year olds who can. Similar comparisons can be made in respect of other topics. For example, the top 15 per cent of 10 year olds in England are able to answer the question "There are 40 children in a class and three-fifths of them are girls. How many boys are there in the class?". By contrast, the bottom 15 per cent of 14–15 year old pupils in Scotland find difficulty in working out $\frac{3}{4}$ of £24; the bottom 15 per cent of 14 year olds in Australia find difficulty with the comparable question "Mixed concrete costs $24 per cubic metre. What would $\frac{3}{4}$ of a cubic metre cost?". There is little evidence to show the attainment of the most capable 11 year olds, because large scale tests do not usually include very many items which will extend these children and so provide the necessary evidence. However, one American study found that there were a number of 12 and 13 year old pupils in Baltimore who performed at the same level as the top 10 per cent of 17 year olds on a mathematics test designed to reveal potential for college study.

343 We believe it is clear from the preceding paragraphs that it is not possible to make any overall statement about the mathematical knowledge and understanding which children in general should be expected to possess at the

end of the primary years. However, the test results which we have quoted, and others which we have studied, indicate that even in the primary years the curriculum provided for pupils needs to take into account the wide gap in understanding and skill which can exist between children of the same age.

344 The study of test results provides evidence only of the achievement of children under test conditions as a result of the curricula and the teaching methods which have been used. They cannot indicate what might be the results given different curricula or different teaching methods. However, the fact that the overall picture is similar in different English-speaking countries suggests that any improvements in teaching are likely to produce slow change rather than rapid results. Even if the average level of attainment can be raised, the range of attainment is likely to remain as great as it is at present, or perhaps become still greater, because any measures which enable all pupils to learn mathematics more successfully will benefit high attainers as much as, and perhaps more than, those whose attainment is lower.

Attitudes

345 During every mathematics lesson a child is not only learning, or failing to learn, mathematics as a result of the work he is doing but is also developing his attitude towards mathematics. In every mathematics lesson his teacher is conveying, even if unconsciously, a message about mathematics which will influence this attitude. Once attitudes have been formed, they can be very persistent and difficult to change. Positive attitudes assist the learning of mathematics; negative attitudes not only inhibit learning but, as we discussed in Chapter 2, very often persist into adult life and affect choice of job.

346 By the end of the primary years a child's attitude to mathematics is often becoming fixed and will determine the way in which he will approach mathematics at the secondary stage. He may thoroughly enjoy his work in mathematics, or he may be counting the days until he can stop attending mathematics lessons. He may have learned that mathematics provides a means of understanding, explaining and controlling his environment, or he may have failed to realise that it has any relevance outside the classroom. He may have learned the importance of exploration and perseverance when tackling a problem and have experienced the pleasure which comes from finding its solution, or he may regard mathematics as a series of arbitrary routines to be carried out at the teacher's behest, with no opportunity for initiative or independent thought. He may be well on the way to mastering some of the mathematician's skills, or he may already see mathematics as an area of work which he cannot understand and in which he always experiences failure.

347· In the previous paragraph we have set out extremes of attitude in order to stress the importance of doing all that is possible to develop positive attitudes towards mathematics from the earliest days at school. At the age of 5, children usually show an uninhibited enthusiasm and curiosity; school is enjoyable and they learn rapidly and with interest as they encounter a great variety of new experience. The challenge for the teacher is to present mathematics in a way which continues to be interesting and enjoyable and so allows understanding to develop. In the course of our visits to primary schools we met a number of teachers who were succeeding in presenting mathematics in this way and whose pupils were clearly enjoying their work. The notes which we made after the visits contain comments such as "the level of work and presentation were most impressive ... pupils went about their work in a quiet, business-like yet enthusiastic manner", "a most enjoyable visit ... a happy and conscientious staff working together to achieve common aims". However, not all the classes we visited were achieving these levels of involvement and the importance of creating positive attitudes to mathematics did not always seem to have been realised. Even in schools in which the general atmosphere was lively and supportive, the need of children to work at mathematics in practical ways had not always been realised, so that attitudes to mathematics in some classrooms contrasted strongly with attitudes towards other work. In some classrooms children were working at abstract calculations with numbers beyond their experience; their need to measure, weigh and pour water, to count real things and to learn about hundreds, tens and units with apparatus and with money had gone largely unrecognised.

The organisation of teaching groups for mathematics

348 In most primary schools children work in mixed ability classes which usually contain children of only one year group. The teacher in charge of the class is normally responsible for the greater part of the work of the class, including mathematics. This arrangement allows flexibility in the organisation of work in mathematics, as in other areas of the curriculum. It is not, for example, necessary for children to do mathematics at a fixed time or for a given length of time, nor for all children to do mathematics at the same time. It is therefore possible for the teacher to work with part of the class while the remaining children are engaged in activities which require less immediate attention from the teacher. Because the same teacher is with the class for most of the week, there is also maximum opportunity to relate work in mathematics to work in other curricular areas (see paragraph 325). However, the quality of the mathematics teaching inevitably depends largely on the strength and interest of the class teacher. If this teacher lacks enthusiasm for mathematics and confidence in teaching it, the children in the class will be disadvantaged. Even though the membership of a class may not change very much from one year to the next it is common for there to be a change of class teacher at the beginning of each year. It is therefore possible for some children to be taught mathematics by seven different teachers during the primary years, with consequent problems of ensuring continuity. An arrangement whereby the same teacher remains with a class for more than one year is likely to improve continuity but may not improve the teaching of mathematics if the teacher lacks the necessary expertise.

349 'Vertical grouping', in which children of two or more year groups are placed in the same class, is quite often used in the infant years and is a necessity in some small primary schools. Although an arrangement of this kind may lessen problems of continuity, there will almost certainly be a greater spread of attainment in mathematics among the children in the class and a consequent increase of difficulty for the teacher in matching levels of work to

*"There is clear evidence from the survey that the performance of children in classes of mixed age can suffer." *Primary education in England*. A survey by HM Inspectors of Schools. HMSO 1978.

the needs of the children.* We do not therefore consider that this form of grouping offers any advantages for the teaching of mathematics.

350 Although most primary teachers group the children in their classes according to attainment for some part at least of their work in mathematics, there are some junior and middle schools in which this practice is extended further so that children from several classes are rearranged for mathematics into groups based on attainment. This enables the range of attainment in any one group to be reduced and so makes it easier for teachers to match the levels of the work appropriately. However, it is necessary to realise that even when children are grouped in this way, considerable differences will exist within each group. This is illustrated by a submission which we have received from a teacher in a middle school: "The middle group itself could have been divided into three, so great was the disparity of understanding". Furthermore, even when grouping of this kind is used, the teaching still has to be shared among the teachers of the classes from which the groups have been formed and so some groups may be taught mathematics by teachers who lack interest in the subject or confidence in teaching it. Nevertheless, rearrangement of this kind is likely to make it easier to provide appropriately for higher-attaining pupils.

351 Some of our members who visited Denmark were able to observe mathematics teaching in a *Folkeskole*, a comprehensive school for children aged 7 to 16. From the age of 7, children in this school were in classes whose teaching was shared by three teachers, and were usually taught by the same small team of teachers for several years. This not only ensured continuity of mathematics teaching but also enabled teachers to make use of their particular strengths. We do not suggest that this form of organisation would necessarily be appropriate in this country but we believe that it suggests the need to examine the advantages of some form of team teaching, perhaps by means of two or three classes working in association with each other, so that continuity of mathematics teaching could be maintained for two or three years. If a teaching team contains a teacher with enthusiasm for mathematics and with some specialist knowledge, this teacher is able to lead the work in mathematics of a much larger group of children than would otherwise be the case and also to assist the other members of the team. A few schools already work in this way with teams of two to four teachers working together. Each teacher takes a major responsibility for one area of the curriculum and follows the lead of the other teachers in the remaining areas. An arrangement of this kind enables children to experience greater continuity of teaching and still to be well known to their class teacher. It also provides for teachers the opportunity of observing the mathematical development of children over a longer period than a single year.

Primary education in England. A survey by HM Inspectors of Schools. HMSO 1978.

352 The report of the National Primary Survey* draws attention to the need for schools to consider how best to deploy staff in order to make the best use of the strengths of individual teachers. We believe that all schools should examine the extent to which the form of organisation which they are using

enables the best use to be made of the mathematical strengths of their staff and provides continuity of teaching in mathematics for the children in the school.

Time allocation for mathematics

353 On average, junior classes devote about five hours a week to mathematics, but this figure conceals wide variations from less than three hours to more than six. If, as we believe should be the case, mathematics plays a part in many areas of the curriculum, children may do quite a lot of mathematics outside the time specifically allocated to it. We do not therefore believe that such specific allocation should exceed five hours per week. On the other hand, we do not consider that the time allocation should fall substantially below four hours. So far as is possible, time for mathematics should be flexible, so that an interesting discussion can be followed to its conclusion or a piece of practical work completed; mental mathematics is usually best done for a fairly short period of time. Even older children should not normally work at mathematics for more than one hour at a stretch.

The mathematics co-ordinator

354 The effectiveness of the mathematics teaching in a primary school can be considerably enhanced if one teacher is given responsibility for the planning, co-ordination and oversight of work in mathematics throughout the school. We shall refer to such a teacher as the 'mathematics co-ordinator'.

355 In our view it should be part of the duties of the mathematics co-ordinator to:

- prepare a scheme of work for the school in consultation with the head teacher and staff and, where possible, with schools from which the children come and to which they go (we discuss this further in paragraph 363);

- provide guidance and support to other members of staff in implementing the scheme of work, both by means of meetings and by working alongside individual teachers;

- organise and be responsible for procuring, within the funds made available, the necessary teaching resources for mathematics, maintain an up-to-date inventory and ensure that members of staff are aware of how to use the resources which are available;

- monitor work in mathematics throughout the school, including methods of assessment and record keeping;

- assist with the diagnosis of children's learning difficulties and with their remediation;

- arrange school based in-service training for members of staff as appropriate;

- maintain liaison with schools from which children come and to which they go, and also with LEA advisory staff.

356 It would not have been difficult to extend this list further by going into greater detail and by making specific mention of a number of other duties which are included by implication. It is, for example, necessary that the mathematics co-ordinator should keep in touch with current developments in mathematical education; and it will be necessary to pay particular attention to the needs of probationary teachers, of teachers new to the staff and of teachers on temporary supply as well as of teachers who lack confidence in teaching mathematics. The overriding task must be to provide support for all who teach mathematics and so improve the quality and continuity of mathematics teaching throughout the school.

357 Good support from the head teacher is essential if the mathematics co-ordinator is to be able to work effectively, and some modification of the co-ordinator's teaching timetable is likely to be necessary in order to make it possible to work alongside other teachers. Appropriate in-service training for the mathematics co-ordinator will also be required; we discuss this further in paragraph 723.

358 There is at present a great shortage of teachers who are suitably qualified to become co-ordinators but we believe that every effort should be made to train and appoint suitably qualified teachers in as many schools as possible. **We consider that, in all but the smallest schools, the responsibility should be recognised by appointment to a Scale 2 or Scale 3 post, or by the award of additional salary increments**; we discuss this further in paragraph 662.

Mathematics guidelines

359 About half of the LEAs in England and Wales have issued mathematics guidelines. These are documents which provide guidance to teachers about the content of the mathematics curriculum and sometimes also about teaching method. The majority of guidelines relate to mathematics for pupils up to the age of 11 or 13, though some relate to mathematics for infants or for middle schools. Many LEAs have sent us copies of their guidelines. We have noted that those which have been produced most recently reflect an increasing concern with assessment. Some LEAs have produced record sheets and assessment materials related to their guidelines which can be used by teachers.

360 Most LEA guidelines have been produced by groups of teachers working under the leadership of an LEA adviser. In many cases the adviser and some of those who have helped to prepare the guidelines have then introduced them to groups of teachers at local meetings. We believe that some procedure of this kind should always be followed, so that the thinking which underlies the guidelines can be explained and discussed. If the guidelines are issued without explanation and discussion, we consider that they are likely to lose much of their effectiveness. Such initial introduction should be followed by discussion of the guidelines in each school.

361 **It is essential that guidelines are kept under review and revised regularly**. At present, for example, few which we have seen offer guidance about the use

of calculators within primary mathematics. We believe also that many guidelines may need adjustment in the light of the information which is now available in the reports of the Assessment of Performance Unit* so as to place greater emphasis on the differences in attainment which exist between children of the same age.

*Assessment of Performance Unit. *Mathematical development. Primary survey report No 1* and *No 2*. HMSO 1980 and 1981.

362 One of the purposes of issuing guidelines is to ease transfer to secondary schools or other primary schools within the LEA. Therefore, where LEA guidelines exist, they should be taken into account when preparing a school's scheme of work; we discuss this further in the following paragraph.

Schemes of work

363 A scheme of work is essential as a basis for the teaching of mathematics in a school. The responsibility for its preparation lies with the head teacher but is likely to be delegated to the mathematics co-ordinator. A carefully planned scheme of work can assist greatly in maintaining continuity both of syllabus content and of approach as children move from class to class. In addition to setting out the progression of work in mathematics which should be followed, the scheme of work should provide guidance about the resources, including both practical equipment and books or work cards, which are available; it should also make suggestions as to the ways in which they can be used for the different topics in the syllabus. It should outline the approaches to be used in the teaching of particular topics and give guidance about such matters as assessment and record keeping.

364 The preparation of a scheme of work takes a considerable time. Wherever possible all members of staff should collaborate in the task. This not only provides an excellent form of in-service training but makes it easier for teachers to implement the scheme in their classrooms because they will be aware of the intentions which underlie it and of the discussion which has taken place during its preparation. As we have already pointed out, the scheme should take account of LEA guidelines where they exist; it is desirable, too, that there should be consultation with the mathematics adviser or advisory teacher and with the schools from which the children come and to which they transfer. It is essential that the scheme of work should be appraised and revised regularly in the light of the experience of the teachers who have been using it in their classrooms.

Small schools

365 The teaching of mathematics in small schools can give rise to a number of problems. The small number of teachers makes it less likely that there will be a member of staff with mathematical expertise; the head teacher may therefore have to act as mathematics co-ordinator. The task of preparing a scheme of work is as great in a small school as in a larger school but there are fewer teachers to share in the work. Because the head teacher is likely to have responsibility for a class, the operation of the scheme of work may be difficult to monitor. The age range of the children in each class is likely to be wide and, because small schools are often in somewhat isolated situations, the teachers in them may lack the support which can come from professional contact with other teachers.

366 For these reasons, some of the suggestions which we have put forward are likely to be harder to carry out in a small school than in one which is larger and an increased degree of support from outside the school is therefore likely to be required. We have been told that in some areas an additional teacher has been assigned to a group of small schools. This teacher is a part-time member of the staff of each school and acts as mathematics co-ordinator for all of them, working in each school in turn on a regular basis. We commend such initiatives. Advisory teachers can also give help of a similar kind and enable those who teach in small schools to become aware, for example, of the different kinds of resources which are available to assist in the teaching of mathematics and of the ways in which they are used.

Aims and objectives

367 It is common for schemes of work in mathematics to begin with a statement of aims. Such a beginning should not be regarded merely as a necessary formality but should be a statement of intent which has been discussed, developed and accepted by those who teach mathematics in a school. The aims of mathematics teaching in primary schools should be closely related to the general aims of primary education. The primary years are a time when children are not only acquiring the skills of language and number but are also experiencing a variety of methods of learning; they are learning to think, to feel and to do, to explore and to discover.

368 Aims are conceived at different levels and are of necessity expressed in general terms; they are realised as much through the methods of teaching which are used as through the topics which are being taught. Amongst the most general can be aims such as developing a good attitude to mathematics and 'opening doors' onto a wide variety of mathematical experience. Others which are rather more specific can relate to the development of spatial awareness, of ability to solve problems, of ability to use and apply mathematics, of ability to think and reason logically, of ability to use mathematical language and to the acquisition of certain skills. We believe that few would dissent from these aims, but the emphasis they receive may vary. An excellent set of aims is set out in *Mathematics 5-11**. This is developed from the statement "We teach mathematics in order to help people to understand things better—perhaps to understand the jobs on which they might later be employed, or to understand the creative achievements of the human mind or the behaviour of the natural world". It concludes "Finally, there is the overriding aim to maintain and increase confidence in mathematics . . .".

*HMI Series, Matters for discussion 9. *Mathematics 5-11*. A handbook of suggestions. HMSO 1979.

369 For classroom purposes some aims can be translated into more explicit and precise objectives but others, such as those concerned with the development of positive attitudes and appreciation of the creative aspects of mathematics are less tangible and so not easily expressed in terms of objectives. For this reason they can sometimes fail to receive their due attention. The way in which a particular topic is to be taught and a particular objective achieved needs to be considered in relation to the aims of the course as a whole and especially of those aims to which it can make a direct contri-

bution. Some topics, for example, offer particular opportunities to develop appreciation of space or pattern, others to help in the development of logical thinking, others to develop persistence in sustained work. It is therefore necessary when working at a particular topic in the classroom to have in mind both the need to relate it to other work which has been covered and also to consider ways in which it can contribute to broader aims. Unless both of these needs are held in mind, it is possible for certain topics to become ends in themselves rather than means through which wider mathematical understanding can develop.

370 It is necessary, too, to ensure that approaches which are used in the classroom do not conflict with the aims which have been agreed for teaching mathematics in the school. For example, one of the aims set out in *Mathematics 5 – 11* — "to develop an understanding of mathematics through a process of enquiry and experiment"—will not be achieved if the methods of teaching which are used do not allow or encourage children to work in this way. Because it is easy for long-term aims to become overlooked as a result of the day-by-day pressures of the classroom, all teachers need to review the aims of their teaching regularly in order to discover whether these aims are being fulfilled within the classroom or whether they are giving way to other more limited and unintended aims.

371 It must be for each school to develop its own aims and objectives for teaching mathematics in the light of its approach to the curriculum as a whole; it must be for each class teacher to seek to achieve these aims and objectives by the provision of suitable activities for the children in the class. It is essential that support is available to help teachers in their task; in the provision of such support the head teacher and the mathematics co-ordinator should play a major part.

7 Calculators and computers

372 We devote a separate chapter to electronic calculators and computers because we believe that their increasing availability at low cost is of the greatest significance for the teaching of mathematics. As a result of the fact that in the mid 1970s the price of electronic calculators started to fall very rapidly, very many adults now possess calculators of their own. So, too, do considerable numbers of school pupils; and many pupils of all ages and abilities who do not have a calculator of their own have access at home to a calculator belonging to some other member of the family.

373 At the present time a similarly rapid drop is taking place in the price of small computers, now more usually referred to as micro-computers, and a Government programme has been announced which is intended to ensure that at least one micro-computer will be available in every secondary school by the end of 1982. Most micro-computers display their input and output on a television-type screen; some also have a printer which will provide a permanent record if one is required. As yet the number of families who possess a micro-computer is still relatively small but it seems clear that this number is likely to increase very considerably within the next few years. We are therefore in a situation in which increasing numbers of children will grow up in homes in which calculators and micro-computers are readily available, in which there is access to a variety of information services displayed on domestic television sets and in which the playing of 'interactive' games, either on micro- computers or by means of special attachments to television sets, is commonplace.

374 **These developments have very great implications for the teaching of many subjects in schools. So far as the teaching of mathematics is concerned, we believe that there are two fundamental matters which need to be considered.** The first concerns the ways in which calculators and micro-computers can be used to assist and improve the teaching of mathematics in the classroom. The second concerns the extent to which the availability of calculators and microprocessors should change the content of what is taught or the relative stress which is placed on different topics within the mathematics syllabus. We have not listed separately the question of the relationship between the use of calculators and the development of fluency in mental and written calculation because we believe that, although important, it arises as one aspect of these fundamental matters and needs to be considered in that context.

Calculators

375 Among the submissions which we have received there are many which make reference to the use of electronic calculators either in school or at work and the range of views which is expressed is very great indeed. For instance:

> Exercise of the basic skills should not depend upon use of calculators: these should be limited to higher education.

> Mathematics must be a compulsory subject, taught to a reasonable standard, using one's loaf and not a calculator.

> The members were agreed on the tremendous part electronic calculators and machines now play in employment. Accordingly, it would seem sensible for the teaching of mathematics to introduce pupils to machines and to familiarise them with their capabilities and use.

> Calculators have revolutionised computation and barely numerate students can overcome their weaknesses with these.

376 It is clear that many of those who have written to us assume that the use of calculators in schools is very much more widespread than is, in fact, the case. Although increasing numbers of pupils in secondary schools, as well as some in primary schools, now possess their own calculators, we believe that in only a minority of secondary schools are sufficient calculators provided for it to be possible for all the pupils who are being taught mathematics at any one time to have one for individual use; this is a matter to which we return in paragraph 393. There still appear to be many teachers of mathematics in the secondary years who discourage or forbid the use of calculators by their pupils. In very many primary classrooms no use is made of calculators at all.

377 It is also clear that there is widespread public concern about the use of calculators by children who have not yet mastered the traditional pencil and paper methods of computation. It is feared that children who use calculators too early will not acquire fluency in computation nor confident recall of basic number facts. These fears are understandable and should not be ignored. However, such research evidence as is at present available suggests that there may be advantages which more than compensate for any possible disadvantages. In recent years a considerable number of research studies carried out in the United States of America have compared the computational performance of groups of pupils who have used calculators with that of groups of pupils who have not. Some of these studies have reported improvements among those who have used calculators in attitudes towards mathematics, in personal computational skills, in understanding of concepts and in problem-solving; other studies have found no differences which are statistically significant between the performance of those who have used calculators and those who have not. **From all the studies the weight of evidence is strong that the use of calculators has not produced any adverse effect on basic computational ability.** We believe that this is important and should be better known both to teachers and to the public at large. Nevertheless, it remains incumbent upon those who teach mathematics to ensure that the development of appropriate skills of mental and written calculation is not neglected. Nor should a school

overlook the need to make parents aware of its policy in regard to the use of calculators by pupils.

378 **We wish to stress that the availability of a calculator in no way reduces the need for mathematical understanding on the part of the person who is using it.** We have already explained that, for example, knowing how to multiply and knowing when to multiply involve distinct aspects of teaching and learning. A calculator can be of no use until a decision has been made as to the mathematical operation which needs to be carried out and experience shows that children (and also adults) whose mathematical understanding is weak are very often reluctant to make use of a calculator. We believe that this is a crucial point which is not always appreciated, especially by those who criticise the use of calculators in schools.

379 There can be little doubt of the motivating effect which calculators have for very many children, even at an early age. This is illustrated by the following extract from one of the submissions we have received from parents.

Following professional advice of mathematical colleagues we kept calculators away from our children until their late teens. But the youngest at age 6 got hold of a calculator to help out his 'tables' and found it such fun that he has been much more mathematically inclined since. So perhaps it would be wise to introduce simple calculators at an early age.

380 Pupils often learn to operate calculators by making use of them in the first instance to check calculations which have been carried out mentally or on paper. Although this should not be regarded as a major use of calculators in the classroom, their use in this way nevertheless offers two important advantages. The first is that feedback is immediate so that pupils are able to check their work frequently. This enables them to seek help if they obtain a succession of wrong answers or to take steps themselves to locate occasional errors. Pupils who lack confidence can also receive encouragement as a result of being able to satisfy themselves that they are obtaining correct answers. The second advantage is that the calculator is 'neutral' and does not express disapproval or criticism of wrong answers. This can be a very great help to some pupils; furthermore, the use of a calculator in this way can provide a motivation and, in some cases, a determination to succeed in order to 'beat the calculator' which would not otherwise exist.

381 It is, of course, easy to make mistakes when using a calculator. These can occasionally arise from faulty functioning of the calculator itself but are much more likely to be the result of faulty operation by the person who is using it. For this reason it is essential that pupils should be enabled to acquire good habits in the use of calculators so as to guard against mistakes. Pupils must learn that calculations should be repeated, if possible by entering the numbers in a different order; that subtraction can be checked by addition and division by multiplication. It is also necessary to stress the importance of checking answers by means of suitable estimation and approximation. Discussion of all these matters in the classroom provides opportunity for increasing pupils' understanding of mathematical concepts and routines and so contributes to their progress.

382 It is also important that pupils should realise that not all calculators operate in exactly the same way; and that they should be able to 'explore' a calculator which is unfamiliar in order to discover whether it has particular characteristics which need to be taken into account when using it. It follows that we cannot regard the fact that the pupils in a class may possess different models of calculator as being a reason for avoiding their use. Comparison of the routines needed to perform the same task on different models of calculator can be very instructive.

383 Before we discuss more specifically some of the ways in which calculators can be used in the primary and secondary years we wish to draw attention to one further matter which we believe to be of importance. This is the use of calculators for the purpose of encouraging mathematical investigation. This can start at the primary stage, as we describe in the following paragraph. At the secondary stage, pupils should be encouraged to undertake exploration of the capabilities of the calculator itself; for example, of the largest and smallest numbers which can be entered, of what happens when the result of a calculation becomes too large or too small for the calculator to accommodate, of the use which can be made of memories and constant facilities and of the most economical methods of carrying out a variety of calculations. Investigations of this kind are appropriate for pupils of all levels of attainment and assist in developing awareness of the range of facilities which calculators provide. We suspect that many adults who possess calculators and use them regularly are not fully aware of the capabilities of the machines which they own; for example, they may not be able to say what would happen if, in given circumstances, they pressed the 'equals' key twice in succession. Calculators can also be used for more general investigations of many kinds. Work of this kind is especially valuable as a means of extending the mathematical insight of pupils whose attainment is high.

The use of calculators in the primary years

384 Many primary teachers feel some uncertainty about introducing calculators into their classrooms until more guidance is available than is the case at present. However, the increasing availability of calculators in many homes means that many children are likely to have access to a calculator at an early age and, by the age of 6, will have started to experiment to see what they can do. We therefore believe that primary teachers need to be able to use a calculator themselves, and that they should ensure that some calculators are available in the classroom for children to use. These can be used as an aid in discovery and investigational work. For example, a number of additions of the form $4 + 6, 14 + 6, 24 + 6, \ldots\ldots, 4 + 16, 14 + 16, 24 + 16, \ldots\ldots$ can be carried out with a calculator to enable a child to find out the result of adding numbers ending in 4 and 6. This can be followed up in subsequent mental work and other similar number patterns explored.

385 There are other ways in which a calculator can be used as an aid to teaching. For example, by entering the number 572 and then asking a child to change the display to 502, to 5720 or to 57.2, in each case by means of a single arithmetical operation, it is possible to reinforce understanding of place value. If the child cannot carry out the necessary operations the teacher is enabled to locate the area in which lack of understanding exists and to provide

appropriate additional experience. Again, many children find difficulty in writing down the number which is, for example, one more than 6399 or one less than 6500. The calculator can assist with problems of this kind and provide a visual display which can be reconciled with an approach to the same question by means of the use of structural apparatus.

386 Children who make use of calculators are likely to meet decimals and negative numbers earlier than is usual at present. This provides the teacher with new opportunities of discussing these topics and the context in which they arise; the fact that some calculators record whole numbers as, for example, 5. or, less commonly, 5.000000 may also lead to questions. Because calculators differ in their capabilities and in the format in which they display their results, teachers need to be aware that some types of calculator are more suitable than others for use at the primary stage. Calculators also enable children to deal with larger numbers than would otherwise be possible. This means that investigations of such things as number patterns can be extended further. A calculator also makes it possible to deal with 'real life situations'; for example, it is possible to find the average height or weight of a whole class rather than of only a small number of children.

387 There is as yet little evidence about the extent to which a calculator should be used instead of pencil and paper for purposes of calculation in the primary years; nor is there evidence about the eventual balance to be obtained at the primary stage between calculations carried out mentally, on paper, or with a calculator. However, it is clear that the arithmetical aspects of the primary curriculum cannot but be affected by the increasing availability of calculators. In our view, it is right that primary teachers should allow children to make use of calculators for appropriate purposes, while remembering that, for the reasons which we have discussed in Chapter 6, it will remain essential that children acquire a secure grasp of the 'number facts' up to 10 + 10 and 10 × 10 and the ability to carry out, both mentally and on paper, straightforward calculations which make use of these facts.

388 It is likely that the use of calculators may bring about some change in the order in which different parts of the primary mathematics syllabus are introduced. We have already referred to the earlier introduction of the concepts of decimals and negative numbers. Decimals will assume greater importance in relation to fractions than is the case at present. Some development work on the use of calculators in the primary years is going on at the present time. **In our view, more is needed both to consider the use of calculators as an aid to teaching and learning within the primary mathematics curriculum as a whole and also the extent to which the arithmetical aspects of the curriculum may need to be modified. We believe that priority should be given to this work and to providing associated in-service training for teachers.**

The use of calculators in the secondary years

389 **We believe that there is one over-riding reason why all secondary pupils should, as part of their mathematics course, be taught and allowed to use a calculator.** This arises from the increasing use which is being made of calculators both in employment and in adult life. We believe that calculators are also likely to be used increasingly in other curricular areas. We have already

drawn attention to the necessity of establishing good habits when using a calculator so that mistakes can be avoided. The prime responsibility for doing this must lie with the mathematics department. If the necessary instruction is not given as part of the mathematics course, it is unlikely to be given at all. Furthermore, the necessary skills cannot be established in a few short lessons; like all other skills, they need to be practised and to have time to develop. As we pointed out in paragraph 381, the development of these skills provides opportunity for developing understanding in other ways as well.

390 Although training in the use of calculators must not be allowed to interfere with the acquisition of appropriate skills of mental and written computation, we believe that their availability should influence the complexity of the calculations which pupils are expected to carry out with pencil and paper and also the time which is spent in practising such calculations. We believe that it is reasonable to expect that most secondary pupils should be able, without using a calculator , to multiply by whole numbers up to 100, even thought they may often use a calculator to carry out calculations of this kind. Some pupils will, without difficulty, be able to multiply by larger numbers and they should not be dissuaded from acquiring this skill. Pupils should also be able to divide by numbers up to 10. However, 'long division' has always presented greater difficulty than 'long multiplication', and we suspect that many adults who were taught this process whilst at school would be unable to explain why the method works. We believe that it is not profitable for pupils to spend time practising the traditional method of setting out long division on paper, but that they should normally use a calculator. Once again, pupils who are interested in acquiring the ability to perform this calculation 'long-hand' should not be dissuaded. We believe that lower-attaining secondary pupils should use a calculator for all except the most straightforward calculations. For these pupils emphasis should be placed on using the calculator correctly and employing suitable checking procedures.

391 The availability of calculators makes it necessary to consider also the extent to which mathematical tables will continue to be used in the future. The purpose of using logarithm tables is to avoid having to carry out heavy calculations with pencil and paper by making use of a method which is easier and quicker; and also, in some cases, to perform calculations, such as those involving fractional indices, which it is very difficult to carry out with pencil and paper. **A calculator provides a means of carrying out such calculations in a still more straightforward way and we believe there can be no doubt that calculators should replace logarithm tables as the everyday aid to calculation.** The more sophisticated 'scientific' calculators enable the values of trigonometrical functions such as sine, cosine and tangent to be obtained without recourse to tables. We have been told that many pupils already possess calculators of this kind. As the availability of these calculators increases we believe that the use of trigonometrical tables will decline and eventually cease; we would encourage this.

392 Calculators can be used to assist in the teaching of a number of topics, including work with operations, functions, exponents, polynomials, square

roots and problem solving. However, relatively little published material is yet available which illustrates ways in which calculators can be used as a means of developing mathematical understanding; much of the material which does exist is contained in the publications of the professional mathematical associations. We suspect that very few of those who teach mathematics at secondary level are at present making use of calculators as an aid to teaching; and that most of those who encourage their pupils to use calculators regard them solely as a way of avoiding tiresome computation. This can, of course, be very helpful when teaching a topic such as trigonometry, which usually involves a considerable amount of computation in its early stages. A reduction in the time spent on computation means that there is greater opportunity to develop the underlying concepts. Examples involving the use of Pythagoras' theorem to calculate lengths are also much easier if a calculator is available. However, these are not examples of the use of calculators in the way to which we referred at the beginning of this paragraph. **There is an urgent need for an increase in the limited amount of work which is at present being undertaken to develop classroom materials designed to develop understanding of fundamental principles.**

The availability of calculators

393 Throughout the preceding paragraphs we have assumed that calculators will be available for pupils to use. We conclude our discussion of the use of calculators in the secondary years by considering two matters which are related and which are likely influence the policy about the use of calculators in the classroom which is adopted by individual schools. The first is the extent to which calculators are available for the use of pupils in secondary schools, the second the question of their use in public examinations. At present there is no common policy among examination boards concerning the use of calculators in mathematics examinations. Some boards do not permit calculators to be used at all, some allow them in certain examinations or certain papers but not in others, a few permit calculators to be used freely.

394 One reason which is often given for not allowing calculators to be used in public examinations is that candidates who do not possess calculators will be at a disadvantage. This argument assumes that a school is unlikely to be able to provide a calculator for the use of each pupil who is attempting a mathematics examination and who does not have a calculator of his own. We do not accept that this should be the case. At the present time the cost of providing some 200 calculators of a type suitable for use by pupils in the 11-16 age range would not exceed £2,000 at a generous estimate. This number of calculators would enable all the pupils in a 6-form entry 11-16 school (approximately 180 pupils in each year group) who were likely to be studying mathematics at any one time to have the use of a calculator, even if no pupil possessed his own; it would also provide some spares. This number would also enable all those likely to be attempting a public examination in mathematics on any one day to be provided with a calculator. We have been told by one LEA that the approximate initial cost of equipping a specialist teaching area for science is £17,000. for woodwork £14,000, for metalwork £25,000, for cookery and housecraft £16,000, for needlework £7,500. This means that the

total cost of equipping specialist areas for the teaching of science, craft and home economics in a 6-form entry school is unlikely to be less then £150,000. We have also been told that the total cost of equipping six rooms of the kind usually provided for the teaching of mathematics is about £10,000. We do not believe that it is unreasonable to expect that one-seventieth of the difference between these sums should be spent on providing calculators and **we recommend that, when equipping new secondary schools, sufficient calculators should be provided to enable each pupil to have the use of one during mathematics lessons.** We recognise, however, that at a time of increasing financial pressure some of those schools which do not yet have sufficient calculators may find difficulty in providing money from their capitation allowances for the purchase of additional calculators. **We therefore recommend that steps should be taken, perhaps by means of a scheme similar to that for providing micro-computers in secondary schools, to ensure that the necessary calculators are available in secondary schools as soon as possible and in any case not later than 1985.**

395 It follows from this that **we believe that examination boards should design their syllabuses and examinations on the assumption that all candidates will have access to a calculator by 1985.** Because it will still be necessary to ensure that candidates are able to carry out straightforward calculations without recourse to a calculator, it may well be appropriate that the use of calculators should not be permitted in certain papers. It will, however, be equally important that candidates should also demonstrate their ability to use a calculator effectively.

Computer Studies

396 Computer Studies was first introduced into the curriculum of some secondary schools during the 1960s. It was developed largely by teachers of mathematics and in many schools there would at the present time be no computer studies in the curriculum at all if mathematicians did not undertake the teaching. For this reason, many people both within the education system and outside it have assumed that computer studies should be regarded as part of the responsibility of mathematics departments.

397 At an early stage of our work we sought advice from a number of people whom we knew to be expert in the field of computer studies in schools, so that we might decide whether computer studies (as opposed to the use of a computer as an aid to teaching mathematics) should be considered as part of the 'mathematics in schools' to which our terms of reference require us to give attention. **Their view was unamimous that computer studies should not be regarded as part of mathematics but should ideally exist within a separate department.** It was pointed out that the teaching of computer studies within the mathematics department was liable to give a misleading view of the subject, with too great a stress on programming and numerical operations and too little on the much wider fields of data processing and social implications; and that training as a mathematics teacher was not of itself sufficient to teach computer studies adequately.

398 However, at the present time the number of specialist teachers of computer studies is very small indeed and, although teachers of subjects other than mathematics are gradually becoming involved, the process is a slow one. **It therefore seems inevitable that for the time being, and probably for some time to come, a significant proportion of the teaching of computer studies will be undertaken by mathematics teachers, who will in consequence have less time available to teach mathematics.** It is probably also the case that mathematics teachers who have displayed the initiative to develop computer studies courses are likely to be among the more enterprising and effective teachers and so the loss to mathematics teaching is correspondingly greater.

399 It is not easy to quantify the amount of time which mathematics teachers devote to the teaching of computer studies. At the present time some 36 000 pupils each year attempt public examinations of one kind or another in computer studies or computer science, but not all of those who follow courses in computer studies necessarily attempt a public examination in this subject. However, it seems likely that the time which mathematics teachers devote to the teaching of computer studies is already equivalent to the normal full-time teaching commitment of at least 600 mathematics teachers, and probably of rather more than this number. We believe that this raises the question of the extent to which computer studies should be taught as a separate subject, if it is done at the expense of good mathematics teaching.

400 There is an associated demand on the time of mathematics teachers of which it is necessary to be aware, because it is a demand which is likely to increase as micro-computers become increasingly available in schools. This is the amount of time which can be taken up in responding to the requests of colleagues on the staff to write computer programs for use in the teaching of other subjects. It can take many hours to write even a short program and ensure that it runs satisfactorily. This means that although the purpose of a program may be clearly defined, its preparation is likely to involve a great deal of work, perhaps to the detriment of the mathematics teaching of staff who are already hard pressed.

401 Computing also makes demands on the time of mathematics advisers. Although a few LEAs now have advisers or advisory teachers who are responsible for computing, it is more usual for this responsibility to be assigned to mathematics advisers. Some mathematics advisers have told us that work in this field already occupies a disproportionate amount of their time. **There is clear danger that, unless LEAs make suitable additional provision, not only will the time which mathematics advisers devote to the teaching of mathematics be eroded still further as more schools acquire micro-computers, but also insufficient help will be available for the development of the effective use of micro-computers themselves.**

Computers as an aid to teaching mathematics

402 There can be no doubt that the increasing availability of micro-computers in schools offers considerable opportunity to teachers of mathematics both to enhance their existing practice and also to work in ways which

have not hitherto been possible. In particular, the availability of a visual display offers many possibilities for the imaginative pictorial presentation of mathematical work of many kinds.

403 Nevertheless, we feel it right to point out at the outset that, although these possibilities exist and are at the present time being exploited by a very small number of teachers, we are still at a very early stage in the development of their use as an aid to teaching mathematics. The amount of work which needs to be done before micro-computers are likely to have any major effect on mathematics teaching is very great indeed and, in the paragraphs which follow, we can do no more than make suggestions based on the limited amount of information which is available to us.

404 The fact that a school possesses one, or several, micro-computers will not of itself improve the teaching of mathematics or of any other subject. It does no more than make available an aid to teaching which, if it is to be properly exploited, requires teachers who have the necessary knowledge and skill and who have been supplied with, or have had time to prepare, suitable teaching programs. It does, however, also provide a valuable resource of which individual pupils can make use and from which some are likely to derive considerable benefit; we discuss this further in paragraph 412.

405 We have already drawn attention to the fact that relatively little advantage is yet being taken of the possibilities which electronic calculators offer as an aid to mathematics teaching. Even in schools in which computer studies courses are well established and are taught by those who teach mathematics, the use which is made in the mathematics classroom of the computer facilities which are available seems often to be very limited. It is, of course, sometimes the case that programs have to be sent away to be processed and that interactive working with the computer is not possible; nor may printout in graphical form be available. Nevertheless, even in these circumstances opportunities exist, especially in respect of work at higher levels, for using the computer to assist mathematics teaching.

406 We mention this matter not because we wish in any way to decry or discourage the provision of micro-computers in schools but in order to underline the extent of the changes in classroom practice which their successful use will require and the small amount of progress in the use of other aids which has so far been made by many teachers of mathematics. For example, experience shows that teachers of mathematics do not always use the graphical and pictorial potentialities of overhead projectors to maximum advantage. We welcome the Micro-Electronics Education Programme which the Government have set up. If good use is to be made of the materials which will be produced, a very extensive programme of in-service training and of follow-up support for teachers in schools and teacher training establishments will be required.

407 There is at present relatively little software (ie prepared programs) available which can be used to assist mathematics teaching. Furthermore, it

has been pointed out to us that much of the limited amount of software which is available is of poor quality, with programs which are badly written and documented, sometimes inaccurate and sometimes merely 'gimmicks'. There can be no point in producing software to teach a topic which can be taught more effectively in some other way. The fundamental criterion at all stages must be the extent to which any piece of software offers opportunity to enhance and improve work in the classroom. Even though the price of micro-computers may continue to fall, the cost of producing software will not. Those who purchase software for use in the classroom need therefore to make sure that it is of good quality.

Use in the primary years

408 Members of our Committee visited two primary schools which were known to possess micro-computers. The software which was in use had been prepared by teachers in these schools and had been designed to suit the needs of children of different ages. In a reception class of 4 and 5 year olds, a program in the form of a game was in use which was designed to develop recognition of figures and letters. With 9 year old juniors an interactive program, which required a group of children to locate a 'hidden' star, was being used to teach the use of co-ordinates. This generated valuable discussion as the group worked out the best strategy to use and also provided opportunity for the development of logical thinking. A third program provided practice in arithmetic skills for 7 and 8 year olds. In every case, when using these and other programs, the children were working with great concentration and the motivation which use of the micro-computer provided was very evident.

409 We believe that it is by means of programs of this kind that micro-computers are likely to make their greatest contribution to mathematics teaching in the primary years. The programs can be stored on cassettes and used by children singly or in groups, as well as by the teacher. There is clearly an urgent need for programs to be written for use by children in the primary years and **we hope that some of the money allotted to the Micro-Electronics Education Programme will be used for this purpose**. We believe that special attention should be paid to the development of programs for mathematical activities which will encourage problem solving and logical thinking in a mathematical context.

Use in the secondary years

410 Among the secondary schools which we visited were two which were making regular use of computers in their mathematics courses. In one school a computer-assisted learning scheme was in use for first and second year pupils. In the other school the computer was used as a means of introducing various algebraic concepts to first year pupils of above-average attainment.

411 At the secondary stage we believe that there is a special need to develop the potentiality of the high-resolution graphical display which is now available on many micro-computers. This enables work to be done on graph-plotting and, at a higher level, can be used to provide a visual presentation of basic ideas in calculus and of the use of iterative methods to solve equations. Many geometrical properties can also be demonstrated in ways which have hitherto only been possible by using cine-films. Moreover, the interactive

nature of work with a micro-computer offers opportunity for pupils to develop greater understanding of many of the mathematical concepts which they will meet. Once again, however, **we wish to emphasise the need to produce programs which are not just 'extras' but which can contribute to the main-stream mathematical work of the school.**

Individual use by pupils

412 We suggest that, in very many secondary schools, the most fruitful results of the availability of micro-computers are likely in the first instance to arise from their use by individual pupils. The motivation provided by access to a micro-computer can be even greater than that provided by a calculator. Experience shows that pupils are happy to 'investigate' computer systems to which they are given access and it is important that their ready acceptance of technological innovation should be fully exploited. Many examples already exist of individual pupils who, often largely self-taught, have developed extensive and effective programs of many kinds. We therefore urge that all secondary schools which possess micro-computers should make them available for use by individual pupils to the greatest extent which is possible. Such access should be afforded both during lesson times, if the machine is not being used for other purposes, and by organising computer clubs which operate outside normal classroom hours. Pupils frequently learn to write programs from one another, and we are aware of schools in which older pupils give time to teaching younger ones. Even teachers who are experienced in the use of micro-computers may find that they are able to learn a considerable amount by observing the activities of their pupils.

413 Although children of primary age will not normally reach the stage of writing their own programs, teachers need to be aware that there are a few children of high attainment whose interest and understanding are such that they wish to attempt work of this kind. In the schools which we visited, one 10 year old had started to write programs in BASIC as a result of studying programs published in magazines and watching his own teacher working on the micro-computer. An 8 year old, although not yet at this stage, was beginning to ask questions about programming. Children who show interest of this kind should be encouraged.

8 Assessment and continuity

Assessment

[handwritten: Intro]

[handwritten: discussion with pupils]

[handwritten: group active]

414 Assessment is an essential part of the work of all teachers; it needs to be carried out in a variety of ways and for a variety of purposes. Much assessment is based on the marking of written work and on information which the teacher gains as a result of comments made and questions asked during discussion with an individual pupil, a group or a class. Assessment can also be rapid and informal—perhaps a brief conversation to discover whether a point has been understood or a glance at a piece of work which is in progress in order to see whether a difficulty has been overcome. When pupils are engaged in practical activity, it is necessary for the teacher to watch them at work in order to discover whether skills such as counting, measurement or the use of drawing instruments are being mastered. Assessment in all of these ways is necessary in order to enable the teacher to monitor the development and progress of pupils.

Marking

[handwritten: Marking]

415 The form of assessment which is most immediately apparent to a pupil is the marking of written work; this may be routine class work or a more formal test. Such marking needs to be both diagnostic and supportive. A cross is of little assistance to a pupil unless it is accompanied by an indication of where the mistake has occurred, together with some explanation of what is wrong or a request to consult the teacher when the work has been returned. This style of marking also enables the teacher to become aware of the kinds of mistakes which are being made and to prepare later lessons in the light of this knowledge.

416 In the course of our visits to both primary and secondary schools we found a very wide variation in the amount of time and care given by teachers to the marking of pupils' work. In some cases marking was of the diagnostic and supportive type which we have already described. On the other hand, we found classrooms in which, although every piece of work had been marked, the result was merely page after page of ticks or crosses with little or no indication of where an error had occurred or what was wrong. There were classrooms in which marking had been so infrequent that pupils had continued to repeat the same mistake because a mathematical concept or routine had not been understood. We were also aware of instances in which, in our view, too little work was being set for pupils in secondary schools to do out of school because the teacher was not able to find the time to mark it.

417 Work in mathematics generates a great deal of marking and it is not usually possible nor, in general, desirable for a teacher to mark every piece of work which is done. A proper balance needs to be maintained between time spent on marking and time spent on preparation of lessons; however, work which has been marked should be returned promptly. Guidance about methods of marking and ways of resolving this problem should form part of a school's scheme of work in mathematics. In a well run class it should not be necessary for a teacher to mark every piece of work in detail. Exercises involving routine practice or applications can well be marked by pupils themselves, either from a list of answers or, in the case of computational exercises, by checking with a calculator. Pupils can then seek help if they are unable to locate and correct their mistakes. We believe that there is value in pupils learning to do this, though it is necessary for the teacher to maintain a general oversight in order to ensure that such marking is being carried out accurately. On the other hand, work involving problem solving or the application of more than a single skill will need appraisal by the teacher. As with so many other things, it is a sensible balance which is required.

Recording of progress

418 Assessment needs to be accompanied by appropriate recording of progress. All teachers carry in their heads a wealth of information about the pupils they teach and it is not always easy to record this concisely on paper. Nevertheless we believe that in both primary and secondary schools it is necessary to maintain a written record of a pupil's progress, not only as a reminder to his teacher and for the information of the head or head of department but also to enable continuity to be maintained when the pupil changes teacher or school. We discuss this further in paragraph 431.

419 It is not easy to decide on the most suitable form in which a record should be kept. Many LEAs provide a standard record card which schools are required to complete and which accompanies a pupil on transfer to another school. However, this may need to be supplemented by a more extensive record. In primary schools it can be helpful for this record to be related to LEA guidelines, though it is necessary to be aware that a checklist of the kind which records the topics which a child has attempted is open to abuse if children and teacher race to reach the end of the list. However, such a list can help a teacher to ensure that all children receive a balanced coverage of topics. In secondary schools the record can be related to the departmental scheme of work. Whatever form of recording is used, some effort should be made to record those qualities which can only be assessed by the professional judgement of the teacher, such as a pupil's persistence in working at a problem, his ability to use his knowledge and his ability to discuss mathematics orally.

Testing

420 **Testing, whether written, oral or practical, should never be an end in itself but should be a means of providing information which can form the basis of future action.** When tests are used in class they should be matched to the level of the pupils who are taking them. In a class which contains pupils whose attainment varies widely, it may not always be appropriate to give the same test to everyone. It is very easy to undermine confidence by using a test which is too difficult and a test which is too easy for those who are taking it

serves little useful purpose, although there can from time to time be a place for a straightforward test designed to encourage pupils to develop mastery of a specific skill. It is important that tests used in the classroom should be followed by opportunity for pupils to explain the thinking, whether correct or not, which has led to the answers they have given and for discussion of difficulties and misunderstandings which may be revealed. Teachers need also to be aware of the difference between 'learning something for the test' and learning in such a way that what has been learned is assimilated and retained for use in the future. Learning of this latter kind is not necessarily revealed by the result of a single test.

Standardised tests

421 Some testing has a wider purpose than the teacher's day-to-day assessment. The head teacher or the LEA may wish all pupils to be tested using nationally validated tests. Results of these tests give some guidance about the ability range and level of attainment of the pupils compared with national norms and should always be made available to the teacher. We are aware of increasing pressure in recent years from some members of the public and from some local councillors for the introduction of 'blanket' testing in mathematics for all pupils. We have been told that at the present time about one-third of all local authorities administer some form of mathematics test to their pupils, in some cases at more than one age. However, a remark which has been drawn to our attention—"no one has ever grown taller as a result of being measured"—underlines the fact that testing does not of itself lead to learning or to the raising of standards.

422 **If LEA testing of this kind is carried out, the tests which are used need to be selected carefully so that they will provide as much information as possible; they should not concentrate exclusively on particular aspects of the mathematics curriculum**. The results of the tests should be capable of use within each school to help teachers to diagnose, and if possible remedy, deficiencies and weaknesses. They should also be capable of use by the LEA as an aid in the identification both of successful practice and of schools which may be in need of additional support.

423 **It is important to realise that standardised tests measure only some aspects of mathematical attainment**. They do not test such aspects as attitude and perseverence nor, very often, the ability to apply mathematics in unfamiliar situations. We therefore recommend that, when the results of standardised tests are used as part of a transfer procedure between the different stages of schooling, these results should not be used by themselves but in conjunction with an assessment of these wider aspects provided by the pupil's previous school.

Evaluation

424 The results of assessment and testing should be used for the purpose of evaluation. The teacher should use the records of individual pupils to evaluate the progress of the class and compare this with the aims and objectives set out in the school's scheme of work. Evaluation of this kind also provides opportunity for the teacher to appraise the results of his or her own teaching, and

critical Evaluation

perhaps, in collaboration with the mathematics co-ordinator or head of department, to modify the scheme of work in some way. For example, if a topic has caused difficulty it may be necessary to reconsider the context in which it was introduced or the teaching approach which was used. In some cases, it may be felt that in a future year the topic should be deferred for comparable groups of pupils; on the other hand, there may be other topics which it is decided to introduce at an earlier stage. Mathematics co-ordinators, heads of department and head teachers need to evaluate the work of groups of classes and of the school as a whole. We believe that evaluation of this kind is not always carried out on a sufficient scale nor on a sufficiently regular basis. It is only as a result of such evaluation that it is possible both to identify successful practice and also to identify areas in which help is required.

Assessment of Performance Unit

425 The mathematics testing of pupils aged 11 and 15 which is being carried out by the Assessment of Performance Unit (APU) has yet another purpose. Its aim is to develop methods of assessing and monitoring the achievement of children at school and to seek to identify the incidence of under-achievement. It makes use of a sampling technique which does not provide information about the performance of individual pupils, classes or schools. The information which is being assembled by APU seems to us to be a potential source of very valuable information for those who plan mathematics courses at all levels, produce text books and other classroom materials and draw up examination syllabuses, as well as for all who teach mathematics. We hope that ways will be found to enable maximum use to be made of this information. **We recommend that in the near future an overall appraisal should be prepared of the educational implications of the mathematics testing which has been carried out so far**.

426 **A novel aspect of the work of APU is the practical testing which is being undertaken**. The examples of this which have been released show clearly the potential of work of this kind in the classroom and we believe that all teachers of mathematics should be made aware of this aspect of APU's work.

Continuity

427 We have already referred to the need for schemes of work to be prepared which will give adequate guidance to teachers; we have also drawn attention to the necessity of maintaining a record of each pupil's progress. If the scheme of work is properly used and the record maintained, it should be possible within a single school to ensure consistency of teaching method and continuity of syllabus as a pupil moves from class to class and from teacher to teacher. The pupil's exercise book will also provide a record of recent work and, unless there have been staff changes, it is possible for a pupil's new teacher to obtain information from the teacher who taught him previously.

428 At the time of transfer to a new school, continuity is more difficult to achieve but no less important to maintain. Transfer is most often to a larger school and many children can be apprehensive both of their new surroundings and also of the prospect of work with new and unfamiliar teachers. In recent years all schools have, we believe, become increasingly aware of the need to

ease problems of transfer. It is common for children to pay a preliminary visit to the school to which they will transfer and for staff of the receiving school to visit the schools from which their pupils will come. However, liaison activities of this kind, while making a major contribution to ease of transfer from a pastoral and social point of view, do not always pay as much attention as we would wish to ensuring continuity of mathematical development. Yet if a pupil is suddenly expected to attempt work which is beyond his capacity or finds himself bored at the outset by having to repeat work which has already been mastered, not only is his mathematical development interrupted but the whole process of transfer is subjected to unnecessary strain. During the first year after transfer, some schools send a report to their contributory schools about the progress of the pupils whom they have received. This is a practice which we commend; if the reports include reference to progress in mathematics, they can contribute towards the development of continuity in mathematics for pupils who will transfer in the future.

429 Problems of continuity on transfer from infant to junior school or from first to middle school are often less great than those which exist on entry to secondary or upper school. This is because infant and junior or first and middle schools are often closer together, their methods of working more similar and the range of attainment which exists among those who transfer less wide than is the case with older pupils. Nevertheless, problems can and do exist even in the case of transfer from infant to junior schools in adjoining premises. However, we believe that the greatest problems exist on transfer to secondary or upper school. These schools often receive pupils from a large number of contributory schools; in some areas, too, children from a single primary school can transfer to two or more secondary schools. As we have pointed out in paragraph 342, the spread in the mathematical attainment of these pupils can be very great and in these circumstances it is not very easy to make sure that pupils continue their mathematical education at a level and a speed which is appropriate. For convenience we refer in the paragraphs which follow to transfer between primary and secondary school; our remarks apply equally to transfer to middle school and upper school.

430 **It is essential that there should be discussion between those who teach in primary and secondary schools in the same area and that such discussion should take place in an atmosphere of mutual professional respect. Both primary and secondary teachers need to take steps to acquaint themselves with the methods and materials which each uses.** We believe that there are many secondary teachers who are unfamiliar with the approach to mathematics which their pupils have been using at the primary stage and also many primary teachers who have not taken steps to discover the type and range of work which those who leave them will undertake during their first term at secondary school. The outcome of such discussions should be overall agreement about the central topics which will have been tackled at the primary stage by pupils of different levels of attainment; where LEA guidelines exist it should be related to these guidelines. We stress that agreement of this kind should take account of the differences in attainment which will exist at the time of transfer. It should therefore be formulated in terms of a 'progression'

of work and not of an agreed list of topics with which all children are expected to be familiar at the time of transfer. However 'reasonable' such a list may seem to be, it cannot be suitable for all pupils. Furthermore, the existence of such a list can produce undesirable pressure on teachers in primary schools to cover all that is in the list even though some pupils will not be ready for some of the work which it includes.

431 We referred in paragraph 419 to the fact that many LEAs provide a record card which accompanies each pupil from school to school, and also to the fact that it is not easy to produce a record card which is sufficiently concise to be completed easily and sufficiently detailed to provide adequate information. However, provided that both primary and secondary schools are committed to the use of record cards, much valuable information about a pupil's range of understanding and skill can be passed on in a way which will help to ensure continuity. It can also be of help to attach to the record card two or three examples of recent work selected to indicate a pupil's general level and style of working.

432 In some cases pupils are given a mathematics test at the time of transfer to secondary school. Such a test is sometimes drawn up by the secondary school in consultation with its contributory schools and sometimes carried out on an LEA basis. **When tests are used, we recommend that they should be conducted before transfer in the primary school in which the pupil feels at home, rather than during the first few days at secondary school in an environment in which the pupil may not yet have settled.** We believe that pupils entering secondary school should not be tested in any formal way until they have had time to settle in their new school and preferably also had time to complete some work on a new topic. Where a transfer test is used it should be related to LEA guidelines or to the 'progression' which has been agreed between primary and secondary schools. This should enable those sections of the test to be used which are suited to a pupil's attainment and avoid the necessity of requiring a pupil to attempt questions on work which has not yet been covered.

433 Some teachers in secondary schools have told us that they take little or no notice of information which they receive from their contributory schools about the mathematical attainment of pupils because they prefer to give all their pupils a 'fresh start'. We cannot accept that it can be justifiable to ignore information provided by schools in which pupils may have spent as long as seven years. It can, of course, happen that a pupil who has not made good progress at his previous school is able at a new school to make much better progress as a result of encountering a different teacher and perhaps a different approach to the subject; the same thing can happen as a result of change of teacher within the same school. It is important that schools should be aware of this possibility and that at all stages the arrangements for teaching mathematics are such as to encourage and support progress of this kind, whatever the age of the pupil. We would also point out that such improved progress is in

any case more likely to occur if the teaching takes account of information which is available about a pupil's previous difficulties and sets out to resolve them.

434 Whatever steps may be taken to avoid categorising pupils prematurely in terms of their mathematical attainment, and to allow for the fact that a change of school may result in greatly improved progress, we can see no justification at all in ignoring and failing to act upon a report from a contributory school which indicates that a particular pupil is good at mathematics. To ignore such information may result in a pupil being given work which is far below the level of which he is capable. This will not only interrupt the pupil's progress but may also lead to the development of undesirable attitudes to mathematics because the pupil feels that his teachers are failing to recognise his ability.

435 Problems of continuity can also arise when pupils transfer to the sixth form. In 11-18 schools the change should present few problems provided that sixth form teachers recognise that some students will need guidance concerning the different pattern of study which is required of them and especially the need to organise private study time effectively. Problems of continuity are likely to be greater when pupils transfer at 16+ to the sixth form of another school or to a sixth form, further education or tertiary college. The need for consultation and transfer of information about attainment in mathematics is just as great at this stage as at the time of transfer from primary to secondary school and those who receive students at 16+ will need to take account of the variety of different backgrounds from which the students have come. Close liaison will be needed to ensure that choices of courses for sixth form study are realistic; pupils should neither embark upon courses for which they are insufficiently qualified nor be debarred from courses which they are likely to be able to undertake successfully.

9 Mathematics in the secondary years

436 As we have continued our work we have become increasingly aware of the implications for mathematics teaching of the differences in attainment in mathematics which exist among pupils of any given age and of the extent to which these differences increase as pupils become older. We drew attention in paragraph 342 to the 'seven year difference' which exists among 11 year olds. If we relate this to work in the secondary years, it means that the mathematical understanding of some pupils who transfer to secondary school at 11 is likely already to be greater than that of some pupils who have just left school at 16. On the other hand, some of those who arrive at the same time may not, while at school, attain the understanding which some of their fellow 11 year olds already possess. When considering work in the secondary years it is also necessary to remember that pupils learn at very different speeds and that, in the sense we have explained in paragraph 228, mathematics is a hierarchical subject.

437 Thus there are three factors—level of attainment at the beginning of the course, speed of learning and the need to obtain a sufficient understanding of certain topics before being able to proceed to others which depend on them—which must be fundamental to our consideration of the content of the mathematics curriculum at the secondary stage. Unless the great differences which exist between pupils are recognised in the provision which is made, those whose potential is high will be denied the opportunity to make the progress of which they are capable and those whose attainment is limited are liable to experience continuing and dispiriting failure. It is from this starting point that we discuss the teaching of mathematics in the secondary years.

438 We start by explaining the differing senses in which we use the words 'syllabus' and 'curriculum'. We use 'syllabus' to denote a list of mathematical topics to be studied but 'curriculum' to include the whole mathematical experience of the pupil; in other words, both what is taught and how it is taught. The curriculum therefore includes the syllabus; it is concerned with the way in which the syllabus is presented in the classroom as well as with other matters which are important. For example, problem solving, logical deduction, abstraction, generalisation, conjecture and testing should play a part in the work of all pupils. However, the kind of work which is undertaken and the methods which are used will of necessity vary with different groups of pupils; nor will the level of sophistication and the depth of understanding which are achieved be the same for all.

439 At the present time the mathematics syllabuses which are followed by most pupils during the secondary years are very strongly influenced by the content of the O-level syllabuses intended for only the top 20 to 25 per cent of pupils in terms of their attainment in mathematics. We believe it is necessary to explain how this has come about. Twenty years ago there were very great differences in the mathematics courses which were followed by pupils in different types of school. There were very few comprehensive schools and most pupils attended either grammar, technical or secondary modern schools. Pupils in grammar and the few technical schools and a small number of pupils in secondary modern schools followed a mathematics course leading to O-level which included arithmetic, algebra, geometry and trigonometry; some of these pupils also studied 'additional mathematics' as a second mathematical subject at O-level before entering the sixth form. Other pupils were usually taught arithmetic only, which sometimes included elements of geometry arising from calculations of area and volume, geometrical constructions, ratio and scale drawing and also simple graphs of a non-algebraic kind; most of these pupils left school at 15 and attempted no nationally recognised examination. Some of this latter group attempted regional or local examinations which varied in scope and status.

440 CSE, which pupils were able to attempt for the first time in 1965, was intended to replace these regional and local examinations and, together with O-level, to provide a public examination system for pupils aged 16 whose attainment in the subject concerned was within that of about the top 60 per cent of the school population. In consequence, many pupils in secondary modern schools, who had hitherto been taught only arithmetic, started to follow the broader CSE mathematics course, even though not all of these pupils stayed at school long enough to attempt the examination*. Since the raising of the school leaving age in 1973, the number of pupils who attempt public examinations, especially CSE, in mathematics has increased greatly; the report of the National Secondary Survey† shows that, in some of the comprehensive schools which were visited, over 80 per cent of the pupils in the fourth and fifth years were following courses leading to O-level or CSE examinations in mathematics.

*By 1968 half of the pupils in maintained schools were continuing at school after the age of 15; by 1973 the proportion was approaching 60 per cent, though these figures conceal considerable regional variation.

†*Aspects of secondary education in England*. A survey by H.M. Inspectors of Schools. HMSO 1979.

441 Success at grade 1 in CSE has always been accepted in lieu of a pass (now grade A, B or C) at O-level for purposes of qualification for entry to further and higher education as well as to many kinds of employment. From the outset, therefore, this has encouraged the use of similar lists of content for both CSE and O-level syllabuses. When CSE was first introduced, relatively few schools offered courses for both O-level and CSE but, as the number of comprehensive schools has increased, so has the number of schools in which it has become necessary at some stage to decide whether a pupil should attempt the O-level or CSE examination. Although in the case of some pupils the choice presents no difficulty, many schools prefer to postpone the choice for as long as possible for some of their pupils and for it to be possible for a pupil to change courses if the first choice proves to be wrong. This has led to understandable pressure that O-level and CSE syllabuses in mathematics

should be 'compatible', that is that they should contain substantially the same mathematical topics, and has lessened still further the differences between the syllabus content of O-level and CSE examinations. Our own study of a selection of CSE syllabuses in mathematics from the earliest years of the examination up to the present day shows that the content of most of these syllabuses has gradually increased. In particular, there has been an increase in the number of more difficult topics of an algebraic kind which have been included and a consequent increase in the conceptual difficulty of the mathematics included in the examination and hence in the courses which pupils are required to follow.

442 We believe it is clear from the preceding paragraphs that the changes in the examination system and in the organisation of secondary schools which have taken place in recent years have influenced the teaching of mathematics in ways which have been neither intended nor sufficiently realised. At the present time up to 80 per cent of pupils in secondary schools are following courses leading to examinations whose syllabuses are comparable in extent and conceptual difficulty with those which twenty years ago were followed by only about 25 per cent of pupils. We have therefore moved from a situation in which, twenty years ago, there was in our view too great a difference between the mathematics syllabuses followed by those who attempted O-level and those who did not to one in which, at the present time, there is far too little difference in the mathematics syllabuses which are followed by pupils of different levels of attainment. Because, for the reasons which we have explained, it is the content of O-level syllabuses which exerts the greatest influence, it is the pupils whose attainment is average or below who have been most greatly disadvantaged.

443 The difference in the level of mathematical attainment between a pupil who achieves CSE grade 1 and one who achieves grade 5 is very considerable yet, in CSE Mode 1 examinations*, both have to attempt the same examination papers. In some subjects other than mathematics it is possible to set questions which candidates of a wide range of attainment are able to attempt because the marks or grades which are awarded to the candidates can be determined according to the quality of their answers. In mathematics examinations of the kind which are most commonly used, an approach of this kind is not usually possible. This is because increase of attainment in mathematics is demonstrated not only by knowledge of additional topics but also by ability to tackle more demanding questions on topics which have been covered earlier in the course. An examination question on an 'elementary' topic which enables candidates to exhibit the level of understanding required for the award of a high grade is likely to be much too demanding for candidates who are able to attain only a low grade; such candidates may not even be able to start the question. Conversely, a question on the same topic which is suitable for candidates who will attain a low grade provides the examiner with little evidence of the capability of those candidates who are able to attain a high grade. It follows that an examination paper in mathematics which is attempted by candidates of a wide range of attainment is bound to contain questions which are too difficult for those who attain low grades. This means that low grades are awarded as a result of a very poor performance in an

*The different modes of examining are described in the note to paragraph 68

examination which draws attention not to the extent of a candidate's knowledge and understanding, but rather to those things which he does not know.

444 As we explained in paragraph 196, CSE regulations state that, in any subject, "a 16 year old pupil of average ability who has applied himself to a course of study regarded by teachers of the subject as appropriate to his age, ability and aptitude may reasonably expect to secure grade 4". It may not, however, be generally realised that the mark required to achieve grade 4 in mathematics is very often little more than 30 per cent on the papers which the candidate is required to attempt and that grade 5—"the performance just below that to be expected from a pupil of average ability"—is likely to be awarded to a candidate who scores little more than 20 per cent of the marks. **We cannot believe that it can in any way be educationally desirable that a pupil of average ability should, for the purpose of obtaining a school-leaving certificate, be required to attempt an examination paper on which he is able to obtain only about one-third of the possible marks**. Such a requirement, far from developing confidence, can only lead to feelings of inadequacy and failure. Although most pupils attempt the CSE examination itself on only one occasion, it must be remembered that during the year leading up to the examination they will almost certainly practise past papers and probably also take some form of 'mock' examination. Those pupils who will achieve the lowest grades or will be ungraded cannot therefore fail to become aware of how little they are likely to be able to achieve on the examination questions which they will be required to attempt. Reference to Figure 6 (paragraph 195) will show how large is the proportion of pupils to whom this applies.

445 We wish now to draw attention to a further difficulty which arises for both examiners and teachers as a result of the fact that CSE examinations are not suitable for many of the pupils who now attempt them. Examiners have a duty to set papers which cover as much of the syllabus as possible. Because they are aware that many low-attaining candidates will attempt the papers, they feel obliged to include within them a number of trivial questions on those topics in the syllabus which are conceptually difficult so that low-attaining candidates may find some questions which they are able to attempt. Teachers in their turn feel obliged to cover as much of the syllabus as possible so that their low-attaining pupils may be able to answer such questions, even though some of the topics which are included are conceptually too difficult for these pupils. This leads to teaching of a kind which, instead of developing understanding, concentrates on the drilling of routines in order to answer examination questions. We therefore have a 'vicious circle' which is difficult to break.

446 Many teachers are aware of this problem but feel unable to do anything about it. Their dilemma was expressed vividly at one of the meetings which we held with groups of teachers. "I know that I should not be teaching in this way and I would much prefer not to do so, but I know also that I have a responsibility to do all that I can to make it possible for my pupils to obtain a CSE grade." However, there can be little doubt that teaching of this kind can lead to disenchantment with mathematics on the part of many pupils and also provide the reason for the comments of young employees, noted during the

*See paragraph 59.

Bath and Nottingham studies*, that some of the mathematics which they had done at school had made no sense to them.

447 Some teachers have endeavoured to resolve their problem by devising limited grade CSE examinations under the Mode 3 procedure. Such examinations usually have a reduced syllabus content and also a maximum grade (commonly grade 3) which can be awarded. This enables the content of the examination to be matched more suitably to the needs of those who are likely to achieve the lower grades. More recently some CSE boards have themselves introduced Mode 1 examinations in mathematics or arithmetic with a maximum of grade 3. We support provision of this kind and regret that there are some CSE boards which do not permit limited grade examinations in mathematics. **We hope that, in the light of the discussion in this chapter, they will reconsider their position and permit examinations of this kind until such time as a single system of examination at 16 + is introduced.** However, even the use of limited grade CSE examinations cannot resolve the problem of providing a suitable curriculum for those pupils for whom CSE is not intended; we return to this point in paragraph 455.

448 We have discussed CSE examinations first of all because it is in the teaching of those pupils who enter for this examination and obtain low grades or are unclassified that we believe the greatest problems to exist. We must, however, emphasise that pupils who achieve grade E at O-level are also unlikely to have been provided with a course which has been suited to their needs. We have been told that schools sometimes find themselves under pressure from parents to prepare pupils whose attainment is not very high for O-level rather than CSE and that the change in the method of recording examination results on O-level certificates, which was introduced in 1975, has increased this pressure.* In our view such pressure is misguided and can be harmful to the mathematical development of the pupils concerned. However, we recognise that it is very difficult for teachers and schools to withstand it in view of the higher status which O-level enjoys and the greater acceptability which is often attributed to it by parents and employers. **As an example of this attitude we instance one employer who told us that he preferred to take applicants who had followed an O-level course but had not even achieved grade E rather than applicants with a good grade in CSE. This is an attitude which we believe to be mistaken and which we regret.**

*Until the beginning of 1975 results at O-level were expressed in terms of simple passes or failures, and no grades of pass were recorded on the O-level certificate. Since the Summer of 1975 results have been expressed in terms of five grades and an unclassified category. Candidates awarded grades A, B or C have reached the standard of the former pass; grade D indicates a lower level of attainment and grade E is the lowest level of attainment judged to be of a sufficient standard to be recorded. Performances qualifying for one of the grades A,B,C,D or E are recorded on O-level certificates but unclassified performances are not recorded.

449 We have considered the question of mathematics examinations and their effect on the mathematics curriculum in some detail because we believe that it is essential to appreciate the way in which existing examination syllabuses have come to influence work in the classroom, especially in the later secondary years, and that this influence can have harmful effects on the mathematical development of pupils. **In our view, very many pupils in secondary schools are at present being required to follow mathematics syllabuses whose content is too great and which are not suited to their level of attainment.** Efforts to introduce pupils to as much of the examination syllabus as possible result in attempts to cover the ground too fast for understanding to develop. The result is that very many pupils neither develop a confident approach to their use of

mathematics nor achieve mastery of those parts of the syllabus which should be within their capability.

450 This situation has arisen because the syllabuses now being followed by a majority of pupils in secondary schools have been constructed by using as starting points syllabuses designed for pupils in the top quarter of the range of attainment in mathematics. Syllabuses for pupils of lower attainment have been developed from these by deleting a few topics and reducing the depth of treatment of others; in other words, they have been constructed 'from the top downwards'. **We believe that this is a wrong approach and that development should be 'from the bottom upwards' by considering the range of work which is appropriate for lower-attaining pupils and extending this range as the level of attainment of pupils increases.** In this way it should be possible to ensure both that pupils are not required to tackle work which is inappropriate to their level of attainment and, equally importantly, that those who are capable of going a long way are enabled to do so.

Courses for 11-16 year old pupils

451 We cannot set out in detail the content of courses for 11-16 year old pupils; we can only indicate essentials and point to certain principles which should govern decisions on the curriculum. We believe that there should be a core of content which should be included in the mathematics course for all pupils; we discuss this further in paragraphs 455 to 458. Beyond that, both the topics to be included and the teaching approach to them must be judged in the light of the needs of particular groups of pupils. **We believe it should be a fundamental principle that no topic should be included unless it can be developed sufficiently for it to be applied in ways which the pupils can understand.** For example, we see no value in teaching and examining, in isolation and as a skill, the addition and multiplication of matrices to pupils whose knowledge of algebra and geometry is not sufficient for them to be able to appreciate contexts within which matrices are of use. On the other hand, some pieces of work may claim a place in syllabuses for higher-attaining pupils because of needs beyond the age of 16. It is, for example, hard to achieve the degree of facility in algebraic manipulation which is required for A-level work in mathematics if pupils have not started to develop it before they enter the sixth form and so this aspect of algebra must receive due attention for such pupils.

452. We therefore consider that there should be both similarities and differences in the mathematics courses which are followed by pupils of different levels of attainment. The similarities will be in the common core of mathematics of which all will have experience and in the approach which underlies the mathematics teaching. The differences will be in additional content, in depth of treatment, in methods of assessment and in the emphases given to different aims. For example, while some pupils will achieve a high degree of competence in simple arithmetic almost incidentally, others will continue to experience difficulty throughout the secondary years and so will need a course which takes account of this. In the paragraphs which follow we discuss the way in which we believe it is necessary to approach the formulation of such a

"differentiated curriculum"; in other words, the provision of different courses to meet the differing needs of pupils in the secondary years.

*Assessment of Performance Unit. *Mathematical development. Secondary survey report No1.* HMSO 1980
†K M Hart (Editor). *Children's understanding of Mathematics: 11-16.* John Murray 1981.

453 Since we started our work, the report of the first Secondary Survey carried out by the Assessment of Performance Unit (APU) * and the report of the Concepts in Secondary Mathematics and Science (CSMS) Project† have been published. These provide valuable information about levels of performance in a wide range of mathematical topics. The CSMS work concentrates in particular on the performance of lower-attaining pupils. Both of these reports illustrate clearly that large numbers of pupils have difficulties in understanding and applying many mathematical processes which are commonly thought to be quite elementary and that these difficulties are much greater than is generally realised either by the public at large or by some of those who teach mathematics. The very low marks which are attained by many pupils who attempt CSE examinations reinforce the evidence which is provided by the APU and CSMS reports.

454 Failure to understand a mathematical topic can result from the fact that the mathematical concepts involved are too difficult for the stage of mathematical development which a pupil has reached. It can also result from teaching which pays too little attention to practical experience or which moves ahead too rapidly so that understanding does not have time to develop. If pupils are following a syllabus which is too large and too demanding, both reasons can contribute to their poor performance. **We therefore believe that, when designing a curriculum which is suitable for lower-attaining pupils, the syllabus should not be too large so that there is time to cover the topics which it contains in a variety of ways and in a range of applications**. We believe that it is by approaching the teaching of mathematics in this way that understanding will best be developed and pupils enabled to achieve a reasonable degree of mastery of the work which they have covered, so that they will develop sufficient confidence to make use of mathematics in adult life. We realise that there may be some who fear that a reduction in the content of the mathematics syllabus will lead to an undesirable narrowing of the curriculum but this is a view which we do not accept. If the approaches to teaching are as varied as we have recommended in Chapter 5, we believe that such a reduction can enable a wider mathematics curriculum to be provided. Teachers will have more time to develop methods which evoke the best response from their pupils. Pupils will be freed from pressure to cover too much ground and will be able to undertake revision and consolidation on a regular basis to the extent which is necessary.

Foundation list

455 **We have therefore decided to set out a 'foundation list' of mathematical topics which, while it should form part of the mathematics syllabus for all pupils, should in our view constitute by far the greater part of the syllabus of those pupils for whom CSE is not intended, that is those pupils in about the lowest 40 per cent of the range of attainment in mathematics**. This group includes those who at present attempt CSE but achieve minimal or no success. We recognise that the content of the list which we propose is considerably less than the content of the syllabuses which many low- attaining pupils now

follow. We believe, however, that the smaller syllabus content will make it possible for the teacher to develop the mathematics curriculum for these pupils in the way which we have set out earlier.

456 Some people have urged us not to attempt to put forward such a list in view of two possible dangers. One is that members of the public will expect that all pupils will be able to master everything which is in the list. The other is that some teachers will regard the list as comprising the total mathematics syllabus for lower-attaining pupils and will teach only those things which are specifically mentioned in it. We are aware of both these dangers but feel that we should nevertheless suggest a list of topics which we expect to be within the capability of most pupils and which we believe that all pupils should attempt so far as they can. We would not wish that any section of the list should be omitted completely, even for those whose attainment is very low, though it must be accepted that such pupils may not be able to tackle all the items in each section.

457 We have not found it easy to draw up the list nor do we suggest that it should be regarded as definitive in every detail. However, we believe it to be a sound basis on which to build; as with all lists of content it will, in any case, need to be reviewed regularly. In preparing it we have started from the premise that mathematics should be presented and taught in a context in which it will be applied to solving a variety of problems. Once a skill has been identified as worthwhile, it is appropriate to practise it. However, such practice should be carried out in order to enable the skill to be used in the solution of problems and not as an end in itself. It is for this reason that in drawing up our list we have added comments (in italics) which both amplify the list and also draw attention to ways in which we believe the various topics should be presented. These comments concentrate especially on the needs of those pupils for whom the foundation list will constitute the greater part of the mathematics syllabus. **We do not intend that they should in any way be seen as limiting the range of work which should be attempted by pupils whose attainment is higher.**

Foundation list of mathematical topics

458 *Throughout their mathematics course, pupils should*

- *read, write and talk about mathematics in a wide variety of ways;*

- *carry out calculations in a variety of modes—mentally, on paper and with a calculator;*

- *associate calculation with measurement in appropriate units and become familiar with the relative size of these units.*

At all stages, pupils should be encouraged to discuss and justify the methods which they use.

NUMBER
Count, order, read and write positive and negative whole numbers and use them in context; eg what is the rise in temperature from $-3°C$ to $10°C$?

Understand place value in numbers of up to 4 digits.

Possess confident recall of addition and multiplication facts up to $10 + 10$ and 10×10 and the related subtraction and division facts.

Understand inter-relationships of the kind $13 \times 8 = (10 \times 8) + (3 \times 8)$.

Be able to select the appropriate operation (addition, subtraction, etc) for use in the solution of a variety of practical problems.

Pupils should possess some reliable method (however unconventional) of carrying out calculations without the use of a calculator when the numbers are small, and with a calculator when larger numbers are involved. There is a need to develop and encourage intuitive methods of both written and mental calculation.

Understand and use the decimal system in practical situations and problems. Add and subtract decimals involving up to two decimal places in the context of measurement (including money).

Pupils should appreciate the implications of movement of figures relative to the decimal point as a result of multiplication or division by a power of 10.

Use the language and notation of simple fractions in appropriate contexts; eg $\frac{1}{2}$ mile, two-thirds of the class.

Pupils should be able to perform calculations involving the word 'of' such as $\frac{1}{3}$ of £4.50.

Be able to add and subtract fractions with denominators 2, 4, 8 or 16 in the context of measurement.

Know the decimal equivalents of $\frac{1}{4}$, $\frac{1}{2}$, $\frac{3}{4}$, $\frac{1}{10}$, $\frac{1}{100}$ and also that $\frac{1}{3}$ is about 0.33. Be able to convert fractions to decimals with the help of a calculator.

The emphasis should always be on the use of number skills in everyday situations. Fluency in computation using methods whose basis is understood should be developed by short and frequent spells of practice; prolonged spells of routine practice are liable to be counter-productive.

MONEY
Recognise coins and notes and know that 100p = £1. Handle money with confidence.

Carry out simple transactions, performing necessary calculations either mentally or on paper.

Add and subtract small sums of money without a calculator.

Multiply or divide a sum of money by a single digit without a calculator.

Perform more complex calculations involving money using any appropriate method, including ready reckoner or calculator.

The emphasis in teaching should be on the use of money in everyday activities such as shopping, leisure activities, 'do-it-yourself', and on budgeting and household accounts.

PERCENTAGES

It is clear that very many pupils find great difficulties with the concept of percentage. We recommend that teaching of percentage should be based on the idea that 1 per cent means "1p in every pound" or "one in every hundred", and not on the use of particular formulae.

Calculate a percentage of a sum of money. Increase or decrease a sum of money by a given percentage.

Appreciate the use made of percentages in everyday life.

The use of percentages should be linked to activities of the kind listed under 'money'. The examples used should be realistic and relevant to the pupils concerned. Emphasis should also be given to the way in which percentages are used both for comparative purposes in many everyday situations and also as a numerical measure based on a 100-point scale of reference.

USE OF CALCULATOR

Use a calculator efficiently to add, subtract, multiply and divide, and to convert a fraction to a decimal.

Appreciate the need for careful ordering of operations when using a calculator.

Be able to select from the calculator display the number of figures which is appropriate to the context of the calculation.

Simple calculations should in general be carried out mentally or on paper and not with a calculator. Emphasis should be placed on using appropriate procedures for checking the reasonableness of the answer which has been obtained. It is not intended that the calculator should be used for calculations such as 3.52×7.04 which are unrelated to a practical context.

Substitute numbers in a simple formula expressed in words and evaluate the answer; eg

gross wage = wage per hour × number of hours worked;
total cost = number of units used × cost per unit + standing charge.

When the numbers involved are straightforward, pupils should also be able to carry out calculations of this kind without using a calculator.

TIME
Be able to tell the time and understand times expressed in terms of 12-and 24-hour clocks.

Be able to calculate the interval between two given times, and the finishing time given the starting time and duration.

Use a bus or train timetable.

Solve simple problems involving time, distance and speed.

Emphasis should be placed on mental calculation where appropriate and on the use of both digital and traditional 'clock' display.

MEASUREMENT
Measure length, weight and capacity using appropriate metric units.

Understand the relationship between millimetres, centimetres, metres, kilometres; grams, kilograms, tonnes; millilitres, centilitres, litres; know that 1 litre is equivalent to 1000 cubic centimetres.

Have a 'feel' for the size of these units in relation to common objects within the pupils' experience.

Use the following imperial units: inch, foot, yard, mile; ounce, pound; pint, gallon; and know their approximate metric equivalents, ie that 3 feet is about equal to 1 metre, 5 miles is about 8 kilometres, 2 lb is about 1 kilogram, 1 gallon is about $4\frac{1}{2}$ litres.

Understand and use simple rates; eg £ per hour, miles per gallon.

Read meters and dials of various types.

Emphasis should be given to practical activities involving measurement and estimation. Opportunity should be taken to relate teaching to measurement of the kind encountered in other curricular areas such as science, geography, home economics, craft and sport.

GRAPHS AND PICTORIAL REPRESENTATION
Organise systematically the collection and tabulation of simple data.

Read and interpret simple graphs and charts and extract specific information from them; construct them in simple cases.

Extract information presented in tabular form; eg cost of making a telephone call.

Pupils should be introduced to a wide variety of forms of graphical representation and use should be made of published information. Pupils should be encouraged to discuss critically information presented in diagrammatic form, especially in advertising. The drawing of graphs should wherever possible be based on information collected as a result of practical activities.

Be able to interpret a simple flow-chart.

Work with flow-charts should include discussion aimed at developing the logic used in mathematical arguments, such as 'if . . . . then . . . . . .'.

SPATIAL CONCEPTS

Recognise and name simple plane figures; understand and use terms such as side, diagonal, perimeter, area, angle.

Understand and use terms relating to the circle; centre, radius, diameter, circumference, chord.

Appreciate the properties of parallelism and perpendicularity; be able to measure angles in degrees.

Draw a simple plane figure to given specifications.

All pupils should learn to use drawing instruments such as ruler, compasses, set square, protractor. For some pupils this may require considerable practice. Plane and solid shapes should be available which pupils can handle and measure, and use for building.

Recognise and name common solid shapes; cube, rectangular block, sphere, cylinder, cone, pyramid.

Attention should be paid to the representation of three-dimensional figures in two dimensions. Pupils should be able to recognise the two- dimensional drawing corresponding to a given solid, and be able to recognise and build solids from such drawings. Use should be made of such things as plans of models, dress patterns, scale drawings, photographs, maps.

Find the perimeter and area of a rectangle.

Find the volume of a rectangular solid.

Understand and use the fact that the circumference of a circle $= \pi \times$ diameter; know that π is a little more than 3.

Calculations relating to plane and solid shapes provide opportunity to practise mental calculation, written calculation and the use of a calculator in a variety of ways and at different levels. Emphasis should be placed on the use of correct units for area and volume and on the use of estimation to ensure that an answer is reasonable.

Appreciate the concept of scale in geometrical drawings and maps, and the use of co-ordinates to locate areas (as on a street map) and points (as on an ordnance survey map).

There is likely to be considerable overlap between this topic and work on graphical and pictorial representation, especially in respect of the interpretation and construction of simple plans and elevations.

Understand the concept of bearings and the ways in which they are measured.

Be able to visualise and understand simple mechanical movement, including the working of simple linkages.

Bicycle gears, car jacks of all kinds, up-and-over garage doors are a few examples.

RATIO AND PROPORTION
Understand the use of ratio as applied to such things as mixtures, eg 2 parts sand to 1 part cement; and recipes, eg work out the quantities required for 6 people from a recipe which serves 4.

This work provides opportunity to discuss ideas such as 'best buy'. Applications to craft work and model making provide overlap with work on scale drawing.

Understand informally simple ideas of direct and inverse variation.

For example, as speed increases, distance travelled in a given time increases; as speed increases, time to travel a given distance decreases. For some pupils it will not be appropriate to attempt detailed calculation.

STATISTICAL IDEAS
One aim should be to encourage a critical attitude to statistics presented by the media.

Appreciate basic ideas of randomness and variability; know the meaning of probability and odds in simple cases.

Emphasis should be placed on the relevance of probability to occurrences in everyday life as well as to simple games of chance. For many pupils it will not be appropriate to undertake work involving combined probability.

Understand the difference between the various measures of average and the purpose for which each is used.

Attention should be paid to the different uses of the word 'average' in newspaper reports, but it is not intended that pupils should necessarily be expected to use the words mean, median and mode.

* * *

Provision for lower-attaining pupils

459 Having set out our foundation list we wish to consider the way in which, in our view, the mathematics curriculum of the secondary years should be developed 'from the bottom upwards' and so we consider first the provision

*Aspects of secondary education
in England. A survey by HM
Inspectors of Schools. HMSO
1979.

of courses for pupils for whom O-level and CSE are not intended, that is those pupils in about the lowest 40 per cent of the range of attainment in mathematics. It is clear from the report of the National Secondary Survey* and also from many submissions which we have received that such provision presents considerable problems in many schools. Many have urged us to stress that mathematics courses for these pupils should be 'relevant to the requirements of everyday life'. However, this aim is easier to state than to achieve. For example, we read in the report of the National Secondary Survey that "it was very common for the schemes of work in the schools visited to refer to the need to relate the mathematics taught to the problems of everyday life, but the convincing realisation of this aim was much more rare".

460 We stated in paragraph 455 that in our view the foundation list should constitute by far the greater part of the syllabus of lower-attaining pupils. In order to complete the syllabus and curriculum for these pupils it will be necessary to add a small number of additional topics and to set out in detail the teaching approaches which should be used; our comments in italics may be regarded as a first step towards carrying out this latter task. The topics which are added may be chosen for a variety of reasons. There are, for example, schools in which the pupils take part in sailing and orienteering, both as a school activity and in their leisure time. Although trigonometry as such is not included in the foundation list, it would be possible in such schools to include work relating to bearings and distances, and to discuss methods by which it is possible to fix one's position. Problems arising from navigation might lead some pupils to the beginnings of trigonometry.

461 The necessity of fulfilling the tasks which we listed at the end of Chapter 1 may also lead to the inclusion of some topics, for example topics which are required by pupils for their studies in other subjects. It may be necessary to introduce certain pieces of work as a vehicle for achieving the curricular aims which we set out in paragraph 438. For example, we believe that efforts should be made to discuss some algebraic ideas with all pupils. For some this will entail little more than the substitution of numbers in a simple formula given in words, discussion of problems of the kind "think of a number . . . ." and the identification of the patterns in a variety of number sequences; other pupils will be able to go further. Some people would maintain that by considering alternative methods of solving the same arithmetic problem one has embarked on the beginnings of algebra and we would also encourage work of this kind. However, as has been pointed out in official publications of various kinds over many years, formal algebra is not appropriate for lower-attaining pupils.

462 We also wish to draw attention to an extract from one of the submissions which has been made to us.

Mathematics lessons in secondary schools are very often not about anything. You collect like terms, or learn the laws of indices, with no perception of why anyone needs to do such things. There is excessive preoccupation with a sequence of skills and quite inadequate opportunity to see the skills emerging from the solution of problems. As a consequence of this approach, school mathematics contains very little incidental information. A French lesson might well contain incidental information about France — so on across the curriculum; but in mathematics the incidental information which one might expect (current exchange and interest rates; general knowledge on climate, communications and geography; the rules and scoring systems of games; social statistics) is rarely there, because most teachers in no way see this as part of their responsibility when teaching mathematics.

We believe that this points out in a very succinct way the need—which is by no means confined only to courses for lower-attaining pupils—to relate the content of the mathematics course to pupils' experience of everyday life.

463 As pupils grow older, the practical applications which are used in mathematics lessons should take account of this fact. It is here that the word 'relevant' becomes important, because illustrations and applications of mathematics which are well suited to 12 and 13 years olds are not necessarily appropriate for 15 year olds; the reverse is, of course, also true. Many of the topics which occur in our foundation list relate directly to topics which are commonly included in courses with titles such as *Design for living* which very often form part of curricula for less able pupils in the later secondary years. Where courses of this kind exist, every effort should be made to relate work in mathematics to the content of these courses. It may be possible to arrange that part of the mathematics teaching takes place within such courses.

464 Especially as pupils become older, a great deal more time is often given to written work than to discussion and oral work. This situation very often arises from the fact that pupils who have become disenchanted with mathematics as a result of lack of success over the years can present problems of control in the classroom which make it difficult to continue oral work for any length of time. However, lack of discussion almost certainly leads to further failure and so the problem is compounded. If the approach which we are recommending is followed throughout the secondary course it is to be hoped that, as pupils grow older, they will not become disenchanted with mathematics because they will have been able to experience success and develop confidence. Problems of control should therefore be lessened.

465 In order to present mathematics to pupils in the ways we have described it will be necessary for many teachers to make very great changes in the ways in which they work at present. This may not be easy; and unless help and support is available both within schools and by means of in-service training, little may be achieved. We therefore welcome the decision of the Mathematical Association to establish a Diploma for teachers of low-attaining pupils in secondary schools. **We believe that there will also be an urgent need for development work leading to the availability of additional classroom materials related to the content of the foundation list and making use of the approach which we are advocating**.

466 At the present time considerable pressure exists for lower-attaining pupils to be provided with some evidence of their achievement in mathematics before they leave school. Since existing public examinations are not suited to the needs of these pupils, schools are increasingly devising schemes of their

own. If we may judge from examples of such schemes which have been made available to us, there appears to have been a considerable increase in the practice during the time for which we have been working. We discuss this matter further in the following chapter.

Provision for pupils whose attainment is very low

467 Before we conclude our discussion of provision for lower-attaining pupils we wish to consider the teaching of those whose attainment is very low indeed. Such pupils are very often withdrawn from normal classes for much of the week and taught in a separate 'remedial' department. Much time is given to developing skills of reading and writing, and the teachers who work in remedial departments are usually experienced in work of this kind. These teachers frequently also teach mathematics to the pupils in their charge but are often less confident in this area; nor is the mathematics syllabus always planned in consultation with the mathematics department. For these reasons syllabuses can sometimes be excessively narrow. It is essential that there should be consultation between the head of mathematics and the head of the remedial department with regard to the mathematics which is taught and the teaching approach which is used. We stress yet again that a great deal of time needs to be given to oral work; mathematics should not be regarded as something which pupils can get on with by themselves while the teacher is hearing other pupils read. We also repeat our remark in paragraph 456 that the mathematics course should be broadly based and related to all the sections of the foundation list.

468 We consider that those who teach mathematics in other parts of the school should from time to time join in with the work of the remedial department, perhaps in a 'non-teaching' period or, preferably, on a time-tabled basis, in order both to assist with the teaching and also to become aware of the problems encountered by pupils whose mathematical attainment is very low. Such an arrangement can also act as a safeguard to ensure that due attention is paid to the mathematical needs of pupils who, although weak in language skills, are not correspondingly weak in mathematics. The needs of such pupils, for some of whom English may not be the mother tongue, are sometimes overlooked.

469 It is relatively uncommon for pupils to remain in a separate remedial department until they leave school. Pupils who have been withdrawn from normal mathematics teaching, sometimes for three years, often find themselves in the lowest mathematics sets in years 4 and 5. Unless there has been effective liaison between the mathematics department and the remedial department this can lead to serious lack of continuity for these pupils. Because of the problems of teaching pupils in bottom sets, such sets are often small and contain some twelve to fourteen pupils. It has been suggested to us that, instead of having two small sets of very low-attaining pupils, it is sometimes better to combine the pupils into one set which is taken by two teachers. We believe that an arrangement of this kind can offer considerable advantages, especially if it is possible to arrange for one of the teachers to be a

mathematics specialist while the other is experienced in dealing with slow learners. We suggest that some schools may wish to consider the possibilities of working in this way.

Provision for pupils for whom CSE and O-level are intended

470 We turn now to the provision of courses for those pupils for whom CSE and O-level examinations are intended. We have already stated that the foundation list, which we have proposed, should be included in syllabuses for these pupils as well as in syllabuses for those whose attainment is lower. However, many will be able to achieve mastery or near mastery of the content of the foundation list well before they are 16 and the syllabuses which they follow should be enlarged accordingly. Such enlargement will both add further topics and also increase the depth of treatment of many of the topics in the foundation list.

471 The syllabuses of the examinations which pupils will eventually attempt should not be allowed to exert too much influence on the curriculum of the early secondary years. We believe that, at the beginning of the secondary course, differentiation of content should not proceed too rapidly; instead, the differing needs of pupils should be met by differences in approach and in depth of treatment. In this way it is possible to provide maximum opportunity for any pupils for whom change of school or change of approach to mathematics results in markedly improved performance. It is necessary also to realise that the content of the examination syllabus and and the content of the teaching syllabus should not be the same; the teaching syllabus should be wider than the examination syllabus. The extra topics which are included should be chosen in the light of considerations of the kind which we have already discussed in paragraphs 460 to 462 and also in order to make the examination syllabus cohere. It is by 'embedding' the examination syllabus in a wider context in this way that mastery of the syllabus is enabled to develop as pupils increase their network of interconnections and so the level of their understanding.

472 **We do not propose to discuss the content of examination syllabuses at 16 + in any detail but instead to indicate two 'reference levels'.** We consider that the content of the examination syllabus for those pupils who at present achieve around CSE grade 4 should not be very much greater than that of the foundation list; that is, of a size markedly smaller than that of many existing CSE syllabuses. There should certainly be some increase in the geometrical content beyond that of the foundation list to include, for example, ideas which underlie trigonometry, such as scale drawing and comparison of right-angled triangles, and the use of Pythagoras' theorem. Experience may show that the availability of calculators will enable some elementary trigonometry to be included. Some simple algebraic work on formulae and equations which involves symbolisation is also desirable. At a higher level, we consider that a syllabus whose extent is comparable to that of existing O-level syllabuses represents a suitable examination target for pupils in the top 20 per cent of the range of attainment in mathematics. **We believe that there is also need for an examination syllabus whose content lies between that of the two reference levels we have indicated;** we discuss these matters further in the following chapter.

473   At this point, **we wish to stress that, in recommending a differentiated curriculum and an associated difference in examination syllabuses, we are in no way advocating any lowering of standards**. A considerable number of people have written to us to say that they did not understand much of the mathematics which they met at school. This has been supported by the evidence from the research studies into the mathematical needs of adult life and of employment. The researchers were told many times that too many topics were covered in mathematics, that the pace of teaching was too fast, that too little time was given to important aspects of the subject. **We believe that if pupils follow a course whose content is better matched to their level of attainment and rate of learning, they will achieve not only greater confidence in their approach to mathematics but also greater mastery of the mathematics which they study. This should contribute to improvements in attainment, attitudes and confidence and so to a raising of standards overall.**

Provision for high-attaining pupils

474   At all stages of the secondary course those whose attainment is high should have opportunity to undertake work which will enable them to extend and deepen their mathematical knowledge and understanding. Although it is to be expected that such pupils will be able to move through the syllabus at a faster rate than many of their fellows, this is not of itself sufficient. Steps must also be taken to develop powers of generalisation and abstraction, of logic and proof, of problem solving and investigation, as well as the ability to undertake extended pieces of work. Pupils should also be encouraged to increase their fluency in routine manipulation and to work fast. It should not be supposed that pupils automatically fulfil their potential or even realise themselves how much they are capable of achieving. Many pupils need help, and sometimes judicious pressure, to discover the 'over-drive' which they possess but of which they are not fully aware.

475   It is especially important that attention should be paid to the needs of high-attaining pupils in the early stages of the secondary course. There is a danger that the fact that they are likely to be coping successfully with the work which they are asked to do may mask the fact that the work is not sufficiently demanding and that these pupils are not being extended. In these circumstances, interest and enthusiasm can be lost and, once lost, may not be easy to regain later.

476   As with the teaching of all other pupils, variety of approach is essential. Although high-attaining pupils may well be able to work profitably on their own for quite long periods of time, they also need the stimulus of regular discussion with each other and with their teacher. They should be encouraged, too, to read books about mathematics and to learn something of the work of the great mathematicians of the past. Pupils of this kind can often appear to grasp new ideas very quickly but practice of the skills associated with these ideas is still required so that the ideas may be assimilated fully. However, the form in which the practice is undertaken should be varied and should not consist only of repetitive examples of the same kind. It is, for example, very often instructive to the pupil to be asked to compose further examples relating to the topic which is being studied. A stipulation that the

examples should lead to particular types of answer can require pupils to consider more deeply the structure which underlies the work which is being done. High-attaining pupils very often relish opportunities to work with computers and work of this kind should be encouraged whenever possible.

477   The provision of suitable work requires careful planning throughout the secondary course. We estimate that between 5 and 10 per cent of all pupils, though not necessarily of the pupils in any one school, are capable of working beyond the limits of existing O-level Mathematics syllabuses by the time they are 16. It is essential that these pupils should be enabled to continue their advance and should not 'mark time' at any stage of the course. It is therefore necessary that, throughout the course, additional provision of some kind should be made for them. There is, however, a fundamental difficulty to be overcome in most schools. Except in very large comprehensive schools or large grammar schools it is likely that the number of pupils for whom such additional provision is appropriate will not be sufficient to form a separate teaching group of normal size. This means that, because of the present shortage of mathematics teachers and the current restraints overall on staffing in schools, these pupils cannot usually be taught as a separate group but must form part of a larger group. This can pose considerable problems for the teacher of such a group, especially as pupils become older and the overall range of attainment within the group increases.

478   Some large schools provide for the needs of high attaining pupils in the fifth year by arranging for them to follow the O-level Additional Mathematics syllabus or prepare for one of the optional papers offered by at least one examination board. An arrangement of this kind presents no difficulty if there are sufficient pupils to form a separate teaching group for the purpose. However, if this is not the case, an attempt to work in this way with, for example, the whole of a top set is likely to result in a course which is too demanding for some of the pupils. This situation is clearly unsatisfactory and can lead to narrow teaching of an 'instrumental' kind with consequent loss of confidence on the part of some pupils which may dissuade them from continuing with mathematics in the sixth form. The situation can be still worse if, in preparation for the Additional Mathematics course, pupils are pushed to take the O-level examination in Mathematics at the end of the fourth year before they are fully prepared for it. On the other hand, if no attempt is made to provide additional work for anyone in the set, those whose attainment is high will be disadvantaged.

479   There are teachers who are able to make provision for pupils to undertake more demanding work—not necessarily for examination purposes—while the remainder continue with the ordinary Mathematics course. However, we believe that there are many schools in which insufficient provision is made for high-attaining pupils. We have therefore considered ways in which it might be possible to provide for the needs of high-attaining pupils in a way which would also offer help to their teachers, so that pupils who are likely to obtain a high grade at O-level will be enabled to extend and deepen their knowledge of mathematics in appropriate ways.

Extra Mathematics
480 We believe that consideration should be given to the provision of an extra paper which could be taken by some pupils at the same time as the existing O-level Mathematics papers (or papers at the corresponding level within the single system of examination at 16 + when it is introduced). The purposes of this paper, to which we shall refer as 'Extra Mathematics', would be

● to develop a deeper understanding of the topics included in the Mathematics syllabus and to take some of them, for example geometry, trigonometry and manipulative algebra, somewhat further;

● to introduce a small number of additional topics;

● perhaps most importantly, to adopt a rather more structured approach towards mathematics, so as to help pupils towards "an appreciation of mathematics as a self-contained logical system"*

Aspects of secondary education in England. A survey by HM Inspectors of Schools. HMSO 1979. page 112.

481 The additional content of Extra Mathematics would be very much less than that of existing Additional Mathematics syllabuses and the level of the paper would be such that it would be appropriate for between 5 and 10 per cent of the pupils in the whole of each year group. Extra Mathematics would not be intended as a course to be followed after the O-level Mathematics syllabus had been completed but as a means of extending the work of some pupils, particularly in the fourth and fifth years. It would be studied alongside and as an extension of O-level mathematics, though without any extra time allocation; it would not at any stage involve the selection of pupils to follow a separate course. We suggest that attainment in the Extra Mathematics paper should be recorded on the O-level certificate by an endorsement at one of two levels of success in a similar way to that which is used in the 'S' paper at A-level; and that a performance which did not achieve one of these two levels should not be recorded. We believe that a paper of this kind could both provide a suitably challenging target for high-attaining pupils and also, by means of its syllabus and the type of questions asked in the examination, give guidance to teachers as to ways in which the work of these pupils could be extended.

Provision for pupils whose attainment is very high
482 A very small number of pupils whose attainment is very high will need special provision over and above that which we have already advocated for high-attaining pupils. These pupils will need to be treated individually and special arrangements made for them. Schools should be aware that they may occasionally receive pupils of this calibre. We have been pleased to note a variety of initiatives for such pupils. Some LEAs, often in association with nearby universities, organise end-of-term, weekend or holiday sessions for these pupils. The Royal Institution has recently completed a pilot series of Mathematics Master Classes for young people, held on ten successive Saturday mornings; further series are planned. There are now regional, national and international mathematical competitions. **We believe that initiatives of this kind should be encouraged and extended.**

Mathematics across the curriculum

483 When considering 'mathematics across the curriculum' it is an over-simplified view to have regard only to the 'service' aspects of mathematics, though these should not be neglected. Those who teach mathematics should be aware of the mathematical techniques which are required for the study of, for example, science, geography, craft or home economics and make provision for them. They should also try to arrange that the mathematics course and the courses in other subjects are developed in such a way that pupils will be familiar with the necessary mathematical topics by the time they are needed in other curricular areas. Furthermore, there should be liaison between teachers so that those who make use of mathematics in the teaching of their subjects do not use an approach or a language which conflicts with that which is used in mathematics lessons.

484 Conversely, those who teach mathematics should be aware of the ways in which mathematics is applied within other subject areas and should ask their colleagues who teach other subjects to provide examples of the applications of mathematics which can be used in mathematics lessons. Too little attention is often paid to this aspect. Similar considerations apply to applications of mathematics which lie outside the school curriculum. Use should be made in mathematics lessons of material gathered from newspapers and other printed sources as well as of information which pupils are themselves able to provide.

485 'Mathematics across the curriculum' is not a direct analogue of 'language across the curriculum' because mathematics is not fundamental to learning and to the development of understanding in the way which is true of language. Nevertheless, because of the ways in which mathematics can be used as a means of communication, it can play an important role in the learning process in curricular areas which may seem to be far removed from mathematics, as well as in areas with which the links are more immediately apparent. The presentation of information by means of graphs, charts and tables, the use of time scales, the use of arrows to denote relationships are only a few examples. **Teachers of other subjects, as well as mathematics teachers, need to be aware of the part which mathematics can play in presenting information with clarity and economy, and to encourage pupils to make use of mathematics for this purpose.**

Organisational matters
Time allocation for
mathematics

486 In recent years the proportion of the teaching week given to mathematics has decreased in most schools as additional areas of study have been introduced into the curriculum. Records available within the DES show that some thirty years ago the usual allocation of time for mathematics in grammar schools was six periods per week, each of about 40 minutes. The survey of fourth year pupils in 150 modern schools, carried out in 1961 and quoted in the Newsom Report*, showed that the average time given to mathematics in mixed schools was 215 minutes per week; the average in boys' schools was 260 minutes and in girls' schools 180 minutes. The present time allowance for mathematics in most secondary schools is the equivalent of five periods of about 35 minutes in a week of 35 to 40 periods for all pupils up to the age of 16—some 175 minutes in all.

*Half our future. A Report of the Central Advisory Council for Education (England). HMSO 1963.

487 There have also been changes in the way in which the time given to mathematics has been distributed through the week. The customary pattern twenty years ago was for all pupils to do some mathematics every day—indeed, a 'daily dose' was thought by many teachers to be essential—but in recent years it has become increasingly common in comprehensive schools to timetable mathematics (as well as a number of other subjects) in double periods. This means that, in schools in which five periods are allocated to mathematics, they are very often timetabled as two double periods and a single period so that pupils are taught mathematics on only three occasions in the week; and these can sometimes occur on consecutive days.

488 We believe that the time allowance which is now usually found—between one-seventh and one-eighth of the teaching week—supported by appropriate homework is adequate time to devote to mathematics but that careful consideration should be given to the way in which this time is distributed within the timetable. This latter point can give rise to conflicts of interest which are not always easy to resolve. If, as we have advocated, pupils are to undertake practical and investigational work, some lessons which last for longer than about 35 minutes are likely to be required. There are some teachers who feel strongly that as many double periods as possible should be provided for mathematics in the timetable, but there are also teachers who feel that shorter and more frequent spells of mathematics teaching are more profitable. Although certain kinds of practical and investigational work require more than a single period for their successful completion, there are other kinds of practical and investigational work which can be completed satisfactorily within a shorter time. It must also be remembered that by no means all mathematics periods are used for work of this kind and that it is very easy for work to lose momentum towards the end of a double period. The number of double and single periods which are provided in the timetable needs therefore to be suited to the methods of working of the mathematics department and to the mathematics curriculum which is being used.

489 Although it is probably most common to divide the teaching week into between 35 and 40 teaching periods, other ways of arranging the teaching week exist which may or may not make it possible to timetable mathematics in either single or double periods. In practice, the way in which mathematics periods are arranged within the timetable must be the decision of each school, bearing in mind the wishes of the teaching staff and the constraints imposed both by the way in which time is allotted to other subjects and perhaps also by the geography of the school. However, whatever the pattern, we believe that it is necessary to ensure that the periods given to mathematics are suitably distributed through the week and also occur at different times during the day so that there is not, for example, an undue proportion of periods at the end of the afternoon.

490 Although we would not normally wish to see the allowance of time for mathematics reduced significantly below the level we have indicated for any pupils, we wish to draw attention to our suggestion in paragraph 463 that part of the mathematics teaching of some lower-attaining pupils could take place

as part of *Design for living* courses. In such cases, the number of periods timetabled as 'mathematics' could appropriately be reduced for these pupils.

The organisation of
teaching groups

491 **When deciding on the way in which the mathematics teaching groups in a school should be organised, we believe that the overriding requirement is to achieve a form of organisation which enables pupils to work at a level and speed which is suitable for them, and also one which enables the teacher to include within his teaching all the elements which we have set out in paragraph 243**. In particular, the form of organisation should be one which enables sufficient discussion and oral work to take place. We also believe it to be essential that the timetable should not be constructed in such a way as to impose on a mathematics department a form of organisation which it regards as unsuitable.

492 The simplest way of achieving such a form of organisation is by timetabling a group of classes from the same year group simultaneously for mathematics. This provides flexibility and makes it possible to arrange the teaching groups in whatever way is best suited to the needs of the pupils; it enables the teacher or method of grouping to be changed during the course of the school year and allows individual pupils to be transferred from one group to another. It also makes possible the rearrangement of the teaching groups for special purposes or to provide cover for teachers who are temporarily absent. If the number of mathematics teachers who can be made available is more than the number of classes which are timetabled simultaneously, the degree of flexibility is greatly increased. It becomes possible, for example, to provide more teaching groups than the number of classes; this not only reduces the average size of the groups but also makes it easier to move pupils between groups other than on an 'exchange' basis. Small groups can also exist without requiring others to be unduly large. Another possibility is to use the additional teacher as a 'floating' teacher who can assist with the mathematics groups in turn or work with pupils withdrawn from their normal teaching group. It should be noted that it is not necessary for a floating teacher to be timetabled for all the periods in which the same group of classes is doing mathematics. However, if a school is short of mathematics teachers, it may not be desirable to gain additional flexibility at the expense of making use of members of staff who are not well suited to teaching mathematics.

Teaching in setted groups
493 The majority of pupils in secondary schools are taught mathematics in groups formed on the basis of attainment. These are most usually 'sets' formed on the basis of attainment in mathematics, but there are also groups formed on the basis of overall ability rather than attainment in mathematics. In some schools mathematics is taught in sets made up within two or more 'bands' formed on the basis of overall ability. It is very important to realise that within any mathematics set (and even more within groups based on general ability) there will still be marked differences in the mathematical attainment of pupils; this is especially likely to be the case if only a small number of sets spans the whole ability range in a small school or in part of a year group in a large school. It is therefore essential that the teaching takes

account of these differences and is responsive to the needs of individual pupils. It should not be assumed that the same teaching approach will necessarily be suited to all in the group, that it will be appropriate for all to do exactly the same work or that pupils should always work as a single group. Teachers should also be aware of the danger that, even if unconsciously, both they and their pupils may lack expectation of what can be achieved by those in low sets and, indeed, also by those in high sets.

494   When mathematics is taught in sets it is important that there should be opportunity for pupils to be transferred between sets if their progress warrants this; and that, when it becomes apparent that a change is needed, it should not be delayed. However, the fact that transfer between sets should be possible does not imply that all sets should attempt to follow the same course. There needs to be common ground between the work of adjacent sets, but there should be differences too so that, so far as is possible, the course in each set is matched to the attainment and rate of working of the pupils in it.

Teaching in mixed ability groups

Aspects of seconday education in England. Supplementary information on mathematics. HMSO 1980.

495   The mathematics supplement to the National Secondary Survey report* shows that half of the schools visited taught mathematics in mixed ability groups during part at least of the first year, about one-quarter during the second year and one-eighth during the third year. However, by no means all of these schools contained pupils covering the full range of ability; for example, some were grammar schools and some were secondary modern schools. Of the comprehensive schools which contained pupils covering the full range of ability, the proportions using mixed ability grouping for mathematics were lower; just under half in the first year, one in five in the second year and one in fourteen in the third year. The use of mixed ability grouping in the fourth and fifth years was very rare indeed—no truly mixed ability teaching across the full ability range was found in any of the schools visited, although it is known that there are a very few schools in which mixed ability grouping is used in these years. It seems, therefore, that it is only a minority of mainly younger pupils who are taught mathematics in groups which span the full ability range.

496   There are undoubtedly some teachers who are able to work in stimulating and effective ways with pupils in mixed ability groups, especially in the earlier secondary years. Where there are such suitable teachers and this method of grouping works well, it clearly provides a form of organisation which is satisfactory and we see no reason to change it. However, we believe that teachers should not be required to work in this way if they are not able to do so successfully. We believe that standards are liable to suffer if mixed ability teaching is imposed upon mathematics departments against their will.

497   We have received much less comment than we had expected about the teaching of mathematics to mixed ability groups, as about classroom organisation generally. A few of the submissions which we have received urge the advantages of this type of organisation—notably that pupils are not 'labelled' and that lower-attaining pupils are therefore likely to make better progress.

The majority of the submissions which refer to mixed ability teaching draw attention to the great differences in attainment which exist between pupils and to the difficulty which some teachers find in coping with these differences within the same class.

498   A major problem to be overcome when teaching mixed ability groups is that of providing sufficient opportunity for oral work and discussion as well as for the practice of mental computation of various kinds. It is very difficult, except perhaps when introducing a topic which is new to all the pupils in the class or an investigation which all can attempt, to work with the whole of a mixed ability class for any length of time. Work with mixed ability classes in mathematics is often made more difficult for teachers because some commercially produced schemes which are commonly used with mixed ability groups are published in such a way that it is almost inevitable that work on several different topics is going on within the class at the same time.* This is in marked contrast to mixed ability teaching in other subjects in which pupils usually work at different levels within the same theme or topic. As a result, it becomes even more difficult for the teacher of a mixed ability class in mathematics to assemble a group of pupils to discuss and consolidate the work they are doing. Pupils therefore work on a largely individual basis; we discuss this method of working in subsequent paragraphs. We recommend that teachers who work with mixed ability groups in mathematics should limit as much as is possible the range of topics on which pupils are working at any one time in order to make class or group discussion possible. We are aware that one scheme which has recently been developed for use with secondary pupils in mixed ability classes has been designed on the basis that all pupils in the class will work at different levels within the same topic.

*This is because the set of materials does not contain sufficient copies of the material relating to any one topic for it to be possible for more than a small number of pupils to use them at the same time.

Individual learning schemes

499   The use of individual learning schemes for the teaching of mathematics has become more common in recent years than used to be the case. Although these schemes are often used with mixed ability classes they can, of course, be used with pupils grouped in any other way. We have received a small number of submissions from teachers who tell us that they use schemes of this kind with success and that the motivation of pupils is increased by working in this way.

500   In our view there are some major problems which need to be resolved when using such schemes. One is that of providing sufficient opportunity for oral work and discussion. Another is the difficulty of devising materials from which all pupils can learn satisfactorily and of ensuring that the necessary interconnections are established between the topic which is being studied and other pieces of mathematics. A third is the necessity for the teacher to have a detailed knowledge of all the material which is included in the scheme. For these reasons the successful operation of an individual learning scheme makes great demands on the teacher, especially in teaching groups of the size most usually found in secondary schools.

501 It is clear that considerable success can attend the use of such schemes in the hands of skilled teachers who are committed to their use, and are able to obtain a similar commitment from their pupils. We would not therefore wish to discourage teachers from working in this way if they are able to do so successfully. However, it should not be supposed that the use of individual learning schemes in mathematics is suited to all teachers or to all pupils. We believe that there are many of each who are able to work more effectively if some form of group teaching is used.

Deployment of teaching staff

502 In most secondary schools the deployment of mathematics teachers presents problems which have no easy solution. Many of these arise from the national shortage of adequately qualified mathematics teachers which exists and which we discuss in Chapter 13. The situation can be made even more difficult if there is a lack of suitably qualified mathematics teachers in a school in which some pupils are studying mathematics at A-level. In such schools there may be only one or two teachers who are able to teach the A-level work and this will deplete still further the availability of adequately qualified staff to teach the younger pupils. It must be for each individual school to deploy its staff in the best way possible in order to minimise the disadvantage to pupils. We believe that head teachers and heads of department are rightly concerned to do whatever is possible to ensure that the same pupils are not taught by inadequately qualified staff for several years running. We do, however, wish to draw attention to certain points which we believe should be borne in mind when assigning members of staff to teaching groups.

503 It seems often to be the case that teachers of other subjects who volunteer, or who are asked, to teach mathematics and who teach mathematics for only a small number of periods in each week are given groups containing pupils of relatively low attainment. In our view this is often a mistaken policy. It is not easy to teach mathematics to low-attaining pupils. We consider that, provided that the member of staff concerned is a competent teacher of his own subject, he is much more likely to be successful with a group of pupils of average or somewhat above average attainment. A group of this kind will be a great deal easier to teach than a group of low-attaining pupils, and, provided that there is suitable support from the head of department and from a well prepared scheme of work, it should be possible for both pupils and teacher to 'learn together'. This should also enable the teacher concerned to develop more quickly confidence and skill in teaching mathematics.

504 There is a further reason why it is undesirable for non-mathematicians who teach only a few periods of mathematics in the week to teach pupils whose attainment is low. When teaching such pupils, it is helpful to be able to provide examples drawn from as wide a variety of mathematics as possible to illustrate the topics which are being studied. It often happens that, within the mathematics which is being taught to other groups of pupils, a straightforward application arises which can be adapted for use with a group whose attainment is low. Those who teach mathematics to several groups of pupils are able to take advantage of opportunities of this kind when they occur. Those whose only mathematics teaching is to a group of low-attaining pupils

have no comparable opportunities and so their already difficult task becomes even harder.

505 Problems of timetabling sometimes make it necessary for one class to be taught by two different teachers. When shared teaching of this kind is unavoidable we believe that it should be timetabled with a group of pupils whose attainment is high. Pupils in a group of this kind can benefit from being able to 'pick the brains' of two teachers and will quickly question any inconsistencies which may appear to exist in the teaching they receive. If, as is probable, they study different topics with each teacher they will also be able without difficulty to keep work in both topics going ahead at the same time. If, on the other hand, shared teaching is timetabled for a low-attaining group, pupils will not receive the continuing reinforcement and revision which they need from period to period and, instead of questioning inconsistencies of approach between their two teachers, are likely merely to become muddled.

506 Shared teaching also provides a method by which an experienced mathematics teacher can provide help and guidance for a teacher who is weak or not mathematically qualified. It may be much better for an experienced teacher and a weak or mathematically unqualified teacher to share the teaching of two sets rather than to assign the whole of the teaching of one of these sets to a teacher who is unequal to the task and who may, in consequence, instil attitudes to mathematics which may not be easy to change at a later stage.

The head of department

507 We believe that the head of department has a crucial role to play in implementing the matters to which we have already drawn attention. Unless he or she provides positive and sustained leadership and direction for the mathematics department it will not operate as effectively as it might do and the pupils will be correspondingly disadvantaged.

508 **In our view, the head of department should be responsible for**

- the production and up-dating of suitable schemes of work;

- the organisation of the department and of its teaching resources;

- the monitoring of the teaching within the department and of the work and assessment of pupils;

- playing a full part in the professional development and in-service training of those who teach mathematics;

- liaison with other departments in the school and with other schools and colleges in the area.

In setting out this list we are aware that it does not, of itself, relate specifically to mathematics and we regard the responsibilities we have identified as being among those which should be undertaken by heads of all subject departments. However, we consider that, among heads of department, the head of mathematics has a task which can be especially difficult and demanding.

Some of the difficulties arise from the central position of mathematics within the curriculum and from its role as a service subject; others arise from the shortage of suitably qualified mathematics teachers. We therefore wish to expand on certain aspects of the responsibilities which we have listed above in order both that heads of department may be helped to appreciate the extent of their task as we see it, and also so that those who hold responsibility at a higher level, whether as heads, governors, advisors or LEA officers, may be aware of the necessity of providing such support and training as may be required.

509 Before we comment on the specific items in our list we wish to draw attention to the need for heads of department to have time to carry out their duties. This is a matter which has come through most strongly in the submissions which we have received from heads of department, as well as in our meetings with groups of teachers and in the discussions which we have held with some of the heads of department who wrote to us. Although it is possible to carry out some of the duties outside normal school hours, certain of them can only be performed while teaching is in progress and the necessary time needs to be provided within the timetable.

510 We put first responsibility for the production and up-dating of suitable schemes of work because it is by this means that the mathematics department makes clear its aims and objectives and provides guidance and help to its members as to the ways in which these may be achieved. The importance of an adequate scheme of work is too often underestimated. The report of the National Secondary Survey* draws attention to the fact that, although in all the schools visited there were written schemes of work of some kind, their quality and usefulness varied greatly. We believe that a suitable scheme of work, in addition to outlining the syllabus to be followed by the different year groups and by the different ability levels within these year groups, should also set out the aims and objectives of the department and give guidance on such matters as teaching method and policies for marking and assessment. It should indicate the teaching resources which are available and state the procedures to be followed for routine matters such as the issuing of text books and stationery. The preparation of such a scheme is a major task and it is one in which other members of the department can and should play a part; but the initiative and the final responsibility for both preparation and implementation must rest with the head of department.

*Aspects of secondary education in England. A survey by HM Inspectors of schools. HMSO 1979.

511 We have already discussed the matters which we believe should be borne in mind when deciding the allocation of staff to teaching groups. In some schools this is almost entirely the responsibility of the head of department; in others the head of department will need to make his own views known to the member of staff who is responsible for preparing the school timetable.

512 We believe that arrangements need to be made for holding departmental meetings on a regular and frequent basis. A regular weekly or fortnightly meeting can provide opportunities for discussion of curricular matters and classroom practice but, if there are only one or two meetings in the term, these

almost invariably become filled with administrative matters. Some schools already provide a timetabled period for this purpose and we hope that more will seek to do so. A record should be kept of matters discussed and decisions taken at departmental meetings. Regular meetings also assist the development of a 'team' approach within the department.

513 We suggest that copies of notes, worksheets and other material prepared by members of the department should be filed systematically so that they become available for others to use. Materials of this kind can provide considerable help to less experienced teachers and to those qualified in other disciplines who are teaching only a few periods of mathematics in the week.

514 We regard it as an essential part of the work of the head of department to be aware of the quality of teaching which is going on within the department and of the various styles and approaches which are being used. He should review the exercise books of pupils in different classes on a regular basis so that he will be able both to monitor the progress of each class and also to check that marking of written work and assessment of pupils is being carried out satisfactorily. In our view it should be normal practice for the head of department to visit lessons given by other members of the department. He should also make it possible for members of the department to see him at work in his own classroom and to see each other at work. Observation of this kind can be of special help to probationary teachers as well as to non-specialist teachers of mathematics.

515 The question of in-service training is considered in greater detail in Chapter 15, but we wish to stress here the very special responsibility of the head of department for giving guidance and support to teachers in their probationary year and also the need to encourage the professional development of the members of the department by delegating specific responsibilities to them. These may, for example, be concerned with the preparation or revision of a unit of work or an organisational task within the department. We draw attention also to the fact that the head of department must not neglect his or her own professional development; it is necessary to take steps to keep up-to-date with developments in mathematical education and to be aware of the journals of the professional mathematical associations and of the articles which appear in them.

516 Because mathematics is a service subject for so many other disciplines, the question of liaison with other departments assumes particular importance. The head of mathematics should not only liaise with the heads of other departments, but should also study their syllabuses and schemes of work. In this way he can become aware not only of the mathematics which will or could be used in the teaching of other subjects but also of the stage of the course at which it will be required. Liaison with the remedial department, if one exists in the school, is also important. As we have already said, it is too often the case that the two departments operate independently of each other, with no awareness of each other's syllabus or teaching methods. Liaison also needs to be established with other schools and colleges in the area, especially

those from which pupils enter the school and those to which they will transfer at 16 + or 18 + .

517 We realise that in many schools the head of department may already be carrying out all the duties we have outlined in the previous paragraphs. Equally, we believe that there are some heads of department who take a more limited view of their role and it is for this reason that we have attempted to delineate carefully the responsibility which should in our view be attached to this post. More generally **we feel it important that responsibility should be properly defined at all levels within the educational system and that responsibility should be linked with authority, accountability and assessment**. Principles of educational management are involved here which go well beyond the brief of our Committee, but we hope that continuing thought will be given to these matters and that, where appropriate, lessons may be learned from good management practice in other fields.

10 Examinations at 16+

518 Throughout our discussion of the teaching of mathematics in schools we have stressed the necessity of matching the level and pace of work in the classroom to the level of attainment of the pupils. We have drawn attention to the pressures which exist for as many pupils as possible to attain some form of accreditation in mathematics by the time they leave school and also to the fact that, for good or ill, the syllabus and papers of the examination which many pupils will attempt at 16+ are dominant factors in determining the content and pace of the work which is done in most secondary classrooms. It follows that **it is essential that the syllabuses of examinations at 16+ and the examination papers which are set should not impose inappropriate constraints on work in secondary schools. Pupils must not be required to prepare for examinations which are not suited to their attainment nor must these examinations be of a kind which will undermine the confidence of pupils.**

519 We discuss first the examinations intended for pupils whose attainment in mathematics is within the top 60 to 65 per cent of that of the school population as a whole. At present these are O-level and CSE, which are shortly to be replaced by a single system of examination at 16+. In the course of our work we have held discussions with representatives of several examination boards and have been permitted to observe the awarding procedures used by some boards which, in preparation for the introduction of a single system of examination, are offering an examination in mathematics which can be attempted by candidates for CSE as well as candidates for O-level. During these meetings we have been impressed by the skill and care with which the officers and examiners of the boards approach their task. We have also been helped by the written submissions which we have received from the examination boards.

520 We explained in Chapter 9 the problems which face those who set examinations in mathematics and the reasons why we believe that CSE and O-level examinations are not well suited to some of those who at present attempt them. We therefore welcome the Government's decision to introduce a single system of examination at 16+ because we believe that it provides an opportunity to introduce examinations of a kind which will assess pupils more appropriately than is the case at present. We also welcome the statement in paragraph 16 of the Government Paper *Secondary school examinations: a single system at 16+*[*].

*Secondary school examinations: a single system at 16+. Cmnd 7368. HMSO 1978.

The educational studies carried out by the Steering Committee led them to the conclusion that in at least some subjects it would be necessary to provide a variety of alternative examination papers and tests, at different levels of difficulty, in order to provide satisfactorily for candidates from the intended wide ability range. This is especially the case where, as in mathematics or modern languages, the range of skills involved is wide or certain concepts are within the grasp of some candidates but beyond the reach of others. The Government accept this view, and consider it essential that the examination system should enable all candidates to demonstrate their capabilities. The assessment procedures must, therefore, provide for the inclusion of items suitable only for some candidates, or required only for some candidates, and in such a way that the curriculum is not distorted for others.

521 **We believe that there are two fundamental principles which should govern any examination in mathematics.** The first is that the examination papers and other methods of assessment which are used should be such that they enable candidates to demonstrate what they do know rather than what they do not know. The second is that the examinations should not undermine the confidence of those who attempt them. Because the syllabuses which will be prescribed and the papers which will be set will be the greatest single factor in influencing the mathematics teaching in secondary schools in the coming years, we believe it to be essential that the examination should provide suitable targets and reflect suitable curricula for all the candidates for whom the examination is intended; and that in order to achieve this it will be essential to provide a number of different papers so that candidates may attempt those papers which are appropriate to their level of attainment.

The proposed single system of examination at 16+

522 We do not believe it is the task of this Committee to prescribe in detail the way in which those who have the responsibility for developing the new single system of examination at 16+ should carry out their task. We believe, however, that it may be helpful to outline a possible approach which we believe would be consistent with the curricular aims for the teaching of mathematics which we have already set out. **We hope that those who have responsibility for devising suitable examinations in mathematics will give this careful consideration.**

523 We outline the possible approach which we are suggesting in terms of the grades which will be awarded in the new single system. Seven grades of success will be certificated. Of these, grades 1, 2 and 3 will correspond to the existing grades A, B and C at O-level, and will subsume the existing grade 1 in CSE; grades 4, 5, 6 and 7 will correspond to the existing grades 2 to 5 in CSE*. We repeat that the new system is intended to cover the range of pupils for whom the O-level and CSE examinations are designed at present, ie about the top 60 to 65 per cent of the ability range in any subject. It is only for this group of pupils that we make the suggestion which follows.

*In particular this means that the existing CSE grade 4—the grade awarded to the 16 year old of average ability—will become grade 6 in the new single system.

524 The scheme of examination which we wish to outline would provide a range of papers which would enable each candidate to attempt a combination of these papers 'focussed' at one of three grades on the scale. For instance, the combination of papers focussed at grade 6 would be one on which a candidate who was awarded grade 6 would be able to obtain about two-thirds of the marks which were available. A rather higher mark would achieve grade 5 and a lower mark (but not, we would hope, below about 50 per cent) would merit

grade 7. It would be possible for a candidate who did exceptionally well and who achieved a very high mark on these papers to be awarded grade 4.

525 We believe that it would be appropriate to provide combinations of papers which would focus at grade 2 (appropriate for candidates expected to gain grades 1, 2 or 3), at grade 4 (for candidates expected to gain grades 3, 4 or 5, though with the possibility of achieving grade 2) and at grade 6. We would envisage that the syllabus for the papers focussed at grade 4 would be more extensive than that for the papers focussed at grade 6 and that some of the questions asked would be of a more demanding type than those included in the papers focussed at grade 6. There would similarly be a further increase in syllabus content and difficulty of question in the papers focussed at grade 2.

526 It would also be necessary to make provision for the award of grades below those suggested as appropriate for the combinations of papers focussed at grade 2 and grade 4. We would, however, expect teachers to advise pupils and their parents as to the combination of papers to be attempted in such a way that only in the most exceptional circumstances would it be necessary to award a grade lower than 4 on the papers focussed at grade 2, and lower than 6 on the papers focussed at grade 4.

527 We have been given to understand that there are some teachers who are expecting that the introduction of a single system of examination at 16+ will remove the necessity of advising pupils and parents as to the papers within the examination which pupils should attempt. However, the whole of our argument for a differentiated curriculum implies that the same set of examination papers in mathematics cannot be suitable for all pupils. If follows that those who teach mathematics must accept responsibility for giving such advice. We believe that the scheme which we are proposing will help teachers in this respect by giving them ample scope for formulating appropriate advice and will not require decisions to be taken at too early a stage.

528 We believe that it would be possible to implement the suggestion which we are making in a number of different ways. One way would be to set a series of four papers graded in difficulty and content so that, for example, papers 1 and 2 could provide an examination focussed at grade 2, papers 2 and 3 an examination focussed at grade 4 and papers 3 and 4 an examination focussed at grade 6. An alternative arrangement would be to provide three pairs of papers, one pair focussed at grade 2, one at grade 4 and one at grade 6, with some questions included in more than one pair of papers if this was felt to be necessary in order to establish comparability between the same grade awarded on different pairs of papers. If our suggestion for the introduction of 'Extra Mathematics' were to be adopted, the relevant paper would be available only to candidates attempting the papers focussed at grade 2.

529 We wish to point out that the provision of a paper which would be taken by all candidates would not accord with the fundamental principles which we have set out in paragraph 521 unless it was suitable for inclusion in a combination of papers which was focussed at grade 6.

530 The two 'reference levels' to which we referred in paragraph 472 would correspond to the syllabuses for papers focussed at grade 6 and grade 2. We have noted with approval the recent moves of some boards to cease to offer both 'traditional' and 'modern' mathematics syllabuses for O-level and CSE Mode 1 examinations and instead to adopt a single syllabus in mathematics. We hope that this policy will be followed by all boards when devising the syllabuses for the new 16+ examination and that alternative 'traditional' and 'modern' syllabuses will not be offered. When drawing up these syllabuses we believe that it will also be necessary to consider whether certain topics which are at present included in many O-level and CSE syllabuses should continue to be included in the new syllabuses. We cite as an example multi-base arithmetic. Although this is a topic which offers opportunity for interesting and often challenging work at a variety of levels in the hands of a skilled teacher, and which can therefore appropriately find a place in some classrooms, we do not believe that a question of the kind 'evaluate 27×3 in base 8' is suitable as an examination question at any level.

Teacher assessment

531 We wish now to discuss a further matter relating to methods of assessment in public examinations. Throughout our discussion of mathematics teaching in both the primary and secondary years we have stressed the importance of presenting mathematics in the classroom in such a way that pupils of all levels of attainment are made aware of the applications of the mathematics which they are studying. We have pointed out that, in order to do this, it is necessary for pupils to undertake relevant practical work, problem solving and investigations.

532 Examinations in mathematics which consist only of timed written papers cannot, by their nature, assess ability to undertake practical and investigational work or ability to carry out work of an extended nature. They cannot assess skills of mental computation or ability to discuss mathematics nor, other than in very limited ways, qualities of perseverance and inventiveness. Work and qualities of this kind can only be assessed in the classroom and such assessment needs to be made over an extended period.

533 It is possible to go further. Not only do written examinations fail to assess work of the kind we have described in the previous paragraph but, in cases in which they comprise the only method of assessment, they lead teachers to emphasise in the classroom work of a kind which is directly related to the type of question which is set in the examination. This means that, especially as the examination approaches but often also from a much earlier stage, practical and investigational work finds no place in day-by-day work in mathematics.

534 We have noted with interest that the first APU report on secondary testing at 15, when comparing the results of its written and practical tests, draws attention to the fact that different methods of testing "can highlight different aspects of pupils' performance, and can assess complementary

*Assessment of Performance
Unit. *Mathematical development.
Secondary survey report No. 1.*
HMSO 1980.

features of their mathematical knowledge".* It therefore seems clear that, **if
assessment at 16 + is to reflect as many aspects of mathematical attainment as
possible, it needs to take account not only of those aspects which it is possible
to examine by means of written papers but also of those aspects which need to
be assessed in some other way.**

535 We believe it is now widely acknowledged that it is appropriate to combine teacher assessment and written papers in the examination of pupils who
are likely to attain the lower grades in CSE; some CSE boards include an element of teacher assessment for all candidates who attempt an examination in
mathematics. Teacher assessment is, however, less common in O-level examinations even though, in some cases, provision exists for this to be included
or for pupils to submit project work which can be taken into account during
the awarding procedure. Because, in our view, assessment procedures in
public examinations should be such as to encourage good classroom practice,
**we believe that provision should be made for an element of teacher assessment
to be included in the examination of pupils of all levels of attainment.** The
proportion of the total assessment which this should represent is a matter for
discussion and might well vary, for example according to the level of attainment of the pupils concerned. We accept, however, that it is also necessary to
provide a method of assessment based entirely on written work in order to
meet the needs of those candidates, in practice often adults, who attempt the
examination independently.

536 If teacher assessment is to be included, it will be necessary for teachers to
develop the necessary expertise in carrying out such assessment and for the
necessary help and guidance to be made available. The aspects of pupil performance which are to be assessed will also need to be made clear; it will not, for
instance, be satisfactory for teacher assessment to be based on testing of the
same kind as is used in the written papers set by the examination board.
Suitable procedures for moderation will also need to be established. Although
some mathematics teachers are already accustomed to assessing their pupils
for the purposes of public examinations, many are not and time will be needed
to develop the necessary skills. For this reason, we believe that it will be
necessary in the early stages for teacher assessment to play a smaller part in
overall assessment procedures than will be possible and desirable as teachers
develop experience in this work. It has been pointed out to us that an increased
amount of teacher assessment within examinations at 16 + will require increased expenditure for training and travel. This, too, may delay the implementation of our suggestion while financial resources are limited.

**Evidence of achieve-
ment in mathematics
for lower-attaining
pupils**

537 We turn now to what we believe to be the more difficult problem of
providing evidence of achievement in mathematics for pupils whose attainment is below that for which O-level and CSE, and the new single system of examination at 16 + , are intended. The choice of 60 to 65 per cent as the proportion of pupils for whom O-level and CSE examinations are designed at present, and for whom the single system of examination at 16 + is intended, is an
arbitrary one: it is based on the recommendation which was made when the introduction of CSE was proposed nearly twenty years ago. Figure 6

(paragraph 195) shows that rather more than this proportion of pupils at present obtain a graded result in mathematics in O-level or CSE (and that the proportion increased from 1977 to 1979), and that a very much higher proportion obtain a graded result in English. If, as a result of following the more suitable courses which we have suggested, more pupils than formerly were to attain the standard at present represented by CSE grade 5 (and we would hope that this might be the case) the proportion of the school population who attained a graded result would increase. It would therefore be necessary either to accept that the examination was appropriate for this higher proportion of the school population or to raise the standard required for the attainment of particular grades in order to keep the proportion of the school population who achieved a graded result to some pre-determined level.

538 It has been suggested to us that, because of the demand on the part of pupils and their parents for some evidence of mathematical achievement, the new single system of examination should cater, in mathematics if not in other subjects, for a larger proportion of the school population than is at present intended by making possible the award of a grade or grades lower than 7. We wish at this stage to make two comments about this suggestion.

539 The first is that we believe it to be most important that the papers focussed at grade 6 should not be distorted and stretched to accommodate pupils whose attainment is below the level at present represented by the award of CSE grade 5. Any such distortion could result either in papers being set which were less well suited to their purpose than they should be or to a lowering of the level at which these papers were focussed. We would not wish either of these to happen.

540 The second is that **very careful consideration needs to be given to the means by which the mathematical achievement of lower-attaining pupils should be assessed. It should not be thought that the existing pattern of timed written papers towards the end of the fifth year, even if accompanied by a substantial element of teacher assessment, is necessarily appropriate for these pupils.** We discuss this further in the following paragraphs.

Schemes at present in use

541 Schemes which provide evidence of achievement on the part of lower-attaining pupils are already in existence in different parts of the country. Several schools have sent us details of 'certificates' which they have devised to provide for the needs of such pupils. In some cases a group of schools in the same area has adopted a common scheme, sometimes after consultation with local employers as we described in paragraph 97; in at least one case the tests which provide evidence of achievement are not only taken by lower-attaining pupils but are also taken by pupils who are expected to obtain a graded result in O-level or CSE. There are also schemes on a somewhat larger scale which have been introduced by local authorities, for example Hertfordshire, which any school in the authority may use. A national scheme is provided by SLAPONS (School Leaver's Profile of Numerical Skills) to which we referred in paragraph 98. One of the purposes of this scheme is to provide tests which can be taken earlier in the year than CSE and O-level examinations. The

results of the tests are then available to the many employers who recruit from the fifth forms of secondary schools in the months before CSE and O-level results are known. It is hoped that in this way the necessity for employers to set their own tests can be reduced. Certificates awarded as a result of other forms of school-based tests very often serve a similar purpose, in addition to providing pupils with tangible evidence of their achievement.

542   There is considerable variation in the ways in which these schemes operate. One pattern is of testing at one level only, sometimes with a requirement that pupils must demonstrate success on more than one occasion. Another pattern provides a selection of tests based on a 'core' together with 'options' from which pupils may choose. A third pattern is of a range of tests at different levels which pupils may attempt in succession. Schools which operate schemes of this third kind have told us that they consider such schemes to increase the motivation of pupils because they provide both incentive and evidence of progress over a period of several terms.

543   There is also considerable variation in the content of the tests which are used. A small number of those which have been sent to us cover a reasonably wide range of mathematical topics but most, including the SLAPONS tests, concentrate wholly or in great part on the testing of computational skills, very often divorced from any context. **We find this a cause for concern because we believe that the use of such tests encourages teaching of the kind which attracted adverse comment in the report of the National Secondary Survey*.** "Lessons seen were often narrowly conceived and in 60 per cent of the schools visited HMI considered that new courses should be developed for the less able pupils . . . . . There is a tendency to restrict the courses provided for the less able to routine calculation divorced from context and to fail to provide a sufficient range of applications of the mathematical ideas within the understanding of the pupils. . . . . . The improvement of these courses is a matter requiring urgent attention."

*Aspects of secondary education in England. A survey by HM Inspectors of Schools. HMSO 1979.

544   We therefore hope that schools which are at present using tests of some kind for their lower-attaining pupils will review the content of their tests so as to make sure that they do not lead to, or result from, a narrow curriculum but instead contribute to its widening in the ways we have suggested. Any schools which may be considering the introduction of tests at this level should consider carefully both the form and content of such tests and also the influence which they are likely to have on the mathematics curriculum which is provided.

Our own view

545   We have considered the principles which should, in our view, govern any tests which are used for lower-attaining pupils in schools. We believe that the principles which we have set out in paragraph 521 in respect of the single system of examination at 16+ should apply to any other form of testing which is used in secondary schools. Any tests should be within the capacity of the pupils who attempt them and should not undermine their confidence. Furthermore, those who succeed in the tests should do so as the result of a good performance on the questions which have been asked. The method of testing

should reflect a curriculum which is suited to the needs of the pupils and, if possible, encourage them to persevere. For this reason we believe that it is likely to be more helpful for lower-attaining pupils to be offered a series of short-term targets, success at each of which provides evidence of achievement, rather than to have to wait for a 'one-off' test when they are about to leave school.

546 In paragraph 455 we stated our view that the foundation list which we set out in paragraph 458 should constitute the greater part of the syllabus for lower-attaining pupils. It follows that the mathematical content of tests should not be limited to computation only but should be based on the foundation list. Thus any tests should include such things as the reading of graphs, charts and tables, mensuration, geometrical representation in two and three dimensions (plans, elevations, nets, etc), the interpretation of flow-charts and of other types of information given in mathematical form, and the use of a calculator. Computation should be tested at appropriate levels but, in our view, this should be done within a series of applications to problems in defined areas such as shopping, travel etc and not by examples testing the four operations in isolation. Practical and oral testing should also be included and a principal aim of any scheme of testing which is used should be to help pupils to acquire the feeling for number and measurement to which we have referred earlier in the report.

547 It would be possible for a school to satisfy the requirements which we have set out in the two preceding paragraphs by using not a single test but a series of tests which pupils could attempt in succession, perhaps from the age of about 14. An arrangement of this kind would make it possible for the tests to be criterion referenced, with each level of test related to the understanding of defined concepts and the ability to apply them to appropriate problems. It would also make possible a requirement that, in order to succeed at any level, it was necessary to achieve a high mark of the order of, for example, 70 per cent. The mathematical content of the lowest level, the number of levels and the size of the steps between them would require discussion. In our view from four to six levels would probably be the greatest number which would be practicable; and we feel that it should be possible within this number of levels to devise a realistic progression in terms of content and depth of understanding. Content could increase from level to level both in breadth and depth but the increase could be gradual; the lowest level might well consist only of oral questions and practical tests such as the measurement and estimation of length, weight and capacity, the reading of dials and very simple tables, and the giving of change.

548 In paragraph 542 we said that schools which operate tests at different levels had told us that such testing provides a valuable means of motivation for many pupils. We feel that this can be the case, especially during that part of the secondary school course when this is often most difficult to achieve, provided that the tests are of an appropriate kind and that the goal of success in a test is perceived by pupils to be possible of attainment in the not too distant future. For many pupils, too, a series of tests is better than one final test of mathematical attainment which allows no opportunity to make another at-

tempt if the result has not been satisfactory. Furthermore, there is no need for a pupil to start at the lowest level of graduated tests nor to attempt every level; pupils can attempt the appropriate level of test when they are ready to do so and can have more than one attempt if this proves to be necessary.

549 The amount of work involved in developing graduated tests which would fulfil the requirements which we have set out in previous paragraphs would be very considerable. It would not be an economic use of resources for a large number of schools to attempt such a task individually or even by working in groups. We therefore feel that it would be preferable for any development work to be undertaken on a larger scale so as to provide a resource, perhaps in the form of a central bank, which any school could use if it wished to do so.

550 Development work of this kind could result in the availability to schools of tests at various levels, together with instructions as to the way in which performance in the tests should be assessed. We believe that, in order to ensure that the amount of testing in any one school should not be excessive, some limits would need to be imposed on the availability of tests; these would have to be discussed. In any case, we do not consider that pupils should be allowed to attempt such tests until they were in the third or perhaps fourth year of the secondary course, assuming the age of transfer to secondary school to be 11, or at the corresponding stage within other forms of secondary organisation. In this way pupils of high attainment, for whom the tests would not be designed, would not be able to enter for them at a very early age.

551 It seems likely that, if graduated tests were to become available, a number of schools might include a record of achievement in the tests as part of profiles which they provide for the benefit of employers and those concerned in further education. It could also be the case that a record of achievement of this kind, which could be made available to prospective employers, would mean that pupils did not have to attempt a succession of mathematics tests set by different employers if they applied for several jobs. Furthermore, if the results of tests from a central source were used in this way, there would be no difficulty in making specimen tests available to employers who wished to inform themselves of the type and level of the questions which were being asked. In this way employers would be enabled to have clear idea of the mathematical achievement of pupils who succeeded at the different levels. Moreover, schools could arrange the timing of the tests in such a way that their results were available by the time at which pupils were starting to apply for jobs.

552 We believe that the availability from a central source of graduated tests based on the principles we have outlined could offer many advantages to schools. It could encourage the provision of better curricula and the motivation of pupils, and assist in the construction of pupil profiles. In particular, it could help to prevent an increase in the use of unsuitable tests; indeed, we would hope for a diminution in the number and use of such tests. Moreover, the availability of a more appropriate alternative for lower-attaining pupils should help schools to persuade these pupils, their parents, and employers that entry for CSE examinations in their present form, or for the examina-

tions which will replace them, might not be desirable. On the other hand, we accept that there could also be disadvantages. The availability of tests, however well constructed, would not of itself lead to the increase in mathematical understanding and to the 'at-homeness' with mathematics which we seek. We are well aware of the dangers of subjecting pupils to more, and more frequent, testing and we accept that the combination of certain kinds of testing and teaching could produce results which are the opposite of those we desire.

553   Nevertheless, we believe that consideration should be given to undertaking development work which could investigate the possibility of providing evidence of the achievement of lower-attaining pupils in a way which would assist and encourage the provision of suitable mathematics courses for these pupils, which could operate under the control of an individual school and which would enable use to be made of material produced outside the school. **We therefore recommend that a study should be commissioned to consider whether it is possible to devise a means of providing evidence of achievement in mathematics for lower-attaining pupils in ways which will support, and not conflict with, the provision of suitable mathematics courses in schools.** It should take account of the views we have expressed in the preceding paragraphs and also of other schemes which already exist. It would be necessary to make some assessment of the costs of any scheme which was suggested and to consider whether any procedure would be required to monitor and regulate the way in which the tests were used. **If it were to be thought that some suitable scheme might be devised in time to make it available to schools no later than the date at which the single system of examination at 16 + was introduced, and that the necessary finance was likely to be available to do so, then the study should be undertaken as a matter of urgency.** Meanwhile, we hope that individual schools which are using tests of their own will examine those tests in the light of the views we have expressed. **We believe, however, that any more general development of tests of the kind we have in mind, at either national or regional level, should await the outcome of the study we are recommending.**

554   If it were decided to make use of a central resource of some kind, we believe that those whose task it would be to set this up should take steps to draw on the experience of teachers throughout the country. The establishment of local working groups which could assist in developing appropriate questions would provide very valuable opportunities for the in-service training of teachers, especially those who were concerned with the teaching of lower-attaining pupils. In this way, the money spent on setting up such a central resource would also contribute to the in-service training of teachers.

555   Our suggestion that any tests which might be devised should reflect the content of the foundation list which we set out in paragraph 458 means that they would cover a large part of the syllabus which we have recommended for 16 + examinations focussed at grade 6. However, we do not consider that in devising the tests any attempt should be made to relate their syllabus content

to that of the 16 + examinations in any specific way. Nor do we believe that initially there should be an attempt to establish any direct correspondence between one or more levels in any graduated tests and specific grades in the 16 + examination. However, we believe it would be appropriate that the levels in any graduated tests should be such that, in general, candidates awarded grade 6 in the new single system at 16 + would also be able to achieve the highest level in the tests; we would expect that some of those who were awarded grade 7 would also be able to achieve this level.

556 If at some stage it were to be decided that it would be desirable, within the single system of examination at 16 + , to introduce the award of a grade or grades for mathematics lower than grade 7, we believe that experience gained in the study we have proposed would assist in the development of suitable assessment procedures.

11   Mathematics in the sixth form

557 Although A-level courses account for the major part of mathematics teaching in the sixth form, several other kinds of mathematics course are also provided in most sixth forms. These include 'service' courses for students whose A-level studies require a knowledge of mathematics beyond O-level standard but who are not studying mathematics at A-level; O-level and CSE courses for students who wish to improve the result which they have obtained in the fifth form; and courses designed primarily for 'new' sixth formers, that is those who are not studying any subject at A-level.

A-level mathematics

558 We start our discussion of A-level mathematics courses by drawing attention to an important change which has taken place in recent years. Thirty years ago, and for many years before that, it was usual to combine the study of single-subject mathematics with the study of two other science subjects, usually physics and chemistry; double-subject mathematics was usually combined either with physics or, less often, with both physics and chemistry. It was rare to combine mathematics with non-science subjects. Although mathematics is still very often combined with physics and chemistry, it has become increasingly common in recent years for students to combine the study of mathematics at A-level with subjects other than these. It is therefore not surprising—indeed, it is to be expected—that additional syllabuses and changes to existing syllabuses have been introduced which reflect the differing needs of those who combine mathematics with subjects other than physics or chemistry. In addition, pressure to avoid undue specialisation at the age of 16 has caused some people to ask whether it is educationally desirable to devote two-thirds of A-level study time to mathematics, as can be the case if double-subject mathematics is chosen; we return to this point in paragraph 586.

*The study was carried out by analysing a 10 per cent sample of applications for university places made through the Universities Central Council on Admissions (UCCA) in 1977 and 1978 by home-based candidates who had attempted mathematics at A-level.

559 A-level courses in mathematics, as in other subjects, are designed as two-year courses for students aged about 16 to 18. We have tried to discover how many of those who study mathematics at A-level go on to further study of some kind. A study* carried out recently for the Standing Conference on University Entrance (SCUE), which has been made available to us, suggests that in 1977 and 1978 just under 60 per cent of those who passed mathematics at A-level were accepted by a university in the United Kingdom; this number represents some 45 per cent of all who attempted A-level mathematics. Some who failed mathematics at A-level also gained a university place, but the information provided by the study suggests that their number was small. It seems, therefore, that between 45 and 50 per cent of those who studied A-level

mathematics in these two years gained university places. Tables 28 and 29 of Appendix 1 show that in 1977 and 1978, as also in 1979, almost 60 per cent of all entrants to universities in England and Wales who had an A-level qualification in mathematics chose degree courses in engineering and technology, the physical sciences or mathematical studies; about half of these chose engineering and technology, which is the largest single 'user' of university entrants with A-level mathematics.

560 Some of the others who studied mathematics at A-level will have been accepted for degree courses in the non-university sector and for other forms of further study, but it has not been possible to determine their number. However, though many of those who study mathematics at A-level go on to degree courses or other forms of higher education, by no means all do so; nor do all of those who proceed to higher education continue their study of mathematics. **It is therefore essential that an A-level course in mathematics should not only provide a basis for further study but also provide a course which is balanced and coherent in its own right and which reaches suitable 'stopping points' for those who will, at least for the time being, go no further.** This is not always easy. Sixth form teachers have told us that, in their view, some A-level syllabuses are over-extensive so that, although the work involved is within the capability of more able students, others find the work over-demanding. As a result, pressure to cover the whole syllabus can lead these students to become confused and to perceive mathematics only as a series of apparently disconnected techniques related to particular examination questions.

The teaching of A-level courses

561 We consider first the aims of mathematics teaching at A-level and the teaching methods which we believe to be appropriate. It should be one of the aims of sixth form teaching in mathematics, as in all other subjects, to enable students to develop the study skills which will prepare them for adult styles of learning; and the need to develop these skills exists equally for those who will go on to higher education and for those who will not. It is very easy for A-level teaching in mathematics to depend too much on exposition by the teacher and for students to adopt passive styles of learning. However, it is as important for students in the sixth form as for pupils of all other ages to develop problem solving techniques, to pursue independent investigations and to discuss and communicate their ideas. It is by working in these ways that students develop the confidence which they will require in order to be able to make use of mathematics in their future studies and careers. A-level mathematics is almost always taught to groups which are a good deal smaller than those in which mathematics is taught to younger pupils and sixth form teachers should exploit the opportunities which this provides to work with their students in a variety of ways.

562 It is also possible for mathematics at A-level to be presented as a very technical and somewhat arid subject with little relation to other school subjects except, perhaps, physics or to the activity of the world at large. It is therefore important that teachers should seek to counteract this impression by making use of opportunities which arise to emphasise the broader role of mathematics. The very varied applications of mathematics should be stressed and illustrations of these applications drawn from as wide a range as possible.

The increasing use of 'mathematical modelling' in, for example, the social sciences provides many possibilities for an enterprising teacher and many more traditional applications are to be found in the physical sciences. Reference to the historical background of some of the topics which are being studied can both help to explain their importance and also add interest and depth to the A-level course. A micro-computer can provide a stimulus to adventurous thinking, very often initiated by the students themselves; the investigative work which can arise in this way should be encouraged. Occasional discussion of some of the assumptions underlying mathematics and of the nature of the knowledge it provides is necessary if students are to be enabled to talk about mathematics in ways which others will understand. **There is at present a lack of teaching materials which assist sixth-form teachers to work in these ways and more are required.**

563 Most mathematics courses in sixth forms require students to undertake far less reading than is the case with courses in other subjects, including science. Much less use is made of libraries and individual reading and study is very often confined to a single textbook. It is not easy to learn mathematics from a book and the necessary ability can take a long time to develop. In order to read a mathematics text it is usually necessary to sit at a table with pencil and paper in order, for example, to fill in the intermediate working which is often omitted; students have to realise this. A helpful 'way in' can be to ask students to make use of the library to look at the way in which a different textbook deals with a topic which has just been studied; the knowledge of the topic which the students already possess will help them to follow the text. A next stage can be to ask students to approach a new topic by studying for themselves the relevant chapter in the textbook which they are using. The chapter can then be re-read and discussed by students and teacher working together so that the students can begin to appreciate the way in which it is necessary to set about reading mathematics. Students should also be encouraged to 'read around' the subject and to acquire knowledge on their own initiative.

564 Throughout this report we have emphasised the necessity of enabling pupils of all ages to achieve as much as they are able. It is especially important that students whose ability is high should be helped and encouraged to extend their work beyond the confines of the single-subject syllabus. Proper use of the library can contribute significantly to this end; it is by no means unknown for mathematically able students, with help from their teacher, to be able to cover the double-subject syllabus in the time which their fellows require to study the single subject.

The applications of mathematics

565 We have referred many times in earlier chapters to the importance of applying mathematics to the solution of problems. **We therefore believe that all A-level mathematics courses should contain some substantial element of 'applied mathematics' so that all who are studying the subject, whether for its own sake or because of its usefulness as a 'service subject' for their other studies, are able to gain a balanced view of mathematics.** It follows that we do not favour single-subject A-level courses which consist of pure mathematics only.

566 It is at this point that we encounter difficult and intractable problems. Thirty years ago no problem arose. The applied element of both single- and double-subject mathematics courses was almost always Newtonian mechanics; some syllabuses also included a certain amount of applied calculus, usually related to problems in mechanics. However, the study of Newtonian mechanics, although related very directly to the study of physics and engineering, is of little relevance in fields such as economics, geography and social studies; in these areas topics such as probability and statistics are of much greater application. The problem, therefore, is to decide how best to provide, often within the same teaching group, for the differing needs of students who may be studying mathematics at A-level for very different reasons. Some may have chosen mathematics because of its value as a service subject in the study of physics and engineering; some may wish to make use of mathematics in their study of, for example, biological or social science; some may wish to continue with mathematics as a principal study in higher education; in many schools there will be some for whom the end of the A-level course marks the end of their full-time education. It is necessary to remember also that, at the age of 16, many students will not yet have decided on their future area of study in higher education.

567 There are strong arguments that, in an ideal situation, all who study mathematics at A-level should have the opportunity to gain some knowledge of Newtonian mechanics as well as of probability and statistics. "Applications of mathematics, such as Newtonian mechanics, are part of our cultural heritage and of the human activity of mathematics. To learn calculus without understanding what led to its development and how it was used by Newton and others, is like learning to play scales at the piano without being shown any compositions."* Equally, the increasing use which is being made of statistical techniques in so many fields makes it highly desirable that those who study mathematics at A-level should have some understanding of probability and statistics.

*H B Griffiths and A G Howson. *Mathematics: society and curricula.* Cambridge University Press 1974.

568. It is the experience of many teachers that the early stages of both mechanics and probability and statistics need to be taught slowly and with great care, allowing ample time for discussion and for the underlying ideas to sink in and develop. Attempts to go ahead too fast can too easily result in failure to develop understanding of the fundamental ideas and so lead to difficulty and failure at later stages of the course. This means that, in practice, the 'ideal situation' which we described in the previous paragraph is very difficult to realise. Within the time available for the study of a single A-level subject it is extremely difficult to cover essential pure mathematics and also to go a sufficient distance in the study of probability and statistics as well as mechanics to reach suitable 'staging points' in each; that is, to go far enough to enable the work which has been covered to form a reasonably coherent whole. Some of the A-level syllabuses which have been introduced in recent years include probability and statistics as well as mechanics, but these attempts to include a wider range of applications have resulted in syllabuses which are very full and hence in courses which can be over-demanding for many students. In consequence, many fail to obtain a secure grasp of either of the applied elements of their course. The same arguments do not apply to those who study mathematics as a

double subject. We believe that all double-subject A-level courses in mathematics should include the study of mechanics and also of probability and statistics.

569 Although, for the reasons outlined above, we would have liked to recommend that single-subject courses should include both mechanics and also probability and statistics, we have concluded with regret that, because such courses make great demands on both teacher and students, it is not at the present time possible for us to recommend that all single-subject courses should be of this kind. On the other hand, we in no way wish to discourage those who are following such courses successfully from continuing to do so. It has been suggested to us that it would be possible to include rather less of both mechanics and probability and statistics within the single-subject A-level syllabus and still provide a coherent course. It has also been suggested that the use of computer simulation might enable essential ideas in mechanics and in probability and statistics to be assimilated more rapidly and in this way enable sufficient of each to be included in a single-subject syllabus. **We believe there is a need for curriculum studies into each of these possibilities.**

570 **We have considered whether we should recommend the study of one area of application rather than another but have decided that it is not possible to do this because of the diversity of students' future needs and interests.** The increasing use of statistical techniques, not only in the study of other subjects but also in business, commerce and manufacturing industry, provides many students with a strong reason for wishing to study probability and statistics. In addition, some who wish to study A-level mathematics may not have studied physics even to O-level standard and may be discouraged from choosing mathematics if the course includes only mechanics. It has been suggested to us that girls are more likely to study mathematics at A-level if they do not have to do mechanics, but we have no direct evidence about this.

571 Some of the greatest pressure for the inclusion of mechanics in mathematics syllabuses comes from those in higher education who are responsible for courses in engineering. However, almost all those who enter engineering courses at this level have also studied physics. Figures supplied to us by the Universities Statistical Record show that, of those who entered degree courses in engineering and technology at universities in England and Wales on the basis of A-level qualifications, almost 98 per cent had an A-level in physics or physical science in both 1978 and 1979. A great deal of the mechanics which is commonly included in A-level mathematics syllabuses is also included within most A-level physics syllabuses. This means that many students who study physics as well as mathematics at A-level cover many topics in mechanics twice, albeit from somewhat different standpoints. It could be argued that such duplication should not be necessary. However, for very many years most of those entering engineering degree courses have studied mechanics as part both of mathematics and of physics. In consequence, teaching for degree courses in engineering is based on an expectation of the level of competence in mechanics which this double study provides. Although some engineering departments accept individual students whose A-level mathematics course has not included mechanics, we have gained the impression that they do so with reluctance, believing that such students are

*T J Heard, *The mathematical education of engineers at school and university*. Department of Engineering Science. University of Durham 1978.

likely to experience difficulty in covering the content of their courses satisfactorily. A recent research study* into the mathematical education of engineers at school and university classifies A-level courses as 'traditional', 'compromise' and 'modern'. It points out that in some syllabuses there is only a restricted coverage of mechanics and that in others it is possible to avoid mechanics entirely. However, no evidence was found that students who had followed one type of A-level mathematics course performed consistently better or worse at university than students who had followed another type of course.

572 In passing, we wish to draw attention to the fact that, so far as we are aware, there is no European country outside the British Isles in which mechanics forms part of school courses in mathematics; it is considered to be part of physics. However, there are also considerable differences in the time required to complete first degree courses in these countries and in the structure of the examinations which are taken by students in preparation for university entrance.

573 The study of mechanics is not only of relevance to those who proceed to higher education courses in engineering; it is also of value to those who study mathematics and physics in higher education. We believe, too, that there are A-level students who, even though they will not continue their study of mathematics after the age of 18, nevertheless enjoy the study of mechanics and derive benefit from it.

574 One further factor needs to be considered. Partly as a result of changes in the structure of some mathematics degree courses in recent years, there are some teachers who find difficulty in teaching mechanics. On the other hand, some teachers, especially those who completed their degree courses some years ago, do not have sufficient knowledge of probability and statistics to teach in these areas effectively. A concentration on the teaching of one aspect of applied mathematics only could, therefore, add to the problems of deployment of mathematics teaching staff in some schools.

575 We realise that our support for either mechanics or probability and statistics or both to remain available in single-subject syllabuses does not resolve problems of choice of A-level syllabus which now exist in schools. In schools and colleges in which there are sufficient students studying mathematics at A-level for it to be necessary to form more than one teaching group, it may be possible to offer alternative courses in order to provide for the differing needs of students. However, it has been pointed out to us by some who teach in such larger establishments that a solution of this kind may fail to take account of the advantages which can accrue from teaching the same course to all students and gaining from the mix of A-level specialisms within the same teaching group which this makes possible. If there are only sufficient students to form a single teaching group, it must be for each school to decide on the course it will offer, taking account of the needs of the students, the capabilities of the mathematics teaching staff, and the mathematical balance of the course which is chosen. We believe that the greatest pressure for a particular type of course to be provided may come from

those who wish to study engineering at a later stage. Since these students will almost certainly be studying physics as well, we do not believe that their needs should necessarily be considered to be paramount when choosing the course to be followed.

A-level core

576 Although we have come to the conclusion that it is necessary for there to be variety in the applied mathematics element of A-level syllabuses we believe that different considerations apply in respect of the range of pure mathematics which is included. In the submissions which we have received there has been a great deal of support for the proposal that a 'core' of pure mathematics should be agreed which would form part of the syllabuses of all GCE boards. The few submissions which have argued against such a core have in general done so because of concern that A-level syllabuses are already becoming overloaded and that syllabus change and development would be inhibited; we believe there is substance in this concern.

577 Although some submissions have given specific support to the 'core' proposed by the Standing Conference on University Entrance (SCUE) and the Council for National Academic Awards (CNAA), many have expressed the view that the content of the SCUE/CNAA core is too large if, as is expected by those who formulated the core, "students should not only have been taught the whole of the core content but should be very well versed in it"*. We are aware that, during the time for which we have been working, the GCE boards have been engaged in consultation among themselves as a result of their study of the SCUE/CNAA proposal. We understand that they are moving towards agreement on a core of pure mathematics which is somewhat smaller than that originally proposed by SCUE/CNAA and which will constitute some 40 per cent of a single-subject syllabus in pure and applied mathematics. **We welcome this move and accept that the 40 per cent suggested forms a reasonable proportion of the total syllabus content. We would not wish to see the core become much larger because of the need to include a substantial proportion of applied mathematics and of the danger that any further increase would prevent the freedom of presentation and development which exists in the best sixth form teaching.**

*Standing Conference on University Entrance and Council for National Academic Awards. *A minimal core syllabus for A-level mathematics.* 1978.

578 **We believe it to be essential that, once a core has been agreed, it should be subject to regular review so that it can be changed as necessary to accommodate current developments.** A suitable body will need to be set up for this purpose. We suggest that such a body should be responsible to the A-level sub-committee of the Mathematics Committee of Schools Council or to such comparable committee as may be formed as a result of the current review of the works of Schools Council.

Variety of A-level syllabuses

579 In the last three or four years, there have been many adverse comments on the apparently very large number of A-level mathematics syllabuses which exist. However, those who have quoted a variety of figures between fifty and sixty have very often failed to point out at the same time that there are eight GCE boards. If each board were to offer only three syllabuses—'pure mathematics' and 'applied mathematics' to provide the double subject and 'pure and applied mathematics' for the single subject— there would be a total

of twenty-four syllabuses. Since it is now necessary to offer the single subject in two alternative forms (so that the 'applied' component is either mechanics or probability and statistics), this figure rises to thirty-two. It is therefore always likely to be the case that the number of A-level syllabuses in mathematics will appear high.

580 In fact, arithmetic calculations of this kind are not very helpful, because there is considerable variation in the way in which the GCE boards draw up and name their syllabuses. For example, some boards include probability and statistics and mechanics as alternative options within the same single-subject syllabus while others provide two syllabuses (with different titles), one of which contains probability and statistics and the other mechanics. We therefore wish to discuss the variety of A-level syllabuses in more general terms and to draw attention to some of the factors which have contributed to this variety.

581 One factor, to which we have already referred, has been the increase in the number of students combining the study of mathematics at A-level with subjects other than physics and chemistry, and the consequent need to provide syllabuses which include probability and statistics rather than mechanics. Another factor has been the world-wide movement towards modern mathematics which started in the 1960s. This has led to the introduction of new courses in many schools and, in consequence, of new A-level examinations which reflect the different content and approach of these courses. In the first place modern A-level syllabuses were associated with the School Mathematics Project (SMP) and Mathematics in Education and Industry (MEI). The SMP and MEI examinations were (and still are) available 'across the boards'; that is, they can be taken by candidates entering through any of the GCE boards. They did not, therefore, in the first instance lead to a great increase in the number of A-level syllabuses since the two sets of examinations served eight boards. However, in due course seven of the GCE boards introduced their own modern syllabuses for single and double subject and the total number of A-level syllabuses increased sharply. We believe that the number of syllabuses reached its highest point in the mid-1970s.

582 The number of different syllabuses is now decreasing. Syllabus changes during the last ten years have lessened the differences between the content of 'modern' and 'traditional' syllabuses and many feel that it is no longer appropriate to attempt to distinguish between them. We have noted with interest that none of the groups which were commissioned some six years ago to propose specimen syllabuses and examination papers in mathematics as part of the feasibility studies for the N & F proposals for examination at 18+ differentiated between traditional and modern mathematics. We support the view that the distinction should no longer be maintained. It should then be possible for the number of A-level syllabuses to be reduced further.

583 However, differences in examinations are not only a question of syllabus content, but are also concerned with the approach to the teaching of mathematics which the syllabus implies and the kind of papers and questions by means of which it is examined. These factors can be very relevant to a

school's choice of syllabus and we believe it is desirable that such choice should exist. Indeed, the difference between some 'modern' and 'traditional' syllabuses is as much in the approach which is used as in the topics which are covered. If, as we hope will be the case, agreement is reached on a core of mathematics to be included in all A-level syllabuses, existing syllabuses will presumably be revised as necessary to accommodate it, and any new syllabuses which may be introduced in the future will include the core. This should reduce the differences between the content of syllabuses and so help to resolve the problems which, we have been told, are caused for both students and their teachers at the beginning of higher education courses by the variations in the content of syllabuses which exist at present. However, it should not necessarily be expected that the introduction of a core syllabus will resolve all problems. Differences in the ways in which A-level courses have been taught may result in greater differences in the performance of students who have followed the same course than result from differences of content between syllabuses.

584 **We do not therefore consider that there are strong arguments for an arbitrary reduction in the number of syllabuses, except on the grounds of duplication.** Indeed, the possibility of introducing new syllabuses is essential if curriculum development is not to be inhibited. There is of necessity a limit to the change which can be made to existing syllabuses in mathematics, as in other subjects, at any one time and it is often easier to reflect curriculum development by the introduction of a new syllabus.

585 **We believe the monitoring of A-level syllabuses which is at present carried out by the A-level sub-committee of the Mathematics Committee of Schools Council is both valuable and necessary.** This includes the review and approval of new syllabuses and schemes of examination proposed by GCE boards or individual projects, and would provide a means of ensuring that any core which had been agreed was included, and assessed appropriately, in any new syllabus. The monitoring process also includes an annual scrutiny of the examinations set by at least two GCE boards; this provides a valuable means of exchange between boards and also of sharing expertise in examining. This work should continue at not less than its present level.

Double-subject
mathematics

586 The fact that mathematics at A-level can be offered as either one or two subjects causes a number of problems. Some have argued to us that double-subject mathematics leads to too great a degree of specialisation and that a more broadly based course, with mathematics as one rather than two out of three A-level subjects, provides a more balanced diet. We recognise the force of this argument, the more so since for many students the choice of double-subject mathematics has to be made at the age of 16 when future plans may be uncertain. Again, except in very large sixth forms, teaching groups for double-subject mathematics are almost always small and so, in comparison with mathematics teaching groups in other parts of the school, not economic in their use of staff. This can be a justifiable cause of concern at a time when, as we point out in Chapter 13, there is a shortage of suitably qualified teachers of mathematics. A further problem can arise from the fact that some entrants to degree courses which require a knowledge of mathematics have taken

double-subject mathematics while others have taken the single subject, so that it is difficult to establish a suitable starting point for all at the beginning of a degree course. However, there are very able students who profit from the double-subject course and we believe that it should continue to be offered to these students in schools and colleges in which the necessary staffing can be made available to teach the course without disadvantage to those who are studying mathematics at lower levels.

587 In our view, too much teaching time is often given to double-subject courses. We do not consider that it is either necessary or desirable to allocate to able students who take double-subject mathematics twice as much teaching time as is allocated to the single-subject course. We accept that those who are studying double-subject mathematics may need the time of two A-level subjects for their own work, but it should not be necessary for their teacher to be present throughout the whole of this time. Indeed, it is arguable that students who require this amount of attention from their teacher have not been well advised to choose the double-subject course.

588 We have already drawn attention in paragraph 180 to the fact that there has been a marked decrease in recent years in the proportion of university entrants to degree courses in mathematical studies who have a double-subject qualification at A-level; in 1979 this figure was 55 per cent. It is most important that schools and colleges should be aware of this and should point out to their students that it is possible in many universities and other institutions of higher education to follow mathematics degree courses successfully on the basis of a good performance in the single subject. It is equally important that those responsible for the early stages of these courses should take account of the fact that many of their students may not have studied the double subject. **We hope, too, that those who select students for admission to higher education will recognise that there are sound educational as well as economic reasons for offering only single-subject mathematics at A-level and will not put either direct or indirect pressure on schools which have only limited teaching resources in mathematics to provide the double-subject course, especially for students to whom it is not well suited.**

589 In recent years the pattern of some double-subject examinations has changed from that of 'pure mathematics' and 'applied mathematics' as separate subjects to that of 'mathematics' and 'further mathematics'. In the latter pattern, 'mathematics' is the single-subject A-level, containing both pure and applied mathematics, and 'further mathematics' consists of more advanced work which presupposes a knowledge of the single-subject syllabus. **In our view the pattern of mathematics and further mathematics is preferable because it does not imply a distinction between pure and applied mathematics, which, as we have already pointed out, we consider to be undesirable at school level.**

The use of formula sheets in A-level examinations

590 We drew attention in paragraph 562 to the need to avoid a style of teaching which concentrates on the acquisition of techniques at the expense of the development of a broader approach to mathematics. However, this does not mean that it is unnecessary for students to acquire fluency in the routine

processes which form part of A-level work. A number of the submissions which we have received from those who teach in higher education have drawn attention to a lack of confidence and accuracy in the routines of algebra, calculus and trigonometry on the part of many who start degree courses. This has been stressed especially by professors of engineering, who have also drawn attention to an over-reliance on the use of formula sheets on the part of many students and a consequent inability to recall simple formulae which are used frequently.

591 In recent years it has become the practice of almost all GCE boards to supply formula sheets which can be used during the A-level examination. There seem to be two main arguments for such provision. Firstly, formula sheets serve as a 'safety net' at a time of possible examination stress which can lead to sudden and damaging lapses of memory, sometimes in respect of quite elementary formulae. Secondly, they are a means of providing candidates with a list of formulae which are either difficult to remember or of relatively infrequent use; these are formulae which, when not working under examination conditions, students would probably obtain from a reference book.

592 We consider both of these arguments to be valid. However, there seems no doubt that the provision of formula sheets has led some teachers to assume, and to allow their pupils to assume, that there is no need to commit to memory any of the formulae which appear on the sheet. This is a view with which we cannot agree. While it is important that students should not be expected to memorise results which have not been adequately derived and discussed, we consider that many students handicap themselves because they do not have rapid recall of certain results which are of fundamental and recurring use in the development of the subject. At all levels understanding must carry with it a degree of remembering and it is our view that, unless students have confident recall* of such results as the trigonometrical addition formulae and the derivatives and integrals of simple algebraic and trigonometrical functions, they will lack the 'building blocks' which they need to develop their study of mathematics satisfactorily. They will also be handicapped in their study of other curricular areas which make extensive use of mathematics both in the sixth form and also in higher education.

*We draw attention to our discussion of memory in paragraph 234. The learning of formulae and standard results should be associated with the use of suitable checking procedures; for example, 'sine is an odd function', or 'put $x = 0$'.

593 It has been suggested to us that because, so long as formula sheets continue to be provided, they are liable to be misused we should recommend either that they should no longer be provided by examination boards or that the more elementary formulae should not appear on them. We do not accept this solution. Quite apart from the difficulty of securing agreement on which formulae should be included and which should not—and a formula which one student may be able to remember with ease may for some reason present difficulties to another—the formula sheet would not serve its purpose as a 'safety net' unless it was complete.

594 **We believe that formula sheets should continue to be provided at A-level and that formulae which relate to any agreed 'common core' should appear on the formula sheets of all boards.** However, we reiterate that the reasons for

this are those which we have already given in paragraph 591 above. The fact that formula sheets are provided should not be regarded by either teacher or student as replacing the necessity for memorising and developing confident recall of fundamental and frequently used results.

I-level mathematics

Examinations 16–18. A consultative paper. DES and Welsh Office 1980.

595 Since our Committee was set up, the Government have published proposals for the development of free-standing intermediate examinations (I-levels)* as a means of broadening the studies undertaken by some of those who currently take full A-level courses. An I-level course would last two years and occupy about half the time normally given to a full A-level course.

596 We support the suggestion that a course of this kind should be available for students who are not studying mathematics as a full A-level subject. We do not, however, believe it will necessarily be easy to design a suitable course and we expect that considerable development work will be required. In our view an I-level course should not be envisaged merely as a replacement for the 'service' courses which are at present provided in some sixth forms, though it would serve some of the purposes of these courses. Nor should it be a course which is comparable to existing AO-level courses in mathematics which would serve the needs of those who find a full A-level too demanding; AO-level courses should continue to exist for this purpose. We believe that the aim of an I-level course should be to develop mathematical ideas and extend previous knowledge without setting ambitious targets in terms of manipulative competence. For example, although calculus would be included, students should not be expected to spend time acquiring facility in the differentiation and integration of complicated functions. The course should illustrate the many ways in which mathematics can be applied and also include some study of the ways in which the subject has developed. We are not aware of any existing course which would be suitable, though we believe that use could be made of some of the ideas which are contained in the *Mathematics Applicable** course and in the N-level study entitled *Mathematical Awareness*. **An I-level course of the type we would wish to see would require skilled teaching and this would have staffing implications for schools and for in-service education.**

*Schools Council Project MA 1601.

597 **We believe that there would also be a place for an I-level course in statistics.** Such a course could serve the needs of many students, especially those who are studying A-level courses such as biology, geography, sociology or economics, in which there is an increasing emphasis on the critical examination and analysis of numerical data. In evidence to us the Royal Statistical Society and the Institute of Statisticians have stressed that statistics is not merely a collection of techniques but is a practical subject devoted to obtaining and processing data; and that the study of statistics should not become separated from the origins of that data. They have also pointed out that statistics in schools frequently ignores the practical situation and concentrates on formal manipulation. Within such an I-level course as we propose there should be time and opportunity to adopt a practical approach and to place emphasis on the application of statistical techniques to data which the students themselves have collected in the course of their own laboratory and field work. In this way it would be possible to demonstrate clearly the applica-

tion of statistics to the analysis of data arising from study in several different areas of the curriculum and to develop a course which did not concentrate mainly on techniques. We believe that in many sixth forms it might be preferable to provide an I-level course rather than a full A-level course in statistics, since such a course would serve the needs of a much greater number of students.

Sub I-level courses

Examinations 16–18. A consultative paper. DES and Welsh Office 1980.

598 The consultative paper* which contains proposals for the introduction of I-level examinations also proposes the development of a pre-vocational course and an associated examination to be taken by students in either schools or colleges at 17 + . This is intended to replace the proposed Certificate of Extended Education (CEE) examination for which some pilot examinations in mathematics have already been developed. **When planning the mathematical component of the proposed new course we believe that account should be taken not only of the pilot CEE courses but also of the findings of the research studies into the mathematical needs of various types of employment which we have discussed in Chapter 3 and of our own proposals for a differentiated curriculum up to the age of 16 + .**

O-level and CSE courses in the sixth form

599 Some have argued to us that the decision not to implement the proposal for the introduction of CEE means that a sixth form examination which, in its pilot form, has been well suited to some students will no longer be available. Some sixth form students who have not achieved O-level grade A, B or C or CSE grade 1 in mathematics in the fifth form have taken CEE pilot examinations in mathematics, on which CSE grades have been awarded, instead of repeating the O-level or CSE examination. Because the proposed pre-vocational course is not a single subject examination but a 'package', it will not provide separate certification in mathematics. Students who wish to improve their O-level or CSE grades will therefore need to take these examinations again.

600 However, pilot CEE courses have not been available in all parts of the country and in practice many students repeat O-level or CSE examinations in mathematics in the sixth form. These 'retake' courses are often among the least satisfactory in a school or college. The time allowance is sometimes meagre, the level of attainment of the students very varied and their examination targets diverse. It is not unusual for students in the same teaching group in a sixth form or tertiary college to be preparing for three or four different O-level and CSE examinations. Furthermore, problems of timetabling sometimes mean that students are not able to be present for all the mathematics periods in the week. In these circumstances the task of the teacher is very difficult indeed and there can be a temptation merely to practise past examination papers in the hope that improved performance will result.

601 We consider that one of the reasons why some pilot CEE courses have been successful is that they have required students to approach the mathematics course in a different, and often more adult, way. In our view a course which includes the introduction of some new work approached in an

adult way is more likely to succeed than one which is based on the practising of past examination papers. Furthermore, if 'retake' courses for CSE or O-level are to be effective, it is essential to provide sufficient time for them, to ensure that students are able to attend all the periods and to consider carefully the approach which is used.

Part 3   12   Facilities for teaching mathematics

602 In the third part of our report we discuss the resources which are required to teach mathematics. We begin by considering the facilities which should be available in primary and secondary schools.

Accommodation

603 In most primary schools each class spends the greater part of the day in its own classroom or class base, moving elsewhere only for activities such as physical education and music. Each class needs facilities for practical work in mathematics but no special accommodation is required. In schools whose accommodation includes areas for practical work which are shared by several classes, it can be convenient to assemble certain pieces of mathematics equipment, for example apparatus for measurement of various kinds, in one part of such practical areas. They are then readily accessible to children from several classes.

604 **We are in no doubt that in secondary schools mathematics should be taught in suitably equipped specialist rooms and that their provision makes it easier for good practice to develop, especially in schools in which there is good leadership from the head of department.** This is underlined in the following extracts from the report of the National Secondary Survey*. "The problems of teaching mathematics are increased by inappropriate accommodation and some of the best practice seen would have been impossible if the teachers had not enjoyed suitably equipped bases in which to work." "Observation showed, and statistical analysis confirmed, that display material, experimental and practical work, the use of realistic material, and the mathematical use of puzzles and games were all more likely to be found in schools in which mathematics was allocated specialist accommodation." However, sufficient suitable accommodation for the teaching of mathematics is not yet available in many secondary schools. In their written submissions, many teachers have drawn attention to the need to provide suitable specialist accommodation for mathematics teaching. Some of the teachers whom we have met during our visits to secondary schools have also stressed this need and have pointed out the difficulties of carrying from room to room all the mathematics equipment which is required.

*Aspects of secondary education in England. A survey by HM Inspectors of Schools. HMSO 1979.

605 About ten to fifteen years ago some of those engaged in mathematical education were advocating the provision of a 'mathematics laboratory' in each secondary school. However, we believe it has become clear that the provision of a single specially equipped room of this kind does not meet the need

satisfactorily. This is in part because of the timetabling problem of enabling mathematics classes to use the facilities which the room provides at the time at which they are needed. More importantly it assumes, in our view wrongly, that only one mathematics class will need facilities for practical work at any one time. **We believe that it is preferable for certain rooms to be designated for mathematics and for these to be grouped together.** If it can be arranged that one of the rooms is somewhat larger than the others, with certain extra facilities such as side benches and a sink, this is an advantage. All the rooms should have tables or desks with flat tops, good blackboard facilities which include a square grid for graphical work, storage facilities for books and equipment, and display boards on the wall. Power points should be available to make possible the use of equipment such as an overhead projector or micro-processor.

606 The provision of grouped specialist accommodation provides many opportunities for the mathematics department in a school. Among these we would include

- opportunity for greater cooperation and mutual support among those who teach mathematics in a school;

- opportunity for those who teach mathematics to observe each other at work;

- opportunity for the head of department to oversee the work of the department more easily and assist its members—this can be especially important in a school which is short of well qualified mathematics staff;

- opportunity to share equipment and other resources more effectively;

- opportunity for better display of pupils' work and other relevant material in circulation areas as well as in classrooms;

- opportunity for flexible use—for example, if one room has certain additional facilities it is possible for classes to exchange rooms at short notice; a set of grouped rooms also enables team teaching techniques to be used if desired, and makes it easier to provide cover for absent staff.

607 We have been told that in some schools pressure on accommodation in recent years has prevented the re-arrangement of teaching accommodation in order to provide a set of specialist mathematics rooms which are grouped together. **Now that secondary rolls are beginning to fall in some parts of the country, we hope that LEAs and head teachers will take advantage of the greater flexibility in the use of accommodation which will result from these falling rolls to provide grouped specialist accommodation for mathematics departments in schools in which it does not yet exist.**

Equipment

608 Throughout our report we have stressed the importance of practical experience at all stages of the mathematics course. **In order to provide such experience the necessary equipment must be available.** Much of what is required is neither elaborate nor expensive but it needs to be available in sufficient quantity and to be readily accessible.

609   In infant classes it is essential to have plenty of small objects which can be used for sorting and counting. There must also be equipment that can be used for measuring, including sand, a sink with a variety of capacity measures and simple balances with appropriate weights. Older children need other measuring devices such as rulers and tape measures, balances and scales of various kinds, and timing devices.

610   Young children need to use simple structural apparatus of various types, such as those in which units can be physically combined into larger pieces or in which collections of units can be represented by larger pieces. At a later stage a different type of structural apparatus should be available which provides a physical representation of units, tens, hundreds and thousands. Apparatus of this kind can assist children to understand the concept of place value. Real or plastic coins should also be available; these can be used both to develop familiarity with handling money and as another form of number apparatus.

611   All children in primary schools need to use a variety of two- and three-dimensional shapes. There should be a supply of simple drawing instruments including shape templates, rulers, protractors and compasses; children also need to use plumb-lines and spirit levels. Paper printed with grids of various shapes and sizes is necessary for some kinds of graphical work as well as for work with patterns and shapes. Its use can also help to develop the concept of area, as can work with 'pin-boards' and 'peg-boards'. Some electronic calculators should also be available. All of this equipment, together with such things as scissors, coloured paper, card, elastic bands, string and glue, needs to be easily available when required and careful thought should be given to the way in which it is organised and stored. It is desirable that a stock of the equipment which is in frequent use should be available in each classroom or class base; it is likely to be more convenient to store equipment which is used less frequently in a place in which it can be easily accessible to all classes.

612   **Practical equipment of similar kinds should also be available in secondary schools.** Cardboard, scissors, gummed paper, glue, plastic counters, small plastic shapes and three-dimensional models are amongst the relatively cheap materials which can assist the mathematical development of pupils. Supplies of paper marked with grids of various kinds in lines and dots should be available as well as equipment which will enable simple statistical investigations to be carried out. In order to undertake the measurement of various kinds, the reading of tables, charts and diagrams, and the geometrical work of a constructional nature which is implicit in the content of the foundation list which we have set out in paragraph 458, very much more practical material will need to be available in secondary classrooms than is commonly found at present. Much of this material, for example a collection of two- and three-dimensional shapes of various kinds and sizes, can be produced by the pupils themselves as part of their work in mathematics and no great expenditure need be involved. However, the 'basic' equipment such as pencils, rulers and compasses which is necessary for work of this kind must be kept in good condition if pupils are to be able to produce work of good quality.

613 We have already advocated that electronic calculators should be available for use by pupils. Some pupils may still need to make use of structural apparatus for number work and a supply of this kind of material should also be available. It is likely to be convenient to divide the storage of practical equipment between classrooms and a central store. Whatever method of storage is used, it is important that all equipment should be readily accessible when it is required.

Display of material and pupils' work

Aspects of secondary education in England. A survey by HM Inspectors of Schools. HMSO 1979.

614 It is usual in primary schools to display pupils' work on classroom walls but much less use is made of display in many secondary classrooms. The report of the National Secondary Survey* says that "mathematical display material was found in only some 40 per cent of schools, and this on a charitable interpretation, including schools with only very limited displays of wall charts or children's work. The proportion of schools in which this material was purposefully used as part of a progressive scheme of work was smaller". Well organised and up-to-date displays of material and pupils' work can provide a valuable resource for teaching whose use should not be under-estimated.

Libraries

615 An increasing supply of books about mathematics which are of general interest has become available in recent years, often in paperback editions. However, copies of these books are not always available in schools nor do some teaching schemes give sufficient encouragement to pupils to read them. In secondary schools, mathematics books of general interest can often remain unnoticed if the classification system which is used to arrange books on the library shelves has the effect of interspersing these books among advanced mathematical texts and similar books which are suitable only for sixth form pupils and teachers. It is desirable that books about mathematics which are of general interest, especially those which are suitable for younger pupils, should be placed together on the same shelf so that they are easily visible. In secondary schools which have a grouped set of mathematics rooms it may be preferable to have a small mathematics library in the mathematics department.

Reference material for teachers

616 **In both primary and secondary schools there should be a supply of reference books for teachers relating to the teaching of mathematics.** These should include some of the publications of the professional mathematical associations as well as of the DES, HM Inspectorate and Schools Council. There should be copies of any teachers' guides which relate to text books in use in the school and also a selection of mathematics text books, other than those which are in general use, which can serve as an additional resource for teachers. In our view it is a proper use of money allocated for mathematics to purchase books of this kind.

Financial arrangements

617 We are aware that very considerable differences exist between schools not only in the provision for teaching mathematics but also in the amount of money which is made available to maintain and improve that provision. **It is essential that sufficient money should be made available to maintain adequate stocks of books and equipment.** It is also essential that flexibility exists in the

way in which money allocated for mathematics can be used. For example, some of the work which we have recommended, especially in secondary schools, will require equipment of a kind which is not readily available from suppliers of educational equipment. It is therefore necessary that arrangements for obtaining supplies should not be such as to make it difficult for schools to obtain from other sources some of the equipment which they require.

13   The supply of mathematics teachers

618 There can be no doubt that the most important resource for good mathematics teaching is an adequate supply of competent mathematics teachers. In this chapter we consider the present situation in schools and methods by which the existing stock of mathematics teachers might be increased. In the chapters which follow we consider the initial training and induction of those who teach mathematics and their subsequent in-service support.

619 The shortage of good teachers of mathematics has been a matter of concern for many years. As we hope will be clear from the earlier part of this report, mathematics is especially vulnerable to weak teaching. "There is no area of knowledge where a teacher has more influence over the attitudes as well as the understanding of his pupils, than he does in mathematics. During his professional life, a teacher of mathematics may influence for good or ill the attitudes to mathematics of several thousand young people, and decisively affect many of their career choices. It is therefore necessary that mathematics should not only be taught to all pupils, but well taught. All pupils should have the opportunity of studying mathematics in the company of enthusiastic and well qualified mathematics teachers."*

*The Royal Society. *The training and professional life of teachers of mathematics.* November 1976.

The present situation

620 The problems in primary and secondary schools are not the same. As we explained in Chapter 6, mathematics in primary schools is almost always taught by the class teacher and only a minority of primary teachers study mathematics as a main subject during their initial training. **The need therefore is to increase the mathematical expertise of primary teachers overall; and also to increase the number of teachers who take mathematics as a main subject during initial training or who, at a later stage, undertake a substantial course of in-service training in mathematics, so that there will be a sufficient supply of teachers who are able to provide leadership and help for their colleagues.** It is, however, necessary to recognise that the proportion of primary teachers who have taken mathematics as a main subject during their initial training is never likely to be large.

621 In secondary schools mathematics is almost always taught on a specialist basis. The shortage of well-qualified teachers of mathematics in secondary schools has increased considerably through the 1970s both as a result of the increasing number of pupils in schools and also of increasing demand for mathematicians in industry and commerce. Many of the submissions which we have received, especially those from LEAs and schools, refer to difficulties

which have been experienced in appointing suitably qualifed teachers of mathematics. Estimates of the extent of the shortage in secondary schools which have been made from time to time have varied considerably. In 1974 and 1976, the DES Secondary School Teacher Shortage Survey suggested that the shortage of mathematics teachers in secondary schools was of the order of 1100. A more recent estimate, given in the report of the National Secondary Survey* and based on visits to secondary schools during the period 1975–1978, suggested that the shortage of suitably qualified mathematics teachers in secondary schools was of the order of 3000. We give our own estimate of the shortage of mathematics teachers in paragraph 631.

*Aspects of secondary education in England. A survey by HM Inspectors of Schools. HMSO 1979.

622 A shortage of teachers of subjects which are at some stage optional in schools can to some extent be overcome by reducing the number of pupils who study the subject, and sometimes also the time which is given to it. However, because it is accepted that all pupils should study mathematics, shortage of mathematics teachers cannot be overcome by reducing the number of pupils to whom the subject is taught nor, other than exceptionally, the teaching time provided; it usually leads instead to the teaching of mathematics by teachers who are neither suitably qualified nor trained to do so. It seems sometimes erroneously to be thought that any qualified teacher should be able to teach mathematics adequately, especially to younger or lower-attaining pupils. Mathematics teaching is therefore especially vulnerable at a time at which rolls in secondary schools are starting to fall. For example, if one of four or five mathematics teachers in a school leaves the staff, there can regrettably be pressure not to replace him but, for the time being at least, to make use instead of teachers trained to teach other subjects, several of whom are likely to have a few teaching periods 'to spare' if the number of pupils in the school has decreased.

1977 Survey of Secondary School Staffing

623 We have therefore tried to obtain information about the levels of qualification of those who teach mathematics in secondary schools. Some information is given in the report of the National Secondary Survey but the only complete source of information on a national scale is provided by a sample survey of the staffing of some 500 maintained secondary schools which was carried out by the DES in November 1977 on the recommendation of the Advisory Committee on the Supply and Training of Teachers. The information collected included details of the qualifications of all the teachers in the schools included in the survey, the subjects which they were teaching, the amount of time given to each subject and the year group (but not level of attainment) of the pupils to whom the subject was being taught. Information was also collected about the curriculum of each school. Sixth form colleges were included in the sample as were middle schools 'deemed secondary' (that is, with pupils up to the age of at least 13) in respect of the teaching of pupils from the age of 11; tertiary colleges were not included.

624 The survey was planned and carried out before our Committee was set up, though analyses of the information which had been obtained did not start to become available until after we had started work. Although the survey was not designed to provide information about mathematics teaching in the detail which would have been possible had it been known that our Inquiry was to

take place, the DES has, in addition to undertaking the analyses which were intended when the survey was planned, carried out at our request a number of further analyses of the data which is available. In this way we have been able to obtain a considerable amount of information. We realise that the figures which we quote are based on information which is now nearly four years old. However, although falling rolls in some areas and enforced economies may have produced some changes, we believe that the picture which emerges is unlikely to have changed significantly. There is, in any case, no more recent information available on a national scale.

Levels of qualification of mathematics teachers

625   When considering the information about the teaching of mathematics which the survey provides, we have classified teachers in terms of their academic qualification. We are aware that the teacher who is well qualified on paper is not necessarily effective in the classroom as a teacher of mathematics; equally, a teacher who has little or no recorded qualification in mathematics may teach mathematics well. However, in the absence of any other information about the effectiveness of the teachers concerned we had no alternative but to proceed on this basis. All those teaching mathematics were assigned to one of four levels of qualification, 'good', 'acceptable', 'weak', 'nil'. The criteria used to assign teachers to these categories are set out in Appendix 1 paragraph A12. They are inevitably arbitrary, but a deliberate decision was made to err on the side of generosity in respect of qualifications which seemed to lie near the boundaries between categories. The overall picture which the analysis suggests may therefore be a little more encouraging than it should be; we do not believe that it is likely to err in the other direction. In our discussion of the outcome of the analysis, we have in some cases combined qualifications in the categories 'good' and 'acceptable' under the heading 'suitable' and qualifications in the categories 'weak' and 'nil' under the heading 'unsuitable'.

626   The results of the survey of secondary staffing show that, in November 1977, 38 per cent of all mathematics teaching in maintained secondary schools was being undertaken by teachers whose qualifications to teach mathematics were either 'weak' (17 per cent) or 'nil' (21 per cent); in other words, almost two-fifths of all mathematics teaching was in 'unsuitable' hands. However, this overall figure conceals very considerable differences between different kinds of school, and also between different schools of the same kind.

627   On a national basis, the information provided by the survey suggests that there were in 1977 some 1500 secondary schools (about 35 per cent) in which at least 70 per cent of the mathematics teaching was in 'suitable' hands; and that these included some 240 schools (mainly sixth form colleges and grammar schools) in which all the mathematics teaching was by teachers with a suitable qualification. On the other hand, the results of the survey suggest that there were nearly 1300 schools (about 30 per cent) in which less than half of the mathematics teaching was in 'suitable' hands and that these included some 150 schools (mainly modern schools) in which none of the teaching was by

teachers with a suitable qualification. In the remaining 1500 schools, between 50 per cent and 70 per cent of the mathematics teaching was 'suitably' staffed.

628 A more detailed analysis is shown in Appendix 1, Tables 15 – 19; we commend these tables to the attention of our readers. We would repeat that, as shown in Table 18, there is very considerable variation between schools of the same type so that, at the time of the survey, there were, for example, at least one modern school in which all the mathematics periods were taught by 'suitably' qualified teachers and at least two comprehensive schools in which none of the mathematics was so taught. The patterns of staffing were not directly connected with the age range of the schools concerned nor were less well qualified teachers only found in particular types of secondary school. The distribution of teaching staff may result, at least in part, from the way in which reorganisation to comprehensive education was carried out in many areas. Even though the range of ability of the pupils in a school changed markedly, there was often relatively little redistribution of teaching staff. Thus, sixth-form colleges and comprehensive schools which developed from grammar schools often found themselves with a high proportion of graduate teachers, whereas comprehensive schools which developed from modern schools sometimes found themselves with many fewer graduates.

Teachers in middle schools

629 It could be argued that, for teachers in middle schools, some of the qualifications classified as 'weak' on the ground that training had been for a younger age group should be regarded as 'acceptable' (see Appendix 1, paragraph A15). However, even if all teachers in middle schools whose qualifications were 'weak' were to be regarded as 'acceptable', 62 per cent of the mathematics teaching of pupils from the age of 11 in middle schools deemed secondary remained in 'unsuitable' hands. This must give particular cause for concern. In secondary schools with pupils only up to the age of 14, a high proportion of the mathematics teaching was also in the hands of 'unsuitably' qualified teachers. However, in these schools, too, it could be argued that some of the qualifications classified as 'weak' should be regarded as being 'acceptable'.

Teaching of younger pupils in comprehensive schools

630 It was thought likely that less well qualified teachers might be used mainly with younger pupils. The hypothesis was examined in respect of the teaching of pupils up to the age of 16 in comprehensive schools with an entry age of 11 or 12; the full outcome is in Appendix 1, Table 19. Some confirmation of the hypothesis was found, but to a lesser extent than might have been expected. In the first year, 47 per cent of mathematics teaching was in 'unsuitable' hands; in the fifth year the figure was 37 per cent. This suggests that there are many schools in which the shortage of mathematics teachers is such that it is impossible to avoid using non-mathematicians with all age groups.

631 Appendix 1, Table 16 shows that, adjusted to a 40-period week, some 178 000 periods of mathematics were taught each week in secondary schools

(excluding middle schools) by teachers with a 'nil' qualification. If we assume the high teaching load of 35 periods per week, this number of periods represents the full-time contribution of more than 5000 teachers. In addition, the number of periods taught by teachers with a 'weak' qualification represents the full-time contribution of almost another 4000 teachers. **These figures give a measure of the shortage of mathematics teachers in secondary schools in 1977, but take no account of the shortage in middle or primary schools.**

Deployment of teachers who are qualified to teach mathematics

632 In our discussion so far we have considered the proportion of mathematics periods taught by teachers with various levels of qualification. It has been suggested that the present shortage of mathematics teachers has been aggravated by the fact that some teachers who are qualified to teach mathematics are not doing so but are teaching other subjects instead. We have therefore investigated this possibility. It is important at the outset to realise that many teachers are equipped, both by qualification and training, to teach more than one subject. Teachers of mathematics, for example, are often equally well, and sometimes better, qualified to teach physics, of which there is also a shortage of teachers. If a teacher of physics were to change to teaching mathematics instead, the effect would be to increase still further the existing shortage of physics teachers and such a change might not be in the interest of the education system as a whole. When considering the question of 'misuse' of mathematics teachers, it is therefore not sufficient merely to count the total number of teachers who are qualified to teach mathematics; it is necessary at the same time to take into account the subjects which they are, in fact, teaching.

633 It proved possible to develop techniques to examine the extent of 'mismatch' in respect of the teaching of mathematics within the schools included in the sample survey. 'Mismatch' was defined as the teaching of mathematics by 'unsuitable' teachers when within the same school, teachers who were 'well' or 'acceptably' qualified to teach mathematics were not teaching either mathematics or a subject for which they were equally or better qualified. The analysis of 'mismatch' shows that in nearly a quarter of the schools in the survey there was no mismatch at all; in the great majority of the remainder the degree of mismatch was small and no greater than is to be expected as a result of the exigencies of timetabling, especially since, as we ourselves have advocated, there are significant advantages in timetabling a number of classes within the same year group to do mathematics at the same time. There were, however, a very few schools in which there seemed to be significant degree of mismatch; several of these were modern schools. **It is therefore the case that attempts to redeploy teachers within individual schools are likely to contribute only minimally to improving the proportion of mathematics periods taught by 'suitably' qualified teachers. The need is undoubtedly to increase the number of teachers who are appropriately equipped to teach mathematics.**

Rates of entering and
leaving teaching

*Aspects of secondary education
in England. A survey by HM Inspectors of Schools. HMSO 1979.

634 We have sought information about the rates at which mathematics
specialists enter and leave the teaching profession at the present time. The information about mathematics specialists in the following paragraphs relates
only to graduate teachers in maintained schools, other than holders of the
BEd degree, with mathematics as the only or first subject of their degree.
These teachers do, however, constitute an important part of the mathematics
teaching force. Figures given in the report of the National Secondary Survey*
suggest that mathematics graduates undertake about one-third of all
mathematics teaching in secondary schools.

635 Figures relating to the years 1975–1979 are given in Appendix 1, Tables
20 and 21. Throughout these years there were some 300 mathematics
graduates working in primary schools. The number of mathematics graduates
in secondary schools increased from about 7100 in 1975 to about 8100 in 1979,
an increase of about 14 per cent. However, during the same period the total of
all graduates working in secondary schools increased by about 30 per cent.
The rate of increase of mathematics graduates was therefore very much
smaller than that of graduate teachers as a whole. Indeed, whereas in 1970
mathematics graduates formed over 11 per cent of the graduate secondary
teaching force, by 1979 the proportion had dropped to about 6 per cent.

636 We have been able to obtain details, by age group, of the number of
graduates in mathematics only or in mathematics combined with non-science
subjects who have left teaching each year since 1974–75, either on retirement
or for some other reason, and the corresponding information for all graduate
teachers (see Appendix 1, Table 22). Comparison of the rates of leaving shows
no significant differences in respect of teachers over the age of 30 nor, until
1978–79, in respect of men aged 25 to 29. However, since 1975–76 rates of
leaving have been markedly higher among men under the age of 25 who are
mathematics graduates than among all male graduates of the same age. These
comparative rates must be treated with caution because the numbers involved
are small; nevertheless, they tend to confirm the impression that young male
mathematics graduates are likely to be especially vulnerable to the attractions
of alternative employment. Opportunities for such employment may well increase when the present recession ends and firms begin to recruit additional
staff.

**Methods of increasing
the supply of
mathematics teachers**

637 We turn now to consideration of methods by which the supply of suitably
qualified teachers of mathematics might be increased. There would seem to be
three ways in which this could be accomplished. The first is to increase the
number of entrants to the teaching profession who are suitably qualified to
teach mathematics. The second is to take such steps as are possible to ensure
that suitable teachers who are already teaching mathematics effectively will
remain within the profession. The third is to improve the quality of the
teaching of some of the under-qualified teachers, who are at present teaching
mathematics, by means of appropriate in-service support and training. We
believe that it is necessary to take action in all three of these ways.

638 Mathematics graduates form a principal source of supply of
mathematics teachers in secondary schools. We have already referred to the

fact that the demand for mathematics graduates is increasing from many sections of industry and commerce and will probably increase further when the present recession ends. Information provided for us by the Universities Statistical Record gives details of the occupations chosen by graduates in mathematical studies who completed their degree courses at universities in England and Wales in 1979; this is shown in Appendix 1, Table 31. These figures show that less than 10 per cent of the graduates in mathematical studies proceeded to teacher training. This figure contrasts with the fact that in 1938 over 75 per cent of newly qualified graduates in mathematics entered the teaching profession. In 1964 the proportion had decreased to 30 per cent, by 1974 to 17 per cent. There are also some mathematics graduates who enter teaching without undertaking teacher training; we return to this point in paragraph 692.

639 In their submission to us, the Association of Graduate Careers Advisory Services (AGCAS) discussed at length the results of an inquiry which they had carried out at our request into the reasons why the number of mathematics graduates entering teaching was small. In their view, the principal factors are pay and prospects within the teaching profession and also perceived problems of discipline in schools; these factors are common to teachers of all subjects. Additional factors relating to mathematics are the fact that mathematics is thought to be a difficult subject to teach and that, because it is a compulsory subject for all pupils up to the age of 16, mathematics classes are likely to be larger and to contain a higher proportion of less well motivated pupils than is the case with many other subjects which are often optional after the age of 14. AGCAS suggest that an improvement in pay and prospects—especially the latter—needs to be an essential part of any attempt to attract more graduate mathematics teachers, together with improvements in the arrangements for induction into the profession in order to allay fears about discipline and allied problems. They also draw attention to the fact that "low pay is seen as a reflection of the status of teaching and its importance to the community and this tends to have an off-putting effect. Students will enter low paid jobs if they are seen as attractive and confer status—eg working in the media. Teaching does not, however, have the same glamour and the low morale of the profession reinforces this view".

640 **We believe it is essential to do much more than is being done at present to improve the public image of teaching, and of mathematics teaching in particular.** Too many potential teachers appear to be put off by widely reported problems of discipline, of the difficulties of teaching or of unhappy schools. Yet the results of a small survey carried out on our behalf by the National Foundation for Educational Research, to which we refer in greater detail in paragraph 671, showed that the group of mathematics teachers included in the survey, who were in their first three years of teaching in secondary schools, were generally happy in their work. A large majority believed that they were well thought of by their colleagues and that they had gained the respect of their pupils. **We hope that both central and local government will respond to our report by affirming their belief in the importance of good mathematics teaching for all pupils, the need to provide good support and facilities for**

mathematics teachers who are already in post and, whatever the overall teacher requirement may be, the need for many more good teachers of mathematics. Good publicity is necessary in order to improve attitudes towards the teaching of mathematics on the part of the general public and, in consequence, in the minds of potential teachers.

641 **In our view it is also necessary to make a much greater effort than is being made at present to recruit mathematics graduates into teaching.** It has been pointed out to us that very many large employers take part in the annual 'milk round' of universities, during which representatives of companies visit universities to inform, advise and interview undergraduates, but that no comparable effort is made by the DES or by LEAs acting in concert. Some careers advisers in universities have told us that students therefore gain the impression that, despite the widely publicised shortage of mathematics teachers, LEAs are not concerned to make a sustained effort to increase recruitment. A further suggestion which has been made to us is that the DES should write to all who are starting the final year of a mathematics degree course pointing out the need for well-qualified teachers of mathematics and the attractions of teaching as a career. Such a letter should include an offer to provide further information. **We believe that the possibilities of making a direct approach to undergraduates reading degree courses in mathematics and of some kind of participation in the 'milk round' should be investigated.**

Mathematics degree courses

*We concentrate on courses in universities because it is from these that most mathematics graduates come. The degree courses in mathematics offered at other institutions are often more varied and less abstract than those offered at universities; not all of our comments may therefore apply to them.

642 Our terms of reference do not extend to higher education and we have not therefore made any extended study of mathematics courses in universities; nor would it be proper for us to make any recommendations relating to this area. However, because a substantial number of teachers of mathematics are, and will continue to be, mathematics graduates, we believe that it is proper for us to consider two matters. The first is the extent to which university mathematics courses in their present form provide a suitable preparation for the prospective school teacher. The second is whether significant numbers of potential mathematics teachers in schools are being lost because university courses are too demanding. We put forward our views on these two questions in the hope that universities* may be prepared to examine and perhaps to adjust the provision which they make.

643 **In our view, the mathematical training provided at university for those who will become mathematics specialists in schools should aim**

- to develop knowledge and mastery of mathematics substantially beyond the level at which they will be teaching and also, where appropriate, provide opportunity to pursue some topic in depth;

- to develop enjoyment of mathematics and confidence in its application;

- to provide an historical perspective of mathematics;

- to provide an appreciation of the relationship between mathematics and other fields of study and application;

- to develop the ability to communicate mathematical ideas both orally and in writing.

It is not clear to us how generally these aims are being fulfilled. We believe that some university mathematics courses contain too much technical material on the grounds that an 'educated mathematician' should have knowledge of this or that particular topic. However, in many cases the pressure of time means that only the rudiments of a topic are covered; although this may be a useful first step for those who will proceed to postgraduate work, it may appear pointless and confusing to others. In fact, the danger of confusing students by too ambitious a course is very real and, for the prospective teacher, can lead to a damaging loss of self-confidence and enthusiasm.

644   There are some degree courses in mathematical studies which combine mathematics and education for those intending to become teachers. While such courses can cater more specifically for intending teachers, they usually require an early commitment to teaching which many students are unwilling to make. The evidence which we have received makes it clear that most undergraduates prefer to keep their options open for as long as possible. **It follows therefore that a mathematics degree course should provide within its structure a set of options which are suited to the needs of those who feel that they may wish to teach but who have not yet made a firm decision to do so.**

645   We do not, of course, suggest that mathematics degree courses should be directed only to the needs of those who will enter the teaching profession. Nevertheless, the supply of an adequate number of well qualified mathematics teachers for schools must be of paramount importance to all institutions of higher education so that the level of mathematical preparation of the students who come to them may be maintained or improved. Furthermore, we believe that, for the great majority of mathematics graduates, the requirements which a mathematics degree course should fulfill are not dissimilar to those which we have suggested are needed by teachers. It is only a small minority of those on mathematics degree courses who proceed to more advanced studies in mathematics after the undergraduate stage. For this small but important minority it is essential that options should be available which prepare them for postgraduate study but we do not feel that it should be necessary for their requirements to dictate the content of mathematics degree courses for the majority of those who attempt them.

646   There can be no doubt that there is a widespread view in schools that university mathematics courses are exceptionally demanding and should only be undertaken by those who are very able. Courses at universities do, of course, differ greatly in their structure and range and we welcome this diversity. Nevertheless, there is a strong tradition, stemming from the older universities, which delineates the level of difficulty of a mathematics degree. It is possible that this is a serious obstacle to the wider recruitment of students who enrol for mathematics degree courses and hence to the provision of a larger pool from which mathematics teachers can come, as well as the mathematics graduates needed by industry and commerce. We drew attention in Chapter 4 to the large increase in the proportion of entrants to degree courses in mathematical studies who have only a single-subject A-level qualification in mathematics. Universities may wish to ask themselves whether broader and

more flexible mathematics courses, designed particularly for single-subject students, might not attract a significantly larger entry. For such a policy to be effective, it would of course be necessary for schools to be made aware of the nature and purpose of such courses so that they could advise their students appropriately.

647 We draw attention to the fact that in recent years the number of girls who have studied mathematics at A-level has been only about one-third that of boys; the number of women who have graduated in mathematical studies at universities in England and Wales has been some 40 per cent of the number of men. On the other hand, among these graduates the proportion of women who have entered Postgraduate Certificate in Education (PGCE) courses to train as teachers has been more than twice that of men, so that the numbers of men and women entering PGCE courses have been about the same. If the number of girls who studied mathematics at A-level were to increase, the pool of potential graduates in mathematical studies would increase and hence the pool of potential teachers of mathematics. The number of potential entrants to BEd courses who would be capable of studying mathematics as a main subject would also be increased. **We believe that active steps to encourage more girls to take mathematics at A-level and then proceed to study at degree level could lead to an increase in the supply of well-qualified teachers of mathematics.**

Entry into teaching from other employment

648 A small number of graduates in mathematics and allied disciplines enter mathematics teaching after some years in other kinds of employment. We have received conflicting evidence as to the extent to which such transfer should be encouraged. Some argue that such entrants to the teaching profession bring with them experience which can be used in the classroom to illustrate the uses and applications of mathematics. On the other hand, we have received adverse comment on the narrow approach to mathematics and the low teaching abilities which have been displayed by some who have transferred to teaching from other occupations. It has also been pointed out that some may turn to mathematics teaching because they have not succeeded in their former jobs. It may have been too easy for graduates to enter teaching from other types of employment, without any suitable training or even any real attempt to assess potential as a teacher. However, at the present time of cutback in many areas of employment, we believe that some who have been very successful in other fields and who would make good teachers are likely to be available and that every effort should be made to recruit them. While we recognise an understandable unwillingness on the part of potential recruits to mathematics teaching from industry and commerce to commit themselves to a year's training with a relatively low income and without certainty of qualification at the end, **we believe that training for such recruits to teaching is just as important as for those who have just graduated. We therefore believe that adequate financial support should be available to remove any disincentive to train.**

649 **There should also be adequate opportunity for those who are considering teaching, and those who might employ them, to be able to make some**

preliminary judgment about their suitability for teaching. We therefore welcome the courses lasting for one term or less which have been arranged by some training institutions and LEAs, in association with major companies, for staff who face the possibility of redundancy. We believe that the availability of one term of 'adjustment to teaching', followed by a further two terms of training for those whose experience of the first term convinces both them and their future employers of their potential as teachers, could lead to a desirable increase in recruitment to mathematics teaching. We also welcome the scheme operated by the Inner London Education Authority which enables mature graduates whose degrees include a substantial mathematical component and who have industrial or commercial experience to be appointed to the authority's temporary staff and paid at the Scale 1 rate for a qualified teacher whilst on secondment to a PGCE course. Those who are recruited under this scheme work in schools within the authority for those parts of the school year which are outside college terms. A further scheme at one training institution allows holders of higher national certificates or diplomas, or some equivalent qualification, to complete a BEd degree in five terms; we have been told of other institutions which would be willing to adapt their courses in order to cater for such entrants if they were forthcoming in sufficient numbers. **We believe that initiatives of all these kinds should be encouraged and that they should be supported by adequate publicity.** It would be helpful to establish a 'clearing house' which could offer advice on the availability of courses.

650 Some improvement in the supply of mathematics teachers has been brought about since 1977 by the Government scheme for training and re-training to teach priority subjects which is operated by the Manpower Services Commission on behalf of the DES. Under this scheme special training awards are available for suitably qualified men and women over the age of 28 who wish to take a one-year course of initial training to teach shortage subjects, including mathematics. Between 1977 and 1980 more than 400 mature entrants with a suitable qualification in mathematics have been funded under the scheme while taking initial training courses to become teachers. In addition, more than 500 teachers of other subjects have taken re-training courses during this period to enable them to teach mathematics. We welcome this contribution to the mathematics teaching force and also the fact that the scheme has now been extended to include one-term or one-year courses of further training for existing teachers of mathematics.

Employment of primary trained teachers in secondary schools

651 There is one source of recruitment to secondary mathematics teaching of which we have become aware which we do not believe to be desirable, nor to be in the interests of the education system as a whole. We have been told of a number of recent cases in which teachers who have taken mathematics as a main subject in BEd degree courses and who have trained to teach in primary schools have been offered, and in some cases urged to take, posts teaching mathematics in secondary schools. Not only are such teachers not suitably trained for this work but, if they enter secondary teaching, do not contribute to the increase of mathematical expertise which is so badly needed in primary schools. **We believe it is important that LEAs should recognise this need and ensure that primary trained teachers with mathematical expertise are ap-**

pointed to primary schools so that the teaching of mathematics in primary schools can be strengthened. As secondary rolls begin to fall we believe that every effort should also be made to enable and encourage mathematically qualified teachers who are teaching in secondary schools, but who have been trained for primary work, to transfer to primary schools.

Grants during training

*The Education (Teacher Training Scholarships) Regulations 1981. Statutory Instruments 1981 No 1328.

652 We welcome the pilot scheme which is about to be introduced to provide national scholarships for intending teachers of mathematics*. These scholarships will provide a flat-rate payment to selected entrants to PGCE courses who are considered to be likely to become teachers of high quality. They require a commitment on the part of the recipient to enter teaching on the satisfactory completion of the PGCE course and offer a guarantee of employment as a teacher. It is, however, possible that some of these scholarships will be gained by students who would have entered teaching in any case, so the net increase in entrants to mathematics teaching may not be large. Nevertheless, this scheme demonstrates recognition at national level of the urgent need for more mathematics teachers.

653 It has been represented to us that the recruitment of mathematics graduates to PGCE courses might increase if the existing system of means-tested grants for these students were to be abolished. We accept, however, that such a change could lead to anomalies in respect, for example, of students in the final year of BEd honours courses and that the likely 'knock-on' effect may, for financial reasons, rule out such a change for several years. **We believe that a system of flat-rate payments to graduates training to teach mathematics (and other shortage subjects) would prove more straight-forward to operate;** such payments should be conditional on the possession of suitable mathematical qualifications (see also paragraph 683) and an undertaking to enter the teaching profession and remain as a teacher for a specified minimum time provided that training has been completed successfully. Grants under similar conditions could be given to fourth year students on BEd honours courses who were taking mathematics as a main subject. A scheme of this kind could be discontinued if the shortage of mathematics teachers ceased, though we believe that a continuing incentive for BEd mathematicians to enter primary schools would be valuable. It is, however, our view that improvement in training grants of any kind is not likely to result in significant increase in recruitment unless future prospects for pay and promotion are seen to be attractive.

Financial incentives

654 We referred in paragraph 639 to the fact that evidence which we have received from the Association of Graduate Careers Advisory Services points out that "better pay and prospects are probably the most important potential factors for increasing the intake" of teachers. We have therefore endeavoured to compare the financial rewards of a career in teaching with those in other occupations which recruit graduates. Such enquiries as we have been able to make point not so much to the difference in initial reward as to the contrasts in perceived prospects at a later stage. This seems to be the case despite the advantages which are sometimes put forward of greater security,

assured and index-linked pensions and longer holidays which teachers enjoy. The newly qualified teacher, with a good honours degree and a PGCE, starting at a salary of £5,547, can be assured of regular though not large annual increments to a total of £7,869 after 10 years service but advancement to a higher salary scale is dependent on promotion opportunities which are controlled by the Burnham salary structure. These opportunities are related to the numbers and ages of the pupils in a school. At the present time school rolls are declining nationally and we have been told that, under the existing regulations, opportunities for promotion beyond Scale 2, the top of which is currently £8,208, are likely to become less. As rolls fall, job security may also be reduced.

655 It is very difficult to quantify the salaries which can be earned by mathematics graduates who do not enter teaching or to generalise about their very varied conditions of employment. Some information about salaries is provided by a survey of its members carried out in 1980 by the Institute of Mathematics and its Applications (IMA) to which some 3600 members replied; one-eighth of these were teachers in schools. The results of the survey show that, for the age range 26–35, the salaries of those members of IMA who were teachers were on average between £1,000 and £2,000 per annum lower that those of members of IMA employed in central government, private and nationalised industry, and commerce. It has also been pointed out to us that, in industry and commerce, the size of a salary increase is very often dependent on the competence and commitment which an employee displays, as revealed and recorded by regular assessment procedures. On the other hand, a teacher's salary advances according to a pre-determined scale which, even in the case of transfer to a higher scale post, limits the total salary increase which can be gained in any year, no matter how committed and effective a teacher may be.

Additional payment to teachers of mathematics

*House of Commons. Fifth report from the Education, Science and Arts Committee. *The Funding and Organisation of courses in higher education*. HMSO 1980.

†Central Policy Review Staff Report. *Education, training and industrial performance*. HMSO 1980.

656 This raises the question of the extent to which additional payment should be made to teachers of mathematics. We have noted that the Fifth Report of the Education, Science and Arts Committee of the House of Commons* issued in September 1980, which deals with the funding and organisation of courses in higher education, recommends that "a higher rate of maintenance grants for students and higher salaries for teachers in certain subjects, such as mathematics, should now be considered". The report of the Central Policy Review Staff,† published in May 1980, say that "while recognising the difficulties we believe that without a pay advantage it will be very hard to overcome these chronic shortages [of teachers of mathematics, science and practical subjects] and we recommend that the nettle be grasped now".

657 The evidence which we have received on the question of additional payment to teachers of mathematics is conflicting. The teachers' unions and the professional mathematical associations do not support additional payment. On the other hand we have received support for such payment in submissions from individual teachers as well as from outside the teaching profession. **We have discussed this matter at length and are agreed that additional funding in some form is necessary if the present situation of acute shortage is to be alleviated.** We suggest two approaches which we believe would be possible.

658   The first is to make greater use of the flexibility which already exists, but is little used, within the Burnham framework and also to introduce additional flexibility. LEAs are already able to offer Scale 2, 3 or 4 posts for reasons which they consider fit; merit as a teacher can be one such reason, but we believe that teachers are not promoted on these grounds as often as we would wish. We have been told that a recent suggestion to LEAs from the joint secretaries of the Burnham Committee for a limited increase in the proportion of Scale 3 posts in schools as a means of relieving the shortage of teachers of mathematics and physics appears to have received little support from LEAs; we welcome the intention behind this suggestion and regret that, perhaps because of current financial pressure, it has not been supported.

659   **We wish to make a somewhat different proposal.** We believe that LEAs should have discretionary power to make appointments in secondary schools at, and promotions to, incremental points above those which are defined nationally. This power exists in local government outside teaching and in many other forms of employment. The case for granting such discretion is that it could be used to encourage ambitious and competent teachers to continue as specialist teachers of mathematics rather than seek promotion by moving to other responsibilities. Such additional payment* would improve the pay of mathematics teachers relative to outside competition. In order to improve prospects it may be necessary for extra payment to be possible above the maxima of existing scales.

*At the present time about one-eighth of teaching time in secondary schools is given to mathematics. The figures which we have quoted from the 1977 survey of secondary staffing suggest that only about 60 per cent of this teaching is in 'suitable' hands. We estimate that the cost of providing an average of two additional increments on the Burnham salary scale for half of the 'suitable' teachers who were teaching a full timetable of mathematics would amount to between 0.2 per cent and 0.3 per cent of the total salary bill for secondary teachers. (This estimate is based on the payment of such increments to $\frac{1}{32}$ of assistant teachers in secondary schools.)

660   A major weakness of incentives which are related to the Burnham scales is the difficulty of ensuring that a beneficiary does not retain his advantage if for any reason he ceases to meet the conditions for which the award was made, for example by ceasing to teach mathematics; it should, though, be possible to devise methods of dealing with this problem. If, however this problem cannot be overcome, we would suggest a scheme which would associate the payment of a specified additional allowance with the tenure of a particular post in a particular school. If a teacher in receipt of such an allowance changed his school or assumed other responsibilities, the allowance would automatically cease.

661   Both of these schemes could also be used as a means of attracting good mathematics teachers to schools whose mathematics departments were inadequately staffed. We suggest that the schemes should apply only to mathematics specialists who were teaching mathematics for at least a given proportion of the timetable and whose teaching was competent; we do not believe that additional payment should be made during the first two years of teaching.

662   **In order to encourage good mathematics teaching in primary schools we believe that there would be merit in allowing LEAs some incremental discretion for teachers in primary schools, especially those in which numbers of pupils limit the availability of posts above Scale 1.** An arrangement of this kind could assist in the appointment or retention of mathematics co-ordinators of good quality.

663 It might be argued that our suggestions would have little hope of success because those who were competing for the services of mathematics graduates would respond by increasing the salaries which they offer. We are not convinced that this would necessarily happen because the proportion of mathematics graduates who enter teaching is in any case small.

664 We recognise that the suggestions we make are open to objection, especially in respect of the problem of establishing competence or the lack of it. We are aware, too, that some may see a threat to relationships between teachers in proposals which they may see as divisive. We also appreciate the financial restraints which exist at present. Nevertheless, we are convinced that our suggestions would greatly help the recruitment and retention of good mathematics teachers without which the improvement in mathematics teaching which we believe to be necessary will not be attained. Any who condemn our suggestions should suggest alternative remedies to the continuing shortage of well-qualified mathematics teachers.

The need to employ newly trained mathematics teachers

665 We believe that a concerted campaign to attract more mathematics graduates into the teaching profession, backed by the financial incentives which we have proposed in the preceding paragraphs, might succeed in its objective. The prospects for success are considerably improved at the present time because the economic recession has reduced competition from other sources for mathematicians. Indeed, the immediate future may present an unrivalled opportunity to achieve a significant improvement in the size and quality of the mathematics teaching force—an improvement which we believe to be essential. Although falling rolls in schools and financial constraints on LEAs are inevitably leading to a reduction in the recruitment of teachers, **we wish to emphasise the vital necessity of maintaining, and where possible increasing, the recruitment of mathematics teachers despite present financial difficulties.** We believe, too, that the need for mathematics to be taught by well-qualified teachers should take priority over the need to redeploy staff. It will be of little use to mount a campaign to encourage more mathematics graduates to enter teaching if LEAs are not prepared to employ them. **We therefore feel that measures should be taken which would ensure that, in the next few years, newly trained mathematics teachers will be able to obtain teaching posts.** It may be that local education authorities should be subsidised, perhaps on a limited and temporary basis, by the DES so that all available suitable mathematics teachers are employed. We believe that the annual cost to central government would be small, that the benefits would be direct and immediate, and that the autonomy of local authorities to appoint their staff would be unimpaired. In addition, newly trained mathematics teachers and those who were considering training as mathematics teachers would feel convinced that their services were in demand. Measures of this kind could be reviewed annually in the light of changing circumstances and could be terminated without difficulty if and when the mathematics teaching strength in schools had reached a satisfactory level.

14   Initial training courses

666 At the present time most of those who teach in schools qualify for entry to teaching in one of two ways. The first is by following a course leading to the degree of Bachelor of Education (BEd). This provides both an academic qualification and also professional training as a teacher. The ordinary degree course lasts for three years, the honours course for four years. The second way is by undertaking a one-year course of professional training leading to the award of the Postgraduate Certificate in Education (PGCE) after gaining a graduate qualification other than BEd. Courses leading to entry by a third method, the non-graduate Certificate in Education, which provided academic and professional training by means of a three-year course, are now being phased out and it is no longer possible to enrol for courses of this kind. A few teachers qualify for entry to teaching in other ways but in this chapter we confine our discussion to entry to teaching by means of BEd or PGCE courses.

667 Initial teacher training in England and Wales takes place in both the university and non-university sectors. BEd courses are offered in some eighty non-university institutions, which include polytechnics and other establishments of higher education; a few are offered in universities. The structure of BEd courses varies from institution to institution. However, most BEd courses include, in addition to professional training and the study of education theory, opportunity to study one or more subjects in depth; we discuss these 'main' courses in paragraph 695. PGCE courses are offered in some thirty universities and in some sixty establishments in the non-university sector.

Entry qualifications to initial teacher training courses

668 All entrants to BEd courses are now required to satisfy the normal requirement for entry to first degree courses; that is, to have passed in at least two subjects at A-level or to have obtained an equivalent qualification. This was not the case with the Certificate in Education courses and so the minimum academic level for entry to teacher training is now higher than has been the case hitherto. A further requirement has recently been introduced in respect of entrants to teacher training who expect to become eligible to join the teaching profession in or after September 1984. These entrants are required to have obtained O-level grade A, B or C or its equivalent in both English and mathematics before starting their training course. This means that all entrants to BEd courses are now required to have obtained this qualification; it will apply to entrants to PGCE courses from September 1983. Reference to Figure 6 (paragraph 195) suggests that many intending teachers are likely to find it

more difficult to gain the necessary qualification in mathematics than in English.

669 The decision to require entrants to initial training to have an O-level qualification, or its equivalent, in mathematics has been welcomed in many, though not all, of the submissions we have received which make reference to it. In our view it has been right to introduce this requirement. However, as is pointed out in DES Circular 9/78, in which the decision to introduce the requirement was announced, "formal qualifications do not necessarily guarantee possession of competence". In addition, "possession of a pass in mathematics at O-level of GCE or the equivalent may, of course, sometimes offer little indication of a student's potential ability to teach the subject"*; we discuss this matter further in paragraph 679. The present reduction in the numbers being accepted for entry to teacher training means that it is not yet possible to assess the effect of the new requirements on recruitment to the teaching profession.

*HMI Series. Matters for discussion 8. *Developments in the BEd degree course*. HMSO 1979.

Recent changes in teacher training

670 The last ten years have been a time of very considerable change in the non-university sector of teacher training. The reduction in the number of training places has led to the closure of some colleges of education and the merger of others with institutions which provide higher education courses of other kinds. It has also been necessary to plan and introduce new courses to take account of the phasing out of the Certificate in Education and also of the reorganisation of colleges; in many cases reorganisation has been accompanied by the transfer of validation from a university to the Council for National Academic Awards. Reductions in staffing resulting from reorganisation have meant that there have been few appointments to the staff of training institutions in recent years; this means that the proportion of staff who have recent experience of full-time teaching in schools is decreasing. Because of the very great changes which have been taking place, much of the comment which we have received about training to teach mathematics relates to courses which either no longer exist or which have been changed very considerably, or to training which took place at a time of uncertainty or rapid change. Many comments, too, are anecdotal and, understandably, many of those who have written to us appear to have knowledge of only one training institution. We have not therefore found it easy to decide on the extent to which either criticisms or compliments which we have received relating to teacher training remain valid at the present time. We have, though, to record that we have received a good deal of comment which has been critical.

Views of recently trained teachers

671 We learned that the National Foundation for Educational Research (NFER) had recently been studying the progress through sixth form and higher education of a group of students who subsequently became teachers. The final stage of the study was the administration of a questionnaire to ascertain the reaction of these teachers to their first year of teaching and to their initial training as seen in retrospect. However, the group contained very few specialist teachers of mathematics and so, in an endeavour to obtain the views of mathematics teachers, the DES, at our request, commissioned NFER to extend their study by administering a slightly amended version of their

questionnaire to a sample of teachers of mathematics who were in their first three years of teaching in secondary schools. This survey was carried out in October 1980 and replies were received from 198 teachers of whom 48 were in their first year of teaching.

672 The reactions of these teachers to a number of statements about their present teaching post revealed, as we have already stated in paragraph 640, a picture of a group of teachers who were happy in their work. More than 90 per cent said that the statements 'I enjoy teaching mathematics very much', 'I am very happy teaching mathematics in my present school', 'I am very satisfied with the duties of my present job' applied at least moderately to them; about half the teachers said that these statements applied strongly.

673 In response to a group of statements about the adequacy of their initial training, these teachers felt, in general, that their training had prepared them better for the subject content of their mathematics teaching and for classroom management and organisation than for dealing with discipline problems, assessing pupils' progress or dealing with the whole range of ability, especially in mixed-ability groups. Those who had followed PGCE courses were more satisfied with the balance between the different elements of their courses than were those who had followed BEd or Certificate in Education courses. Among those teachers who commented on the balance of the course, there was a general wish for a greater emphasis on the more practical elements of the course, such as methods of teaching and classroom management, practical teaching and observation in classrooms. Those teachers who had asked themselves how time was to be found for a greater emphasis on these elements almost all suggested that the time given to education theory should be reduced. Many teachers felt that the relevance of education theory had not become apparent by the end of their initial training course.

674 Some teachers commented that their initial training courses had failed to provide adequate preparation for the teaching of mathematics to very slow learners. In our view it should not be the task of initial training courses to provide preparation for such teaching; nor do we believe it to be practicable, though students should be made aware of the variety of special needs which they may meet in their pupils. In order to be able to appreciate the special difficulties of very slow learners, as also of children with other special needs, it is necessary first of all to gain experience of teaching mathematics to pupils who do not have problems of this kind. **We therefore consider that training to teach pupils with special needs should be provided by means of in-service courses undertaken after teachers have had opportunity to gain classroom experience.** In our view new entrants to teaching should not be required to teach mathematics to classes of such pupils.

Professional training

675 Professional training within BEd and PGCE courses is related to the age of the pupils—primary, middle or secondary—whom the student intends to teach. Two of the elements of professional training are courses of instruction in methods of teaching particular subjects, commonly called 'curriculum

courses' in BEd and 'method courses' in PGCE, and school experience. We discuss both of these.

Curriculum and method courses

676 We start our discussion of curriculum and method courses by drawing attention to two matters which we believe to be of importance whatever the age range to which the course relates. We consider that the need to take account of these will require changes in a number of courses.

677 **The first is the need within these courses to stress the various elements which, as we have pointed out in paragraph 243, should be included within mathematics teaching and in particular the place of oral work, discussion and practical work.**

678 **The second is the need to emphasise the relationship between mathematics and other areas of the curriculum.** "The contribution of a subject to the pupil's education is weakened if it is not perceived by the teacher and presented to the pupil in terms of its relation to the rest of the curriculum. There are two associated aspects of this notion. One is the extent to which the teacher sees his subject as part of the curriculum as a whole and as contributing to it. The other is the explicit making of links between subjects in such a manner that the skills and ways of thinking in a particular subject are given opportunity for development in others."* The more integrated approach to the curriculum which is used in the primary years may perhaps make it easier for intending primary teachers to view mathematics in this way, provided that they have been able to develop confidence in their own knowledge and use of mathematics. It is no less important, though it is likely to be more difficult to achieve, that those who teach mathematics in secondary schools should also be led to consider the relationship of mathematics to the curriculum as a whole.

*Teacher training and the secondary school. An HMI discussion paper. DES 1981.

Preparation for teaching mathematics in primary and middle schools
679 Because nearly all primary teachers and many of those who teach in middle schools are required to teach mathematics, all students who are training to teach at these levels take a curriculum course in mathematics. Most of them will not be studying mathematics as a main subject and many may well not have enjoyed mathematics at school. Even though the introduction of the O-level or equivalent requirement for entry to teacher training may lead to some improvement in overall competence and also in the attitude of some students to mathematics, it cannot be expected that all those who intend to teach in primary and middle schools will necessarily be looking forward to teaching mathematics; indeed, some may have given up their studies in mathematics with a sense of relief after gaining their O-level qualification. It is not therefore surprising that some students will start their training with fears about the teaching of mathematics, and that training institutions should have difficulty in giving to some of their students the positive attitude to the subject and the confidence which are necessary if these students are to be able to teach mathematics well. However, the teaching of mathematics will probably occupy about 20 per cent of the time of those who teach in primary

schools. **It must therefore be the major task of those who train these students to establish positive attitudes, to consolidate and deepen the students' knowledge of mathematics—a process which may involve filling some gaps—and to provide them with a firm basis from which to start teaching mathematics.** Our discussion of primary mathematics in Chapter 6 indicates how much is involved in this task.

*In 1979 rather more than 25 000 teachers in primary schools (ie about 13 per cent of all primary teachers) had a graduate qualification; of these only about 300 had mathematics as the first subject of their degree.

680 Only a minority of those who are now teaching in primary and middle schools have trained by means of a PGCE course*. Indeed, there are many who feel that one year is too short to provide adequately for all the needs of PGCE students training to teach in these schools. Their mathematical needs are those which we have already described but the problems of providing within a PGCE course for students who have a sense of inadequacy in mathematics are likely to be greater, both because of the short time which is available and also because many may not have studied mathematics for at least five years. **In our view the long-term solution lies in lengthening the PGCE course for primary teachers.** In any event, we consider that training institutions should not offer places on PGCE courses for intending teachers in primary or middle schools to students whom they know to be mathematically weak.

*As defined in Appendix 1, paragraph A12.

Preparation for the specialist teaching of mathematics

681 Although a large proportion of those who take mathematics as a main course in BEd become specialist teachers of mathematics, the majority of mathematics specialists in secondary schools train as teachers by taking a PGCE course with mathematics as their first method subject. Figures relating to the enrolment on PGCE courses since 1975 of graduates who are taking mathematics as their first method subject are given in Appendix 1, Table 24. A very sharp drop in numbers in 1978 and 1979 has been followed by a considerable upturn in 1980 but it seems that the increase is not necessarily in students who have an honours degree in mathematics. Interim findings from a survey of PGCE courses in university departments of education which is being conducted by Professor G Bernbaum at the University of Leicester show that in 1979-80 there were about 200 men and 200 women who were taking mathematics as a first method subject in PGCE courses at universities. About half of these students had an honours degree in mathematics, about 7 per cent an honours degree in a related* subject and about 13 per cent an honours degree in a subject which was not related. The remaining 30 per cent had joint honours, pass or ordinary degrees but there was no information about the mathematical content, if any, of these degrees. Information relating to PGCE courses in the non-university sector for the year 1980-81, which has been made available to us by the Mathematical Education Section of the National Association of Teachers in Further and Higher Education, suggests that less than 30 per cent of those who were taking first or second method courses in mathematics possessed a mathematics degree; a further 20 per cent had joint degrees including mathematics. It therefore seems clear that at the present time by no means all those who are taking mathematics as a first method subject in PGCE courses in either the university or non-university sectors are mathematically well qualified (see also Appendix 1, Table 25).

682 Two reasons have been suggested to us which may account for the fact that an increasing number of graduates in other disciplines are taking mathematics as a first subject in PGCE courses. The first is that it will be easier to obtain a post teaching mathematics than teaching some other subjects. The second is that, at a time when job opportunities are not as plentiful as used to be the case, it is a useful 'insurance policy' to obtain a teaching qualification in a shortage subject.

683 We therefore believe that the increase in recruitment to PGCE courses in mathematics should be viewed with caution. We welcome the increase in numbers of students training to teach mathematics if their degree courses have contained a substantial mathematical component. On the other hand, we view with some concern the fact that students whose degree subject is completely unrelated to mathematics should be accepted for a PGCE course in mathematics. We agree that "these students are likely to swell the numbers of those whose qualifications are inadequate or inappropriate for the work that they are being asked to do".* We do not think that, within the time allotted to most method courses, they are likely to be able to learn to present mathematics effectively in the classroom. In our view, **students should not be admitted to first method courses in mathematics unless their degree courses have contained a substantial mathematical component**.

Teacher training and the secondary school. An HMI discussion paper. DES 1981.

684 The number of teachers in secondary schools who are expected to teach more than one subject is likely to increase during the next few years because of pressures resulting from falling rolls. In some PGCE courses students preparing to teach in secondary schools take one main subject but in others they follow courses in two subjects. In some cases these courses are equally weighted but often the second subject is given less, and sometimes very much less, time. At the present time mathematics is a popular choice as a second subject. The Leicester survey showed that in 1979-80 there were about 360 students taking mathematics as a second method subject in PGCE courses at universities. Of these students, about 20 per cent had an honours degree in a subject related to mathematics and about 50 per cent an honours degree in a subject which was not related; the remaining 30 per cent had joint honours, ordinary or pass degrees but there was no information about the mathematical content, if any, of these degrees. It is highly probable that many of those who take mathematics as a second subject will be required to teach it. Indeed, it may be the case that some who choose a second method course in mathematics will do so in the hope that some training in the teaching of mathematics will help them to obtain a teaching post. It is therefore essential that these courses are substantial enough to give adequate grounding and that "recruitment to them must be governed by exacting standards".* In our view **students should not be admitted to a second method course in mathematics unless their degree course has contained a reasonable mathematical component or they are mathematically qualified in some other way**.

Teacher training and the secondary school. An HMI discussion paper. DES 1981.

Time allowance for curriculum and method courses
685 From the submissions which we have received it is clear that there is very considerable variation in the amount of time which is given to curriculum and

method courses in mathematics, and to their positioning within BEd and PGCE courses as a whole. A frequent complaint has been that insufficient time is given to curriculum and method work but complaints of this kind are directed at time allowances which are widely different; nor is there any apparent connection between the amount of time allowed and the intensity of the complaints. The time allotted to curriculum and method courses appears to depend on a number of factors; these include the timing of these courses in relation to teaching practices and other school visits, their relationship to other elements within professional studies which deal with classroom method and the amount of private study which is expected of students.

686 A number of submissions suggest that within primary and middle BEd courses a minimum of 90 hours spread over three or four years should be given to a curriculum course in mathematics. This suggestion was supported in oral evidence to us by the Council for National Academic Awards, which is responsible for the validation of many BEd courses. We were told that if the time allocation fell to below 90 hours, it was expected that compensation would be provided in the form, for example, of a larger than usual time allocation for curriculum science. We have also noted that "in HMI's view a total of some 90 hours, deployed suitably in relation to school experience and block practice, would be reasonable provided that students were mathematically ready for the work"*. This implies that some students may need to spend additional time in reaching the necessary state of preparedness. There are, however, many courses which provide considerably less time than this; we have been told of courses which give as little as 40 hours. We wish to express our concern that it is possible to spend so little time on preparation for teaching a part of the curriculum which occupies up to one-fifth of a child's time during the primary years. There is also very considerable variation in the proportion of time which is given to method work within PGCE courses. We have been told of courses in which it occupies rather more than one-third of the total time which is available; on the other hand, there are courses in which the proportion of time given to method work is one-eighth. We refer later to the need to review the great variety of provision which exists for both curriculum and method courses.

*HMI Series. Matters for discussion 8. *Developments in the BEd degree course.* HMSO 1979.

School experience

687 School experience forms part of all initial training courses. It is likely to start with short periods of observation in a school, sometimes in the company of a tutor, and will certainly include at least one extended period of teaching practice during which school staff must assume major responsibility for the student. **It is therefore necessary for the staff of schools which receive students on teaching practice and the staff of the training institutions from which the students come to act together in a well-defined and mutually supportive partnership**; unless such a partnership exists and operates effectively students will not benefit as they should from their time in schools. This means that schools and training institutions need to explore together over a continuing period the nature of the partnership and the contribution which each can make towards it. These discussions should involve not only the senior staff of the schools concerned but also the teachers in whose classes and departments the students will be working. So far as mathematics is concerned, it is to be ex-

pected that the mathematics co-ordinator or the head of department will play a major part in providing support for students on teaching practice. We believe it to be desirable that they should have opportunity to discuss their role with the staff of the training institution and receive any necessary guidance. The members of staff of both school and training institution need also to be clear about their respective responsibilities in regard to the assessment of students on teaching practice.

688 It has been suggested to us that in some cases the partnership between schools and training institutions is not sufficiently well defined, with the result that schools are sometimes not sufficiently aware of the extent of their responsibilities towards the students who come to them nor of the aims, objectives and priorities of the institutions from which they come. Equally, some training institutions may not be aware of the aims, objectives and priorities of the schools to which their students go. At worst, there can be a conflict of aims between school and college in respect of mathematics teaching. If this is the case, it is essential that steps should be taken to discuss and resolve any such difficulties.

689 We have been concerned to be told that in some cases current financial restraints are limiting the choice of schools which can be used for teaching practice, so that institutions are having to send students to schools which are near at hand rather than to schools which would offer greater opportunities but which are further away. **We believe it is essential that such restriction of choice should not lead to the use for teaching practice of schools in which the necessary support cannot be given to students**. The same restraints are said to mean that students on teaching practice sometimes receive fewer visits from their tutors than would otherwise be the case.

690 We recognise that it is not practicable for all students who will teach mathematics in primary or middle schools to be supervised during their teaching practices by staff who are mathematically qualified. **It is therefore essential that training institutions and schools working in partnership should develop reliable methods of identifying and helping those students whose work in mathematics during teaching practice is weak**. For the reasons we have given in paragraph 680, students on PGCE courses may be especially at risk.

691 We draw attention finally to the fact that **special attention needs to be given during practice to students preparing to teach in secondary schools who have taken a second method course in mathematics**. Not only do they need opportunity to teach mathematics in addition to their main subject but their teaching of mathematics needs to be supervised adequately. Since the tutor who visits them may not be a mathematics specialist, the support given by the head of mathematics in the school will be especially important.

Requirement to undertake
initial training

692 Mathematics graduates are at present allowed to teach in maintained secondary schools without taking a PGCE course. A great majority of the

submissions we have received which refer to this matter have urged that this exemption should cease. In our view the ability to obtain a degree in mathematics does not necessarily imply an ability to teach mathematics well without further training. As we have pointed out earlier in this report, it is not easy to teach mathematics well or to cope with the very differing requirements of pupils of different levels of attainment. The number of graduates in mathematical studies who enter teaching directly after completing their degree course has been decreasing in recent years. Information relating to the period 1974–1979 is given in Appendix 1, Table 23.

693 In our view **the exemption of newly qualified mathematics graduates from the requirement to undertake initial training as a teacher should cease as soon as is practicable**. A majority of us feels that, in respect of those who will in future qualify as mathematics graduates, action to remove the exemption should be taken as soon as is administratively possible; and believes that, although this might discourage some from entering mathematics teaching, the number is likely to be small so that the loss of these untrained teachers would not be sufficient to outweigh the arguments in favour of professional training for all future mathematics graduates who wish to enter teaching. We all hope that the withdrawal of the exemption would be accompanied by the incentives to undertake training to teach mathematics which we have advocated in the previous chapter.

694 Even after the withdrawal of the existing exemption, there would remain a number of men and women who would still be eligible to enter teaching in maintained schools without undertaking professional training as a teacher. **We believe that, both for their own good and for the good of the pupils whom they will teach, all those who intend to teach mathematics should undertake suitable initial teacher training and, in the strongest possible terms, we urge them to do so. We believe also that LEAs should be required to make special induction arrangements for all untrained mathematics graduates which would ensure good support and thorough assessment procedures throughout their probationary period**.

Main mathematics courses in BEd

695 We have been told that in 1980 some 500 students chose mathematics as a main course in the first year of BEd out of a total entry to BEd courses of rather more than 5000. The introduction of the two A-level requirement for entry to BEd courses means that main courses in mathematics should now be able to start from a base of previous A-level study in mathematics. In paragraph 643 we listed the elements which we believe should be included in the mathematics degree courses of intending teachers. These apply equally to main mathematics courses in BEd. No matter what the age of the pupils whom they may be teaching, teachers of mathematics must be able to take a wide view of the mathematics curriculum and be able to make informed judgements about the priorities within it. We believe that this aim is more likely to be fulfilled by the provision of a main mathematics course which aims at achievement over a broad front rather than by one which seeks to achieve depth by restricting its coverage to only a few areas of mathematics. **In our**

view this should be a major consideration in the validation of main mathematics courses.

696 There is considerable variation in the size of teaching groups for main mathematics courses. We have been told that in the year 1980–81 there were nine training institutions with first year mathematics groups of less than five students, twelve with groups of between five and ten students, thirteen with groups of between ten and fifteen, and ten with groups of more than fifteen students.

697 The existence of some very small groups raises the question of whether main courses which recruit such small numbers should be allowed to continue. We have received a number of submissions in which it is argued that such courses should be maintained even although they are not economically viable. Among the reasons advanced is the fact that, because mathematics is a 'core' subject of the curriculum, it should be seen to be available as a main course in as many training institutions as possible, rather than in only a few specialist centres. It has also been pointed out that many students wish to train at an institution which is within reasonable reach of their home or, in some cases, at a particular institution and that the absence of a main course in mathematics could result in the loss of potentially good mathematics teachers. Some have argued further that, because many students take up teaching posts in schools near to the institution in which they have trained, a geographical spread of institutions which offer main mathematics courses is necessary. Many of those who are engaged in teacher training have told us that they value the opportunity to teach both main and curriculum courses in mathematics because the interaction between the two kinds of course is beneficial both to them and to their students and allows valuable links to be made between the main course, the curriculum course and the experience of students during their periods of teaching practice.

698 We accept the validity of many of these arguments for maintaining as many main courses as possible and **we do not believe that the decision as to whether or not to allow a course to continue should be made only on economic grounds**; we have been told that on these grounds groups of about twelve are desirable. We believe that in all cases the decision should be taken on educational grounds which take account of the need to make use of the variety of approaches to teaching which we have advocated throughout this report. Groups should not be so large as to discourage discussion and co-operative working but should be large enough to permit a sufficient interplay of ideas. We recognise that there is bound to be some fluctuation in the annual intake to most courses; however, we believe that, if a course regularly recruits less than eight students, its continued existence should be called into question.

699 We recognise that the logic of our argument means that some colleges might lose their main mathematics course and that questions of possible redundancy among mathematics lecturers could arise. We do not believe that this need happen. In our view, time previously given to the teaching of main mathematics courses could with advantage be used to assist with in-service

support for teachers in the locality. In this way systematic contact between lecturers and schools could be extended, perhaps to include a regular teaching commitment in a school. We believe that such an arrangement would be beneficial to the work of the training institution concerned.

Balance within initial training courses

700 Some submissions to us have expressed concern at the apparently low status accorded to curriculum courses. Others have suggested that the introduction of the BEd degree has resulted in an increased emphasis on main subject and education theory courses at the expense of curriculum courses and other professional preparation, so that students are less well prepared than formerly for effective work in the classroom. We recognise the importance of these concerns. In our view **proper professional preparation is of the greatest importance for intending teachers and should be given appropriate status.**

701 We believe that requests for a greater emphasis on the more practical elements of both BEd and PGCE courses, such as those of some of the recently trained teachers who responded to the NFER questionnaire, stem from what is perceived to be a lack of relevance in some parts of the course. **It is clearly important that students should feel that all parts of their initial training are relevant to their needs.** The relevance of courses on methods of teaching and classroom organisation is clear, as is the relevance of school experience, but the relevance of education theory may be less apparent. It has been suggested to us that this may be because there are few education tutors who have a background of mathematics teaching. In consequence, for instance when illustrating education theory with examples from the classroom, they may seldom make use of examples from mathematics but instead choose examples from other curricular areas of which they have greater experience. There is need for close co-operation between education tutors and mathematics tutors so that education theory can be seen by mathematics students to be firmly grounded on practice in schools and in this way to make a valid contribution to their initial training.

Induction

702 Initial training prepares a student for entry to the profession but much of that preparation is likely to be less effective than it should be if it is not followed up and developed during the first year of teaching. "There is no major profession to which a new entrant, however thorough his initial training, can be expected immediately to make a full contribution. The Government share the view of the James Committee that a teacher on first employment needs, and should be released part-time to profit from, a systematic programme of professional initiation, guided experience and further study."* The aim set out in the paper from which these words are taken was for the introduction of a national 'induction' scheme in the school year 1975-76. This was to provide both enhanced support and a lightened teaching load for teachers during their first year of service. Sadly, this intention has not yet been realised.

Education. A framework for expansion. HMSO 1972.

703 Most LEAs make provision of some kind for the induction of probationary teachers. A survey carried out in 1979* on the induction of teachers

See DES Statistical Bulletin 9/80

serving in maintained schools suggested that a high proportion of new entrants to teaching were involved in some kind of induction programme but
that relatively few induction programmes included provision for significant
release from normal classroom duties and/or a lightened teaching load. It is
possible that recent expenditure reductions may have caused a reduction in
the modest provision which existed during the year 1978-79.

704 The results of the NFER survey which we described in paragraph 671
showed clearly that the provision made for the teachers included in the sample
had been uneven and often inadequate. Of those teachers who had been
teaching for at least a year, only a quarter said that they had attended more
than two meetings arranged by the LEA during their probationary year; one-
third had attended no meeting at all. On the other hand, eighteen teachers who
had attended at least ten meetings had evidently taken part in substantial induction programmes. Within schools, eleven teachers seem to have taken part
in formal induction programmes but almost half of the teachers said that they
had not attended any formal meetings for probationers in their schools. This
does not necessarily mean that probationers lacked guidance in their schools;
they may, for example, have been given help and support by the head of the
mathematics department which they did not classify under the heading
'meetings for probationers'. Indeed, over 40 per cent of those who had completed their probation said that they were very satisfied with the guidance and
help which they had received from their schools during their first year of
teaching and a further 50 per cent said that they were satisfied or fairly
satisfied; only 8 per cent expressed dissatisfaction. On the other hand, only 7
per cent expressed themselves as very satisfied with the support given by the
LEA and 30 per cent expressed dissatisfaction; 40 per cent said that they were
fairly satisfied—perhaps being unaware of the level of support which can, and
in our view should, be provided. Average teaching loads during the probationary year were very little lighter than subsequently and some teachers who
had been given a somewhat lightened teaching timetable felt that they had
been imposed upon by having to cover more frequently than their colleagues
for teachers who were absent. We do not have such specific evidence in respect
of primary teachers but the general indications which are available to us do
not suggest that provision for induction is significantly better.

705 It has been pointed out to us that revised patterns of initial training which
have been adopted in many institutions in recent years have been planned on
the assumption that college courses will be followed by a properly structured
induction year. If this is not provided, the training of all who have taken these
courses will be incomplete. In our view **a proper programme of induction,
which should include some lightening of the teaching load, is necessary in the
first year, as is good support from the mathematics co-ordinator or head of
department, the head teacher and LEA staff.**

706 We appreciate the problems which schools can face in providing support
and a lightened teaching load during the first year of teaching. In some cases
probationary teachers are appointed at short notice to replace experienced
teachers and at a stage at which it is very difficult to change arrangements

which have already been made for the following school year. Nevertheless, it is essential that schools and LEAs recognise the need for suitable induction procedures and make every effort to provide them. We have already pointed out that mathematics is not an easy subject to teach; it follows that adequate support is of vital importance during the first year of teaching as well as subsequently. Ideally, probationers should be appointed only to schools in which suitable support is assured. When an appointment of this kind is not possible, LEAs should make every effort to supplement the support which the school is able to give. Teachers in secondary schools for whom mathematics is the second teaching subject can be especially in need of support, particularly in schools in which the standard of mathematics teaching is not high. In the absence of such support, these teachers are liable to find themselves falling back on a narrow style of teaching aimed mainly at containing the class and keeping order, rather than working in the ways which we recommend in this report and which, we hope, will also have been advocated during their training.

Future developments

707 In the course of the chapter we have quoted from *Developments in the BEd degree course*, from *Teacher training and the secondary school*, and from other evidence we have received. All these sources highlight the complexity of existing initial training courses, draw attention to problems of the kind which we have discussed in this chapter and pose questions which we believe should be answered.

708 It is not our task to prepare a report on the initial training of teachers nor is our Committee suitably constituted to do so. We are, however, concerned that there seems to be so little knowledge in the training institutions about the relative effectiveness of, for example, the different forms of curriculum and method courses in mathematics, their length and their positioning within the course as a whole. We are also concerned that little systematic attempt appears to be being made to evaluate the different forms of provision which exist and to distil or discover good practice. *PGCE in the public sector** refers to the 'bewildering variety of practice that exists'. The submissions which we have received make it clear that this is indeed the case and we find it hard to believe that such variety can be justified. **We believe that there is an urgent need to review and evaluate the courses provided for the initial training of all those who will teach mathematics.**

*PGCE in the public sector. An HMI discussion paper. DES 1980.

709 We are disturbed to read that "the problem for teacher training is to know what the newly qualified teacher should be equipped with on emerging from his course and what could be left to induction and in-service training. There is no consensus on this, and even if there were it would produce no ready formula which would fit the circumstances of every institution"*. **We believe that it is essential that efforts should be made to achieve such a consensus.**

*Teacher training and the secondary school. An HMI discussion paper. DES 1981.

710 We have noted that the recent report from the Education, Science and Arts Committee of the House of Commons* recommends that HMI should no longer be concerned with courses in the higher education sector. It is not for us to comment on the general issue but **we feel it to be essential that HMI should continue to be involved with teacher training courses.** Indeed, we find

*House of Commons. Fifth Report from the Education, Science and Arts Committee. *The funding and organisation of courses in higher education.* HMSO 1980

it anomalous that, as part of their duties, HMI appraise initial training courses in institutions in the public sector but have not been given a similar duty in respect of university departments of education which are also responsible for the initial training of teachers. **We are aware that HMI maintain informal links with university departments of education; we recommend that they should be given the duty of appraising the initial training courses which these departments provide.**

711 We have received several suggestions that PGCE courses and three-year BEd courses should be made longer in order to provide a better preparation for the intending teacher. Although we can appreciate the reasons for these suggestions, and although we have been told that the number of three-year BEd courses is decreasing, we doubt whether any overall extension of initial training courses would be practicable in the foreseeable future. However, this is not to say that there should not be changes. We believe, for example, that consideration should be given to the extent to which it would be possible to make use of the weeks which lie outside the existing university and college terms, and especially of those which lie within school terms.

712 We have already stressed the need for good induction programmes; **we believe that consideration should also be given to ways in which initial training and induction might be more firmly linked.**

713 Our report has many implications for the initial training of teachers of mathematics in both primary and secondary schools. We have suggested emphases on the teaching of mathematics which are different from those found in many schools at present. These include a greater differentiation in the curriculum of secondary schools, the use of a greater variety of teaching styles, and a greater emphasis on discussion, practical experience, applications of mathematics, problem solving and investigation. It will be the responsibility of those who train teachers to ensure that students experience a variety of forms of learning for themselves, are aware of the need for variety in teaching methods, and start teaching with an ability, at a beginner's level, to use these methods in their teaching. Many students will not yet be able to observe our suggestions in practice in the classrooms in which they find themselves; but young teachers can influence their schools, as well as be influenced by their schools, and we hope that the staff of initial training institutions will respond to our report and take account of its recommendations in their work with students. Perhaps the most important characteristic with which a new teacher of mathematics can enter teaching in the 1980s is a determination continually to monitor and reappraise his teaching and his curriculum. It must be for those who train teachers to seek to develop a flexible attitude in their students which will enable them to respond positively to curriculum change in mathematics.

15 In-service support for teachers of mathematics

714 In the course of our report we have drawn attention to a number of ways in which we believe that the mathematics curriculum for some pupils should be changed, as well as to ways in which we believe it is necessary for mathematics teaching to develop in the coming years. Whatever the reaction of central or local government to our recommendations, the extent to which those recommendations are implemented in the classroom will depend upon the response of those who teach mathematics and upon their ability to work in the ways which we have suggested. We wish now to consider the various forms of support which can and should be provided to enable those who teach mathematics to develop and extend their professional skill.

The need for in-service support

715 In our view the need for such support is self-evident. Even if greatly increased numbers of teachers who are well equipped to teach mathematics were to enter teaching in primary and secondary schools in the next few years, it would take many years for this to have a significant effect on the overall quality of the mathematics teaching force. It follows that **any improvement in the standards of mathematics in schools must come largely as a result of the efforts of those teachers who are already in post**; they must, therefore, receive all possible support to enable them to improve the effectiveness of their teaching.

716 The analysis of the present teaching force which we have given in Chapter 13 shows that there are many teaching mathematics in primary, middle and secondary schools whose qualifications for this task are weak or non-existent. It should not, however, be supposed that in-service support is needed only to remedy the deficiencies of those who lack suitable qualifications. However good their initial training and induction may have been, **all those who teach mathematics need continuing support throughout their careers in order to be able to develop their professional skills and so maintain and enhance the quality of their work**.

717 The need to provide adequate support for teachers is stressed in very many of the submissions which we have received, including submissions from LEAs, training institutions, professional bodies and teachers themselves. The need for support is not, of course, confined to teachers of mathematics alone and a number of the matters which we discuss in this chapter apply equally to teachers of other subjects. **We are, however, concerned with the specific needs of mathematics and believe that there are a number of reasons which justify**

support for teachers of mathematics on a scale which may not, on financial grounds, be possible for teachers of all subjects at the present time.

718 In the first place, there is the general acceptance that mathematics is an essential part of the curriculum for all pupils in both primary and secondary schools. Secondly, there is the public concern about the teaching of mathematics which has led to the setting up of our Inquiry. Thirdly, there is the lack of suitable qualification, to which we have already drawn attention, of many of those who teach mathematics. Fourthly, the curricular changes which we have recommended in this report will require many teachers to work in ways which are different from those which they use at present. Fifthly, the increasing availability of calculators and computers requires all who teach mathematics to reconsider the content of the mathematics curriculum and the methods which they use in the classroom. Sixthly, other areas of the curriculum which make use of mathematics are at a disadvantage if mathematics is not well taught. Finally, as we have noted in Chapter 13, it is possible that falling rolls in secondary schools in the coming years, and consequent reductions in staffing, will increase the pressure for mathematics teaching to be undertaken by teachers for whom mathematics did not form a significant part of their initial training.

Types of in-service support

719 In-service support for teachers needs to be provided in a variety of ways. We may distinguish in general terms between the support and training which is provided within the school itself, the support which comes from meetings with other teachers, visits to other schools and membership of professional subject associations, the support provided by the staff of local authorities and training institutions, and the support which comes from training courses of various kinds; these categories are not, of course, wholly distinct. There is one other kind of support which, although perhaps not so immediately apparent to many teachers, is provided by research into mathematical education and by the centres which exist to undertake and disseminate work of this kind. We wish to discuss all of these.

School-based support

720 **We are in no doubt that school-based in-service support for teachers is of fundamental importance.** It can be directed specifically to the needs of the school and its pupils, so that those who teach mathematics develop professionally as a result of working together to improve the work of the school. Above all, it can and should be a continuing process which is not limited to the length of a lecture, a discussion or a course.

Every staff group has within it the ingredients of a kind of continuous educational workshop. For it is in the staff group itself that meeting points can be found between student and practitioner, between the young and the middle-aged, between the inexperienced and the experienced, between the enthusiasts and the cynics, the optimists and the pessimists, between the so-called 'pupil oriented' and the so-called 'subject oriented' teachers.*

*E Richardson. *The teacher, the school and the task of management*. Heinemann Educational Books 1973.

721 A description of the kind of day-by-day support which is needed is implicit in the descriptions of the responsibilities of the mathematics co-

ordinator in a primary school and the head of department in a secondary school which we have given in paragraphs 355 and 508. We emphasise at the outset that the effectiveness or otherwise of school-based support and of the professional development of teachers to which it should contribute depends on the ability of the mathematics co-ordinator or head of department to provide the necessary leadership and example. The foundation of school-based support should be a suitable scheme of work which gives guidance not only about syllabus content but also about teaching method, the availability of resources, assessment and record keeping, and necessary administrative procedures; this scheme should be reviewed regularly. There is a need for those who teach mathematics to be given opportunity to observe and work with each other and to share teaching materials and other resources. In our view it is essential that those who teach mathematics should hold meetings on a regular basis. Some of these meetings should be used to discuss the teaching of particular groups of pupils or particular topics; such discussions can assist the development of a common approach and lead to the preparation of teaching materials which all can use. Because meetings of this kind are likely to assist with lesson preparation, the overall demand which they make on the time of teachers can be less than might be supposed.

722 It is important that school-based activities do not become too inward-looking. It is therefore helpful from time to time to invite someone from outside the school to join in meetings and discussion in order to offer new ideas and additional expertise. Such a person might perhaps be a teacher from another school, a mathematics adviser or advisory teacher, the warden of a teachers' centre or a member of the staff of a teacher training institution. Unless this person knows the school well and is well known to those who teach mathematics, a 'one-off' session may be of little value; it is usually more profitable to arrange a series of meetings so that mutual understanding can develop and discussion be based on the perceived needs of the school.

723 **Because the effectiveness of school-based support depends upon the leadership of mathematics co-ordinators or heads of department, it is essential that they themselves should receive support and training**. We believe that such training must be the responsibility of local education authorities and that it is necessary for it to be provided not only for newly appointed co-ordinators and heads of department but also, on a continuing basis, for those who are already in post. Training should emphasise the leadership and organisational role of the co-ordinator or head of department as well as the need to be aware of current developments in mathematical education. In our view **the training of co-ordinators and heads of department is likely to contribute most quickly and effectively to the overall improvement of mathematics teaching and should be given top priority**. Training provided by local education authorities will need to be accompanied by support for the work of the co-ordinator or head of department within each school. This must be the responsibility of the head. We realise that it is not easy in either primary or secondary schools to provide opportunity within the school day for the co-ordinator or head of department to work with other teachers but we believe that it is essential that this problem should be overcome, if necessary by the use of part-time staff to provide cover.

724 Except, perhaps, in small schools the co-ordinator or head of department will need assistance with some of the tasks for which he is responsible. This will need to come from other teachers of mathematics who should be encouraged, singly or in groups, to undertake specific jobs within the school or department; help of this kind will not only assist the co-ordinator or head of department but will also contribute to the professional development of the teachers concerned. The delegation of responsibility in this way also encourages co-operation, both inside and outside the classroom, between those who teach mathematics. There are sufficient examples of co-operative working of this kind to be found in both primary and secondary schools to show that such working is both possible and effective. However, it is necessary to realise that such methods of working have to be developed gradually; perseverance is required over many terms throughout which support for the co-ordinator or head of deparment must continue both inside the school and from outside.

Meetings with other teachers

725 However good the support which is provided within a school, it needs to be supplemented by provision of other kinds. In particular, **it is necessary for those who teach mathematics to have opportunity to meet other teachers**. Local teachers' centres can play an important part in providing facilities for teachers to meet each other. We have been interested to note that very considerable impetus to the establishment of teachers' centres in their present form was provided by the centres set up as part of the Nuffield Mathematics Project in the latter part of the 1960s. These centres were seen as places in which teachers could meet and where the ideas and activities put forward in the Nuffield Teachers' Guides could be 'discussed, elaborated and modified'. We have no doubt that there is need for provision of this kind and regret that, according to reports which we have received, several teachers' centres have recently been closed. We hope that they will be re-opened as soon as circumstances permit and that in the meantime LEAs will endeavour to provide in other ways the kind of support which can come from teachers' centres.

726 Teachers' centres can not only facilitate formal and informal discussion and arrange courses of various kinds but can also act as resource centres. A few authorities maintain one or more teachers' centres which are devoted entirely to mathematics. We have received evidence of the excellent work which is carried out by these mathematics centres and of the support which they provide for the teachers whom they serve. The staff of these centres are mathematicians who are able to arrange mathematical activities in their centres for groups of teachers, for teachers with their classes, and sometimes also for children. They also give help to individual schools and teachers and generally provide a focal point for mathematical activity in their area. In some cases mathematics centres also produce a magazine or newsletter for teachers. **We believe that, provided they are adequately resourced, centres of this kind can play a most valuable part in improving the teaching of mathematics; we strongly support their continuation.**

727 **In areas where mathematics centres do not exist we recommend that efforts should be made to provide at least one resource centre for mathematics**

in each authority. This should have a library of reference books on the teaching of mathematics, a selection of journals, a range of published text-books and a collection of mathematical equipment; these should be available both for use in the centre and for loan to teachers. Arrangements should be made for a mathematics specialist to be available in the mathematics resource centre at stated times so that he can discuss with teachers the classroom use of the materials which are in stock.

Visits to other schools

728 Teachers of mathematics—as, indeed, teachers of most other sub-jects—very often have little idea of what goes on in other schools in their neighbourhood. Time spent observing and joining in with the teaching in another school can provide valuable insight into different forms of organisa-tion and different teaching methods. For this reason **we consider that all teachers should be enabled to visit other schools from time to time**. These visits must of necessity take place during school hours and cover of some kind will have to be provided for the teachers involved. In secondary schools the fact that many fifth form pupils are no longer in school during the second half of the summer term may make it easier to release staff for visits to other schools during this period. Some authorities arrange for a school to be closed for a single day from time to time so that the whole staff may engage in in-service work. An arrangement of this kind can enable all the teachers in a school to spend the day visiting other schools in their own or a neighbouring LEA.

729 Further opportunities of meeting and working with other teachers are provided by membership of examination panels and especially of local work-ing parties. These may be concerned with such matters as the preparation of guidelines in mathematics, the preparation of teaching and assessment materials, arrangements for continuity and transfer between schools or liaison with local employers. Work of this kind is not only of benefit to the mathematics teaching of an authority or group of schools but can also provide a valuable means of advancing the professional development of those teachers who take part.

The professional associa-tions for teachers of mathematics

730 The professional mathematical associations provide yet another means of enabling those who teach mathematics to meet other teachers, both through local branch meetings and national conferences. They also assist pro-fessional development through their journals and other publications. However, a disappointingly small proportion of those who teach mathematics in schools belongs to one or other of these associations. We understand that the joint membership of the two associations which are most directly concerned with mathematics teaching in schools—the Mathematical Association and the Association of Teachers of Mathematics—amounts to some 12 000 but that, if those who belong to both associations are counted on-ly once, the total is nearer to 9000; nor do all of these teach in schools. In com-parison with some 30 000 teachers who teach mathematics in secondary schools and the very many who teach mathematics in primary and middle schools, this is a very small number. Both the Mathematical Association and the Association of Teachers of Mathematics publish journals which are ex-

pressly intended for those who teach mathematics in schools and which contain many articles which offer help and suggestions at both primary and secondary levels. **We believe that every effort should be made to encourage membership of the professional mathematical associations and that the associations themselves should do as much as possible to develop their local activities.**

731 The professional mathematical associations are already playing an important part in the development of mathematical education. We consider that it is vital that they should continue to be able to present informed and independent opinions. If they are to be able to do this effectively it is necessary that serving teachers should play a full part in the working groups and committees of these associations. **We hope that LEAs will give such support as may be necessary so that those of their teachers who are invited to take part in the work of these groups and committees may be able to do so.**

Mathematics advisory staff

*There are differences in the titles used by LEAs. Some have 'advisers', some have 'inspectors' and some have both. We use the terms 'mathematics adviser' and 'mathematics advisory staff' to refer to all those who have responsibility for assessing the quality of the mathematics teaching within an LEA and providing support for those who teach mathematics.

732 **We believe that mathematics advisory staff* have an essential part to play if the kinds of in-service support which we have already discussed are to operate effectively.** The first task of a mathematics adviser must be to monitor the quality of the mathematics teaching in the schools for which he is responsible; in LEAs which do not have an adviser specifically for primary mathematics this must entail close contact with the advisers who work in primary schools. He has then to take such steps as are open to him to improve the quality of mathematics teaching throughout the LEA and to encourage and disseminate good practice.

733 We have already stressed the central role of the mathematics co-ordinator or head of department in providing school-based support for those who teach mathematics. It follows that the mathematics adviser must be aware of the strengths and weaknesses of the mathematics co-ordinators and heads of department in the schools which he visits and be prepared to offer them such help as is possible. He will need, in particular, to make sure that the necessary in-service training is provided for them; in some cases it may be helpful to arrange this on a shared basis with other authorities.

734 In addition to monitoring the work of schools he will also need to maintain regular contact with wardens of teachers' centres and staff of local training institutions, as well as with local employers' organisations. He may need to set up local working parties and should be able to advise on visits of teachers to other schools. He will be concerned with arrangements for the provision of in-service work within the authority and will need to identify teachers and others who are able to help with local courses. He is likely also to be called upon to give advice about the appointment of mathematics staff in schools.

735 Just as it is necessary for teachers in schools to be aware of what is going on in other schools, so it is necessary for the mathematics adviser to be aware of what is going on in other authorities. He must also be aware of current developments in mathematical education both regionally and nationally. In some parts of the country it has become the custom for mathematics advisers

from a group of neighbouring authorities to meet together regularly, perhaps once in each term, to exchange information and ideas, and sometimes also to arrange in-service activities which are shared between two or more authorities. **We believe that consultation of this kind is to be encouraged and that it provides a valuable method of sharing experience and assisting the development of good mathematics teaching**.

736   **The duties to which we have already referred constitute a formidable list, and we do not suggest that it is complete. If the adviser is to be able to do his job effectively and offer the necessary support to others, it is essential that he himself receives support**. We do not believe that all advisers receive adequate training, either on appointment or subsequently. On appointment they should be provided with a proper programme of induction which includes opportunity to see advisers at work in other authorities; there should also be opportunity for further training from time to time.

737   There is considerable variation between authorities in the level of provision of mathematics advisory staff. Some authorities have two or more mathematics advisers and a team of advisory teachers; some have a single mathematics adviser; some have an adviser who has responsibility for mathematics as well as for one or more other subjects; some have no mathematics adviser. It is the responsibility of LEAs to ensure that the teachers whom they employ are sufficient in quality and quantity. **We do not believe that an LEA can ensure that the quality of mathematics teaching in its schools is adequate unless it has within its advisory staff adequate mathematical expertise to carry out the necessary assessment and identify schools which are in need of assistance**. The money required for the appointment of a mathematics adviser is exceedingly small compared with the cost of the education service as a whole or the cost of that part of the teaching force which deals with mathematics. **We therefore believe that more expenditure on mathematics advisers is essential, since without it there are bound to be wide and unacceptable variations in quality and inadequate resources to improve it**. If mathematics advisers are to spend their time to maximum advantage, they need to be provided with adequate secretarial and other administrative support and, where appropriate, with support from mathematics advisory teachers.

738   Some advisers devote their whole time to mathematics while others combine a specialism in mathematics with more general duties. Whichever arrangement obtains within an authority, we believe it to be important that mathematics advisers should be aware of the place of mathematics within the whole curriculum and be able to view the organisation for teaching mathematics in the context of the organisation of the school as a whole. It is to be expected that advisers who are seen to be effective in their own field are likely to be asked for comment and help on other matters and so they need to be aware of developments and sources of help outside their specialist field. We do not therefore support the suggestion which has been made to us in some submissions that mathematics advisers should not also be required to work in more general ways. However, **it must be for each LEA to ensure that suffi-**

cient time overall is available for advisory work in mathematics and to appoint further advisory staff if necessary.

Mathematics advisory teachers

739 Although many LEAs do not make use of advisory teachers for mathematics, some have teams of advisory teachers with well-defined tasks. In some LEAs there are teachers who work part-time in a school and part-time as mathematics advisory teachers. Advisory teachers most commonly work alongside teachers in the classroom. They may spend a period of several days in the same school or visit each of a group of schools on a regular basis. They may perhaps help with drawing up or revising a scheme of work or assist with the in-service programme of the LEA or a school. We believe that advisory teachers can play a valuable role in providing in-service support in a school and in assisting with the introduction of new approaches to teaching. This is perhaps especially the case in primary schools which lack a member of staff who has mathematical expertise; it is likely that regular visits from an advisory teacher who is able to work alongside teachers in the classroom could be a very effective method of improving the level of mathematics teaching within such a school. An advisory teacher may be able to make an especially valuable contribution in rural areas where the distance between schools and problems of transport can make it difficult for teachers to take part in the activities of a teachers' centre or to meet teachers in other schools.

Establishments of higher education

740 Establishments of higher education of all kinds have for many years made a major contribution to the in-service support of teachers. In addition to providing full-time and part-time courses for serving teachers both on a regular basis and in response to particular needs, their staff engage in many other kinds of in-service support. One form of support which we believe to be of great value, even though it is on a relatively small scale, is the provision of 'school-teacher fellowships'. Teachers who are appointed to these fellowships, which are sometimes funded by the institutions themselves and sometimes by outside sources including industry, are enabled to undertake a period of study or research away from the classroom.

741 Those who work in establishments of higher education, and especially in training institutions, often assist with courses organised by local authorities or teachers' centres and serve on local in-service committees. Increasingly, too, staff of training institutions assist with school based in-service work, especially in schools in which they have become known as a result of their visits to supervise students on teaching practice. In some cases, too, their institutions serve as resource centres for local teachers, who are encouraged to make use of libraries and other facilities. Furthermore, training institutions are often able to work across LEA boundaries, which enables teachers from different LEAs to meet and work together.

742 Our attention has been drawn to a matter which is said to operate against the involvement of members of staff of training institutions in in-service work with teachers. Such work is clearly of direct relevance to those who are concerned with the initial training of teachers but, if undertaken, reduces the time

available for academic research. However, there can be no doubt that publication of the results of academic research is considered to be essential for promotion to higher posts in institutions of higher education, especially universities. We have been told that some of those who work in training institutions therefore feel that time spent on in-service work can be to the detriment of their academic careers, because those who have responsibility for making appointments do not value evidence of experience gained during in-service work as highly as evidence of published work. If this is the case, we greatly regret it.

743 'Consultancy' work in schools undertaken by the staff of training institutions is not only of benefit to schools but also enables those who work in training institutions to gain up-to-date and first-hand knowledge of the work which is going on in primary and secondary classrooms. We have been told that the way in which time given to work of this kind is 'costed' can in some cases limit the amount of time which training institutions feel able to give to it, particularly if a significant amount of travelling is involved; nor do the methods which it is intended should be used to 'cost' time given to work in schools appear to be entirely clear. We understand that the Pooling Committee is aware of this problem and is at present seeking ways of clarifying the situation. **We believe it to be very important that any arrangements should be such as to encourage and facilitate consultancy work undertaken by the staff of training institutions.**

In-service courses

744 Although it is not our view that courses should be regarded as the most important form of in-service support for teachers, **we nevertheless consider that they have an essential role to play because not all the forms of support which are required by teachers can be provided within a school**, by meetings with other teachers, by advisory staff or by the professional mathematical associations. Courses provide a means whereby teachers from several schools can come together for a special purpose, either to consider a particular aspect of mathematics teaching or to add to their own knowledge or qualifications. We have, for example, already referred to the need to provide training for mathematics co-ordinators and heads of department; this will almost certainly have to be done, at least in part, by the provision of suitable courses on a local or regional basis.

745 Most short courses, and also some longer courses, operate on a part-time basis, very often with one session each week or fortnight. It is common for these sessions to take place after the end of the school day. However, many have argued strongly in submissions to us that courses held at this time are often less effective than they should be because teachers are tired after a full day in school; in addition, preparation for the next day's work may suffer. They have therefore urged that more courses should be held during school hours than is the case at present; we believe that there is substance in this argument. We know that in some LEAs it has been found helpful to start each session in mid-afternoon, so that half of the session takes place during school hours.

746 An advantage of courses which operate on a part-time basis can be that

teachers are more easily able to make use in the classroom of ideas and activities which are suggested during the course and, if necessary, to discuss these further or obtain any help which they require during later sessions. On the other hand, the interval of a week or a fortnight between sessions may make it difficult to maintain continuity in certain types of course. Full-time courses, whatever their length, take teachers completely out of the classroom and enable them to devote their full attention to the course without the distractions of daily school life. If such courses are residential, there are additional opportunities for reflection and discussion. Some longer courses, for example DES regional courses, often combine regular weekly or fortnightly sessions with occasional periods of full-time work for two or three days. In our view there is need for both part-time and full-time courses.

747 The range of mathematical knowledge and experience which exists among teachers on the same course can often be very wide and it is not easy to ensure that the content and level of a course is suited to all those who are taking part in it. For this reason it is important that, when a course is advertised, there should be a clear explanation of its purpose and of the level of mathematical knowledge and experience which will be expected of those for whom it is intended. In this way teachers, and those who will support their attendance, will be able to judge in advance whether a course is relevant to their needs.

748 **The long term effectiveness of in-service courses, especially those which last only a short time, can be greatly diminished unless there is suitable follow-up**. In the case of courses which are locally based, efforts should therefore be made to arrange one or more follow-up sessions at intervals after the course has been completed. The cost of organising such sessions is likely to be small in comparison with the cost of the original course. It has been pointed out to us that, if several teachers from the same school have attended a course, either together or on successive occasions on which the course has been offered, it may be more effective to organise follow-up work within the school itself. There was strong support for this method of working from teachers and others whom we met during our visit to the Scottish Education Department. In any case, teachers who have been on courses should be encouraged to share their experiences with their colleagues on the staff and to discuss them at staff or departmental meetings. It is all too easy for in-service training courses to result in no long-term improvement because of lack of interest or support when a teacher returns to his school.

749 Full-time courses for teachers lasting either a year or a term are offered by a number of training institutions but recently a considerable number of both primary and secondary mathematics courses have had to be cancelled because they have been insufficiently subscribed. Although we have not been able to obtain complete information, it is clear that some LEAs have reduced the number of teachers whom they have seconded to these full-time courses because of the effects of current financial stringency. **If a sustained effort is to be made to improve the qualifications of those who at present teach mathematics, the number of teachers seconded to full-time courses in mathematics will need to increase substantially**.

Diploma in Mathematical Education

750 A substantial contribution to the in-service education of teachers in primary and middle schools has been made by the introduction of the Mathematical Association's Diploma in Mathematical Education which is intended for teachers of children in the 5–13 age range who have at least two years of teaching experience. Two year part-time courses for this diploma started in 1978 and are now offered at some fifty centres throughout the country. About 600 teachers enrolled for the diploma course in 1978 and much the same number have done so in each following year. Of those who enrolled in 1978, about 400 obtained the diploma in 1980; a similar number are expected to do so in 1981. Teachers who obtain the diploma are entitled to one increment on the Burnham salary scale. **We believe that every support should be given by LEAs to teachers who wish to enrol for a diploma course since it provides an effective means of increasing the mathematical qualification of teachers at a significant rate**.

751 Because diploma courses normally take place in teacher training establishments, there are parts of the country in which it is not possible for teachers to take the course. **We believe that energetic efforts should be made to explore ways in which diploma courses can be made available in areas which are remote from training institutions**.

Open University

752 In recent years many serving teachers have gained mathematics degrees awarded by the Open University or degrees which include some mathematics. We have been pleased to note the extension of the University's work to include in-service support for teachers. A course on *Mathematics across the curriculum* is already being offered and a complementary course on *Developing mathematical thinking* is in preparation. There are also provisional plans to include these courses as part of a mathematics diploma course.

753 We believe that the Open University, which has now developed considerable experience and expertise in the development of 'distance learning' courses in mathematics, could play a major part in improving the quality of the mathematics teaching force. At the present time the course fees for courses such as *Mathematics across the curriculum* are high because courses of this kind which the University offers are required to be financially self-supporting. **We believe that means should be explored of providing financial support for in-service work in mathematics provided by the Open University,** since the provision of suitable courses which could be followed by groups of teachers within LEAs might prove to be an effective method of providing in-service support of good quality on a wide scale.

Radio and television

754 Mathematics programmes designed for use in the classroom are broadcast by both BBC and IBA, who consult to prevent duplication. These provide material which can be used to form the basis of a mathematics course or supplement other work in the classroom; different series of programmes cater for the needs of pupils of different ages and levels of attainment. The programmes are usually accompanied by notes for teachers and often by work books

for pupils. Programmes of this kind provide another form of in-service support for teachers which can not only be used in the classroom but can also serve as a basis for discussion by groups of teachers at a teachers' centre or within a school.

755 In preparation for the raising of the school leaving age in 1973, the BBC broadcast a series of television and radio programmes for teachers which were followed and discussed by large numbers of teachers either at teachers' centres or within individual schools. **We believe that programmes of this kind on different aspects of mathematical education, designed to provide a basis for discussion among groups of teachers and perhaps considering some of the issues raised in this report, could make yet another valuable contribution to in-service support and training in mathematics.** We have been told by the BBC that, in their view, an advantage of broadcast programmes is that they are impersonal, so that the ideas contained within them can be criticised frankly without fear of giving offence; this is not always possible following input from a visiting lecturer or adviser.

Research into mathematical education

756 A considerable number of research studies have been carried out which relate to aspects of mathematical education. These include work in such fields as concept formation and development, mathematics learning, the classroom behaviour of teachers and pupils, and curriculum development and evaluation. Many of these studies are summarised in the *Review of research* to which we have already referred several times. We are pleased to learn that plans are being made for a book based on this review to be published shortly* because we believe that much of the research which has been undertaken remains largely unknown to those who teach mathematics in schools and also to many who produce textbooks and other teaching materials. Even when teachers become aware of the existence of a research study on a topic, they very often find it difficult to appreciate its relevance to their own classroom. One of the reasons for this is that research reports are usually written in a technical style which is not always easy to follow and are very often published in journals which most teachers do not see.

*Information is available from the Shell Centre for Mathematical Education, University of Nottingham; see also paragraph 188.

757 More recently some research projects have published their work by means of articles, which are written in a non-technical style, in the journals of the professional mathematical associations. The project *Concepts in secondary mathematics and science*, to which we have already referred, is an example of this. We suggest that **more use should be made of the educational press both to disseminate the results of specific research projects and also to review and interpret for teachers the state of research on different topics.** Advisers and others who have responsibility for in-service support for teachers should be aware of the need to disseminate and interpret the results of research studies so that teachers can be helped to find ways of making use of these in their own thinking and classroom practice.

Centres for mathematical education

758 There are a few centres whose reputation is international, such as those at the University of Nottingham and at Chelsea College, University of London, which foster curriculum development and research into mathematical educa-

tion and which also provide in-service support for teachers of mathematics. The work which these centres undertake is closely related to mathematics in the classroom and we believe it to be of the greatest value. **We would welcome the establishment of a few further centres of comparable quality; we believe that the creation of such centres could lead to a significant improvement in the quality of mathematical education**. However, a paramount consideration in the establishment of any further centres must be the appointment of staff of suitable quality. In our view any further centres should be suitably situated geographically and be based on existing institutions, whether in universities or elsewhere, which are known to be strong in the field of mathematical education. In order to ensure that the work done by any new centre would be able to be strongly classroom-related, it would be important that the LEAs, whose teachers were likely to be most closely involved, should take part in the discussions leading to the setting up of that centre.

Financial support

759   There are two matters which are fundamental to the provision of almost all kinds of in-service support—finance and time. Unless the will exists to make the necessary financial provision and to provide the necessary time, no appropriate plans for in-service support can be formulated, either within a school or an authority, or on a regional or national basis. It is clear from national surveys, as well as from information which some LEAs have provided for us, that a great deal of in-service work is undertaken by mathematics teachers outside school hours and sometimes at their own expense. This voluntary effort is praiseworthy and to be encouraged. However, the problem of providing support and training for all those who teach mathematics cannot be solved entirely by schemes which depend on the voluntary attendance of teachers. The fact that many teachers undertake in-service work outside school hours shows that they already accept in-service training to be part of their professional commitment; the fact that LEAs provide in-service training courses shows that they accept the provision of training and support for teachers to be part of their duty. We are aware that in some countries it is part of the conditions of service of every teacher that a certain number of days each year should be spent on in-service education and we believe that many local authorities and teachers would wish to see a similar provision in this country. Provision of this kind has been recommended in many of the submissions which we have received. We realise that there are difficulties, not only of a financial kind, in such a proposal but without some such arrangement we do not believe that there can be sufficient opportunity to influence and improve the quality of mathematics teaching. **We are convinced that the curricular changes which we are proposing and the changes in attitudes and perceptions on the part of teachers which they will require would have a far better chance of implementation if such an arrangement were to exist**.

760   There are two elements in the cost of providing in-service training for teachers: one is the cost of providing the training courses themselves and the other is the cost of releasing from their schools the teachers who take part in them. These costs are met from a variety of sources. Part of the cost is borne by local authorities, either individually or jointly; voluntary colleges and

universities contribute through the courses which they provide. Central government contribute through the rate support grant to local authorities and through their funding of voluntary colleges and universities. However, no part of the rate support grant is specifically ear-marked for in-service training; nor is any part of the grant to voluntary colleges or universities ear-marked in this way. Teachers themselves contribute both in terms of the expenses which they incur and also by giving up their own time.

761 There are arrangements whereby LEAs can recover a proportion of their expenditure on certain in-service training costs for teachers, including tuition fees, travel and salaries, from a national 'pool' which is financed by contributions from all LEAs according to an agreed formula. Pooling was at one time restricted to full-time courses included in a national programme and lasting either a year or a term, but the present position is that, in general, full-time courses lasting for more than four weeks and part-time courses lasting for at least sixty hours are eligible for pooling. In addition, the costs of providing courses of in-service training in departments of education in local authority maintained colleges are redistributed between authorities through the Advanced Further Education (AFE) Pool. Many maintained sector departments of education provide direct in-service support in schools; a proportion of the lecturer costs involved in giving this support is also chargeable to the AFE Pool.

762 We have noted two types of training which are not at present eligible for pool support. The first is the cost to authorities of releasing mathematics teachers to gain experience of industry and commerce. The second is the cost of appropriate courses from the Open University Associated Student Programme which are taken by teachers. **We recommend that both of these should be eligible for support.**

763 Most LEAs pay, in full or in part, the expenses of teachers who undertake in-service work, either locally or further afield. We have, though, been told of cases in which unhelpful limitations have been placed on teachers, for example in terms of travel or of refusal to permit attendance at a course which, although suitable and near at hand, is taking place in the area of another LEA. We regret restrictions of this kind and also failure to recompense in full expenses which have been incurred; in our view, **such disincentives to take part in in-service work should not be placed in the way of teachers.**

764 The methods by which in-service support is provided by local authorities are likely to vary considerably. It will be appropriate for LEAs to pay for their own facilities, whether in schools, teachers' centres or elsewhere and to make recompense for facilities provided for their teachers by other LEAs. In some cases joint resource centres may be established, often in conjunction with training institutions, and joint funding will be appropriate. We believe that in any such arrangement there should be an incentive for the training institution to provide what is required. Payment, at least in part, for services rendered would help to ensure that good provision not only survived but was developed, whereas less good provision would cease.

765 The Secretaries of State bear considerable responsibility for the quality of the initial training of teachers; we believe it is also appropriate that they should influence the quality of teachers during their subsequent careers by the provision of direct financial support for programmes of in-service training. We understand that under Section 3(a) of the Education Act 1962 they possess powers to do this which, until the introduction of the National Scholarship scheme (see paragraph 652), have not hitherto been used. Although we have been told that local authorities do not, in general, like the concept of funding for specific purposes, they nevertheless have experience of it and accept it in a number of other fields. The evidence which is available to us suggests that at the present time sufficient money is not being spent on the provision of in-service training and that in some areas the position is worsening. **Unless the Secretaries of State take effective action in this field we do not believe that sufficient resources to improve the quality of mathematics teaching will be made available**. In this matter we believe that the Secretaries of State should work in close association with local authorities as the employers of teachers and also, as appropriate, with teacher training institutions.

766 In the words of the Government paper *Education: a framework for expansion* published in 1972, "expenditure to achieve an expansion of in-service training ... is a necessary investment in the future quality of the teaching force". The scale on which it was at that time proposed to increase the provision of in-service training has not been achieved. We note that the Government's expenditure plans presented to Parliament in March 1981 state that "provision for the release of teachers to in-service training and induction has been held at the current level." However, we believe that in-service support and training for those who teach mathematics needs to be increased and that this need is pressing if the developments and changes in mathematics teaching which we are advocating are to be realised. Not all of our recommendations will cost money but many of them will. It will be for those who control finance to determine whether additional money can be made available or whether money must be found from within the public expenditure limits already determined. In his address to the 68th North of England Conference in January, 1981, the Secretary of State for Education and Science said that "Even within our restricted resources, I believe it to be crucial that a high priority should continue to be given to the right in-service training ... In asking authorities and teachers to give a high priority to carefully-managed in-service programmes ... I am in effect asking that, in their very proper concern to do their utmost for today's children, they should not lose sight of the need to do equally well by tomorrow's". **Unless additional money is spent on in-service support for those who teach mathematics, the improved mathematical education which we believe could and should be provided for children in the future is unlikely to be available to them.**

16 Some other matters

767 In this chapter we discuss four matters for which we have not found place elsewhere in our report; they are not related to each other. These matters are mathematics in middle schools, the teaching of statistics, mathematical education in other countries and statistical information relating to mathematical education.

Mathematics in middle schools

768 The mathematical needs of children are related to their ages, their levels of attainment and their rates of progress, and not to the type of school in which they are being taught. It is for this reason that, in the titles of Chapters 6 and 9, we have referred to mathematics in the primary and secondary years and not to mathematics in primary and secondary schools. Pupils in middle schools span the ages traditionally associated with the words primary and secondary but the mathematical needs of children in these schools are not different from those of children of the same age in either primary or secondary schools. For this reason we have not included a separate chapter on mathematics in middle schools; had we done so we would of necessity have repeated a great deal of what we have written elsewhere in the report.

769 We have received very few submissions which make specific reference to middle schools. However, we have been able to visit middle schools and, in the course of our series of meetings with teachers in different parts of the country, to meet teachers who work in middle schools. In addition, we have been able to discuss the teaching of mathematics in middle schools with a small group of mathematics advisers whose areas contain schools of this kind. We have also known that HM Inspectors have carried out a survey of middle schools for pupils aged 9 to 13 during the school year 1979–80 and a survey of middle schools for pupils aged 8 to 12 during the year 1980–81. We understand that a report on each of these surveys is to be published and that each report will contain information about the teaching of mathematics.

Staffing

770 Mathematics is most often taught by the class teacher for at least the first one or two years of the middle school course, but setting for mathematics is introduced in many middle schools in the later years. The results of the 1977 survey of staffing in maintained secondary schools, to which we referred in Chapter 14 and which included middle schools 'deemed secondary'*, show that there is a very considerable shortage of mathematically qualified teachers in middle schools. In the middle schools included in the survey, 62 per cent of the teaching of mathematics to pupils aged 11 to 13 in these schools was in the

*Middle schools with pupils aged up to 12 are normally deemed primary; those with pupils aged up to 13 are normally deemed secondary.

hands of teachers who had no qualification in mathematics (see Appendix 1, paragraph A15). We have no comparable information about the qualifications of those who teach mathematics to pupils aged 11 to 12 in middle schools deemed primary but we have no reason to suppose that these teachers are any better qualified; indeed, it seems possible that the reverse is the case.

771   Most middle schools have been formed from either primary or secondary modern schools. Staff from the original schools often remained as part of the staff of the new middle schools and so were in some cases required to teach pupils of an age of which they did not have experience or for which they had not been trained. We were, for example, told by the group of mathematics advisers whom we met that former primary teachers who are now teaching in middle schools for pupils aged 9 to 13 can find difficulty in meeting the needs of 13 year olds, especially those whose attainment is high; equally, former secondary teachers can very often find it difficult to work with 9 year olds. As existing middle schools recruit new staff the situation may change, but the advisers reported difficulty in attracting mathematically qualified teachers to work in middle schools. One reason for this would seem to be the present shortage of mathematics teachers in secondary schools, in which the promotion prospects are also likely to be better. Another may be the fact that very few teachers in middle schools are able to work as specialist teachers of mathematics; they are likely instead to have to teach one or more other subjects or take responsibility for the general teaching of a class. Furthermore, because of the uneven distribution of middle schools throughout the country, it is difficult for teacher training institutions to provide initial training courses which focus specifically on middle schools.

The mathematics co-ordinator

772 Because, as we pointed out in paragraph 770, a high proportion of mathematics teaching in middle schools is undertaken by teachers who lack suitable qualifications, **the post of mathematics co-ordinator (or head of department) is clearly of particular importance**. We have, however, been told that there are some mathematics co-ordinators in middle schools who are not themselves mathematically qualified; we find this disturbing. Furthermore, it is often the case that the mathematics co-ordinator has to combine responsibility for mathematics either with responsibility for another subject or with administrative or pastoral responsibility of some kind. This clearly limits the time which is available for the oversight of mathematics teaching within the school. Often, too, the mathematics co-ordinator teaches only the older, and perhaps also the higher attaining, pupils. It is desirable that he should have experience of teaching pupils in each year group but, if it is not possible to arrange this, **the mathematics co-ordinator should at least be enabled to visit the classrooms of all those who teach mathematics and work with these teachers from time to time**.

773 When preparing a scheme of work the mathematics co-ordinator will need to pay special attention to maintaining continuity of progress in mathematics both when pupils transfer from first school and when they move on to upper school. This may not be easy, especially if pupils move to any one of a number of upper schools; we have been told that difficulties can

sometimes arise because of requests from upper schools that, by the time they leave middle schools, pupils should be working from a particular textbook. Because pupils from several middle schools very often proceed to the same upper school, liaison is necessary not only with first and upper schools but also with other middle schools in the area. We consider that mathematics advisory staff are likely to have an important role to play in encouraging and facilitating liaison of this kind.

The teaching of statistics

774 Surprisingly few of the submissions which we have received have made direct reference to the teaching of statistics. In the course of our report we have ourselves made several references to it, either directly or by implication, for instance in the foundation list (paragraph 458) and in Chapter 11. We now consider briefly certain aspects in greater detail.

775 Statistics forms part of the mathematics course in the majority of schools at all age levels, ranging from the collection and representation of data in primary schools to the mathematically oriented material in A-level syllabuses. A survey carried out in 1976 by the Schools Council Project on Statistical Education 11–16 (POSE) indicated that over three-quarters of all secondary schools were teaching some statistics as part of the mathematics courses followed by 11–16 year olds. However, the submissions which we have received from POSE and from the Joint Education Committee of the Royal Statistical Society and the Institute of Statisticians have stressed that, although statistics is commonly taught within mathematics courses, it should not be regarded solely as part of mathematics. The submission from POSE states that "statistics is not just a set of techniques, it is an attitude of mind in approaching data. In particular it acknowledges the fact of uncertainty and variability in data and data collection. It enables people to make decisions in the face of this uncertainty".

776 Statistics is essentially a practical subject and its study should be based on the collection of data, wherever possible by pupils themselves. It should consider the kinds of data which it is appropriate to collect, the reasons for collecting the data and the problems of doing so, the ways in which the data may legitimately be manipulated and the kinds of inference which may be drawn. Work in subjects such as biological science, geography and economics can therefore contribute to the learning and understanding of statistics. When statistics is taught within secondary mathematics courses too much emphasis is very often placed on the application of statistical techniques, rather than on discussion of the results of ordering and examining the data and on the inferences which should be drawn in the light of the context in which the data have been collected. The work can therefore become dry and technique-oriented and fail to show the power and nature of statistics.

777 Many of the ideas of which statistics makes use need time and exposure in order to mature. This means that some of the more elementary ideas should be introduced at an early stage so that understanding can develop and deepen over a period of time. We have been told that in the preparation of many CSE,

O-level and A-level syllabuses the difficulty of some of the statistical concepts which they contain appears to have been underestimated, with the result that the statistical content of these syllabuses is too extensive. This is said to be even more true of the syllabuses of other subjects which make use of statistical techniques.

778 Few teachers, including those whose degree or other courses have included the study of statistics, have received training in how to teach statistics in schools. **There is therefore a considerable need for in-service training courses on the teaching of statistics not only for mathematics teachers but also for teachers of other subjects.** It is possible to identify three levels at which such training is required. The first is training for those who will teach statistics as a subject in its own right or as part of a mathematics course which includes statistics. The second is training for those who teach subjects which make use of statistics and who may themselves have to introduce their pupils to the statistical methods which are used. The third is training to enable other teachers to develop in their pupils a numerically critical approach to data and an awareness of the forms of misrepresentation which are very often to be found in published materials of various kinds. It is desirable that attention at the appropriate level should also be paid to these matters within initial training courses, preferably with groups of students whose main subjects span a variety of disciplines. We were pleased to hear of one PGCE course in which such an inter-disciplinary curriculum option on the teaching of statistics in schools has been introduced; we would support the provision of a similar option within other initial training courses.

779 Because work in subjects other than mathematics can contribute to the learning and understanding of statistics, efforts should be made to ensure that there is co-operation between all those in a school who make use of statistics in their teaching. Such co-operation could be assisted in secondary schools by the nomination of a member of staff who would identify the use which was made of statistics in the teaching of a variety of subjects and act as co-ordinator for the teaching of statistics; such a co-ordinator need not necessarily be a teacher of mathematics.

780 We have been told that there is at present a shortage of suitable books about statistics for both pupils and teachers, and that such text books as are available concentrate on theory rather than practice; it seems clear that **there is need for the provision of further teaching materials which will emphasise a practical approach to the teaching of statistics.** The introduction of electronic calculators has eased problems of calculation and so has made it possible for pupils to make use of 'real-life' data rather than of data which has been artificially contrived in order to avoid heavy calculation. This provides opportunity to emphasise the interpretation of data rather than techniques of calculation. In our view **the increasing availability of micro-computers and the visual display which they provide should also offer opportunities to illuminate statistical ideas and techniques; we believe that development work in this field is also required.**

781 To conclude our discussion of the teaching of statistics we quote from the submission we have received from POSE.

> Statistical numeracy requires a feel for numbers, an appreciation of appropriate levels of accuracy, the making of sensible estimates, a commonsense approach to the use of data in supporting an argument, the awareness of the variety of interpretation of figures, and a judicious understanding of widely used concepts such as means and percentages. All these are part of everyday living. Good statistical teaching can encourage pupils to think in these ways.

We endorse this statement and draw attention to the way in which it complements our discussion in Chapter 2 of the mathematical needs of adult life.

Mathematical education in other countries

782 Our terms of reference have not required us to study the teaching of mathematics in other countries. Nevertheless, among the earliest submissions which we received were some which drew comparisons with other countries and it was clear that both similarities and differences existed in the approaches to mathematics teaching which were used in these countries. We therefore decided to seek information about mathematics teaching in other English-speaking countries as well as in some comparable European countries, while realising that any effort which we were able to make would of necessity provide only very limited information which we could not assume would be representative of the country from which it had come. We were aware that different cultural and social attitudes influenced education at all levels, from central legislation to classroom practice, even in other English-speaking countries, but we wished to see whether our own debate could be enhanced by some knowledge of methods of working in other countries.

783 With the help of the League for the Exchange of Commonwealth Teachers and the Central Bureau for Educational Visits and Exchanges we were able to write to a number of teachers from England and Wales who had recently been teaching on exchange in Australia, New Zealand, Canada and the United States of America as well as to teachers from these countries who were at the time working in schools in England and Wales; we received many helpful replies from these teachers. Through the good offices of the Secretariat of the European Community we received official statements about mathematics teaching in several of the countries within the community, and the British Council provided us with information relating to countries in other parts of the world. Some of our members were able to pay short visits to Denmark, Holland and West Germany and, nearer home, to Scotland. During these visits we were able to talk with teachers and industrialists, as well as with those concerned with curriculum development and teacher training, and to visit several schools.

784 We cannot over-emphasise the diversity of approach which we ourselves have observed and of which we have been told, a diversity which we have also found in the schools in this country which we have visited. This diversity has, on the one hand, provided a valuable stimulus to our discussion but, on the other, has made it clear that, especially on the basis of the very limited information which we possess, we can in no way generalise about mathematical

education in any other country. It is clear that there are no model solutions; in all countries of which we have some knowledge we have been aware of concern to improve the quality of mathematical education.

785 Notable differences exist between countries in the extent to which there is central control of the curriculum. In Scotland, the existence of a single examination board exercises a unifying effect on the secondary curriculum, as does the use of centrally developed materials at both primary and secondary levels. In Holland and the various *Länder* in West Germany, curriculum control is exercised by national or regional government. In France, there is centralised control of the syllabus and of the associated textbooks. However, the existence of nationally or regionally determined curricula in no way removes the requirement for teachers to exercise professional judgement in their mathematics teaching.

786 The part played by public examinations also varies considerably from country to country. In some countries, only those pupils who wish to proceed to higher education are required to attempt a public examination; other pupils are assessed by their teachers, often according to criteria which are specified in detail. In some other countries pupils attempt an examination only at the time at which they leave school. On the other hand, in some countries internal assessment procedures within schools can mean that pupils can be required to repeat a year if their progress is adjudged to have been insufficient. These practices are in marked contrast to the situation in England and Wales in which public examinations exert a strong influence throughout the secondary years, but in which it is unusual for pupils not to be promoted with their fellows.

787 In other countries it is often the case that secondary teachers are required to have a higher academic qualification than those who teach in primary schools. Furthermore, the subjects which secondary teachers are permitted to teach are often restricted to those in which they have a graduate qualification or to subjects which are closely related to these. The restriction which is often placed on the level of qualification of those who teach in secondary schools as well as on the subjects which they can teach is also in marked contrast to the situation in England and Wales. It is clear that at the present time it would not be possible to require all those who teach mathematics at secondary level to hold a minimum mathematical qualification. However, just as we have urged that all who teach mathematics should be professionally qualified, so we believe that they should possess suitable academic qualifications. **We believe, therefore, that there should be an investigation into ways in which such a requirement might be introduced over a period of years and that the first steps towards introducing such a requirement should be taken as soon as possible.**

788 Shortage of mathematics teachers did not seem to be thought a problem in Denmark, Holland or West Germany and in all of these countries teachers appeared to enjoy a high status. In contrast we were told that, in some parts of the United States of America, there is an extreme shortage of mathematics teachers above elementary level. In West Germany salaries of teachers in

secondary schools were said to be comparable with those of 'middle management' in industry and commerce. In Bavaria we were told that teachers were assessed every four years by the head teacher and that the results of this assessment could materially affect the time which was spent by a teacher at a given grade.

789 Our discussions with teachers in other countries, as well as the letters from the exchange teachers to whom we wrote, revealed a very wide range both of teaching styles and syllabus content. It is clear, for example, that both modern courses and traditional courses are sometimes taught in very formal ways. We have received criticism of modern courses taught in this way from those who see such courses as too abstract and difficult for the average pupil. Equally, we have received criticism of formal approaches to traditional courses because such approaches can lead to teaching of a kind which emphasises routines at the expense of understanding. On the other hand, in the course of our visits abroad, as well as from other sources, we have become aware of the high regard which is often accorded in other countries to some of the teaching materials which have been produced in this country and some of the educational research and curriculum development which has been undertaken. Both are seen as more firmly based on classroom practice and less theoretical in approach than is often the case abroad.

790 We have been impressed by the Danish tradition of in-service training for teachers which goes back for very many years. We were told that each year about one-third of all teachers in *Folkeskoler* take in-service training courses organised by the Royal Danish School of Educational Studies. In the United States of America many teachers follow Master's degree programmes in education during the summer vacations and in some areas there is an expectation that teachers will undertake training of this kind. In West Germany we were told that, in some *Länder*, head teachers had a responsibility for staff development which could include a requirement that teachers should undertake appropriate in-service courses. We have also been given to understand that, in some Canadian provinces, the conditions of service of teachers require them to undertake some in-service training each year.

791 The diversity of practice in schools which we have outlined in the previous paragraphs is complemented by the diversity of employers' views of which we have become aware, both through the comments of those engaged in education in other countries and also in the course of the small number of discussions which we have been able to have with employers themselves. No uniform picture has emerged and the mixture of satisfaction, reservation and criticism has been not unlike that which we have obtained from employers in England and Wales. In Denmark and Holland we gained the impression from educationalists that there was little criticism of the mathematical attainment of school leavers. In West Germany we were made aware of the advantages which were seen by those employers to whom we talked to result from the *Berufsschule* system of vocational education, which requires all those who have left full-time education before the age of 18 to continue part-time vocational education up to this age for the equivalent of about one day a week.

These employers expressed enthusiasm for the courses provided by the *Berufsschulen* and words such as 'meaning' and 'relevance' were used frequently in connection with the courses which were provided. Another advantage was seen to lie in the fruitful liaison which developed between industry and schools. We were told that courses for some 15 year olds in parts of West Germany were very much more vocationally orientated than is the case for pupils of the same age in this country and one UK industrialist has made clear to us his regret that similar provision is not made in schools in England and Wales. We have been told that in the United States of America some employers have complained of a lack of ability among school leavers to apply mathematics; on the other hand one distinguished industrialist expressed the view that the differences to be found in educational provision in the United States "reflected to a considerable extent the ability of the different regions to support through direct taxation establishments of education that have the quality in their teaching staff essential to produce the best results".

792 For the reasons which we explained at the beginning of this section, our references to mathematical education in other countries can only be fragmentary and incomplete. However, we believe that the information which we have been able to obtain has been of use in our work. All the points to which we have drawn attention in the preceding paragraphs have been reflected in our own discussions and so have contributed to our thinking and our conclusions; some of them raise more general issues which we believe merit further investigation. We are most grateful to those whom we met in the course of our visits for the help which they afforded us so willingly.

Statistical information relating to mathematical education

793 It has proved very much more difficult and has taken very much longer than we had expected to obtain statistical information about the various aspects of mathematical education. This has not been because of any unwillingness on the part of the DES or of other bodies to provide information but because, for example, much of the information which is collected each year from schools and universities has not hitherto been analysed in ways which provide the kind of information which we have been seeking. Accordingly, the computer programs which were needed to do this did not exist and have had to be specially written. A less important, though not insignificant, problem is that it is not, in general, possible to obtain separate figures relating to computer studies and statistics because they have traditionally been included as part of mathematics for the purposes of statistical analysis.

794 We have tried to obtain information relating to four main areas. The first of these relates to pupils in schools. In addition to making use of information published in the DES *Volumes of Statistics* and by the various examination boards, we have made extensive use of information provided by the annual 10 per cent survey of school leavers carried out by the DES. We understand that, as a result of the work which has been carried out for our Committee, computer programs have been developed which will in future enable the information provided by this survey to be analysed more readily. We welcome this move because we believe it is important that this information should be readi-

ly available for analysis in ways which are subject specific. **We recommend that analyses of the kind which have, at our request, been produced for the years 1977–1979 should continue to be produced for subsequent years. We wish also to recommend, that, so far as is possible, the DES and examination boards should cease to include computer studies and statistics under the general heading of mathematics and treat them as separate subjects for the purpose of statistical analysis. We draw attention to the fact that we have not been able to obtain information about the CSE and GCE results of students aged 16 to 19 who are studying in tertiary and FE colleges. We believe that this information should be available.**

795 The second area about which we have sought information relates to the entrance qualifications on a subject specific basis of those entering courses in higher education. As we explained in paragraph 170, detailed information about students at universities is collected by the Universities Statistical Record but comparable information about students on courses at other institutions of higher education is not at present available. To our regret we have not therefore been able to obtain information about the mathematical qualifications of students entering courses at these institutions nor have we been able to reconcile information about the numbers of students on degree courses in the non-university sector provided by the DES and the Council for National Academic Awards. **We have noted with interest that the recent report of the Education, Science and Arts Committee of the House of Commons* recommends that "the sort of information collected for the Universities Statistical Record should be collected for college and polytechnic personnel" and that "the data bases maintained by the Universities Statistical Record at Cheltenham and the Further Education Statistical Record at Darlington should be made compatible". We support this recommendation strongly.**

*House of Commons. Fifth Report from the Education, Science and Arts Committee. *The funding and organisation of courses in higher education.* HMSO 1980.

796 The third area relates to the mathematical qualifications of those who enter initial training courses for intending teachers. We have not, for example, been able to identify the proportion of students enrolling for BEd degrees, other than at universities, who have an A-level qualification in mathematics. This problem would be overcome if the recommendation of the Parliamentary Committee to which we referred in the previous paragraph were implemented. **We believe that there is a need to identify the kind of information which should be available about the academic qualifications of entrants to BEd and PGCE courses and the way in which this is collected and analysed; for example, it should be possible to discover the extent of the mathematical qualifications of those who choose mathematics as a first or second method course in PGCE or as a main course in BEd.**

797 The fourth area about which we have sought information relates to the qualifications of teachers and the subjects which they are teaching. A major difficulty in identifying teachers who are mathematics specialists arises from the fact that education is recorded as the first subject of qualification of those who hold the BEd degree. This means that, for BEd graduates who are mathematics specialists, mathematics is recorded as the second subject rather than the first, whereas for mathematics specialists with other kinds of

graduate qualification, mathematics is recorded as the first or only subject of qualification. **We recommend that education should not be recorded as the first subject of qualification of graduates who hold the BEd degree since the title of the degree itself provides this information; instead, the subject of the main course should be recorded as the first subject of qualification.** The survey of secondary staffing carried out in 1977 provided a great deal of valuable information. **We believe that surveys of a similar kind should be carried out at least once every five years and that agreement should be reached on the kinds of information which it is necessary to obtain so that any LEA which wishes from time to time to collect information about its own teaching force will be enabled to do so in a way which it knows will be compatible, and therefore comparable, with any information which becomes available as a result of national surveys.**

17   The way ahead

798 Our terms of reference have required us to consider the teaching of mathematics in schools in the light of the mathematical needs of pupils when they proceed to further or higher education or to employment, as well as of their needs as adults. In the first part of our report we have sought to identify these needs, in the second part to discuss the ways in which a balanced and coherent mathematics curriculum can provide for these needs, and in the third part to identify the provision and support which is required to enable good mathematics teaching to take place.

799 Changes of many kinds have led to increasing pressure on teachers during the last fifteen years. These years have been a time of rapid social, technological and economic change. There has been a move to comprehensive education in almost all parts of the country and the school leaving age has been raised; in some areas the reorganisation of local authorities placed additional strains on teachers. The 'bulge' in the birth rate, which reached its peak in 1964 and led to a rapid increase in the teaching force, has been succeeded by the present sharp fall in the number of pupils in many areas and so to uncertainty and reduced opportunities of promotion for teachers. Pressure on mathematics teachers in particular has also arisen from the introduction of new mathematics syllabuses resulting in the need to teach courses of a different kind, often without opportunity to undertake the necessary preparation and in-service training. The introduction of CSE has created examination pressures on pupils and teachers which did not exist earlier; metrication and the introduction of decimal currency led to initial difficulties in terms of the availability of suitable books and other teaching materials in many schools. More recently the rapid increase in the availability of electronic calculators, without sufficient guidance as to the use which could and should be made of them in the classroom, has presented further problems to many teachers. Finally, throughout these years there has been a continuing shortage of mathematically well qualified teachers in both primary and secondary schools.

800 Alongside these changes, there has been increased public criticism of the education system, and especially of mathematics teaching, expressed by parents and employers as well as by many in political and public life. In these circumstances it would be surprising if the morale of some of those who teach mathematics had not suffered. Yet these years have also been a time of mathematical development during which many teachers have given willingly of their time to join, for example, in local working groups, in curriculum

development and in other kinds of in-service work. In many schools good work is being done in the teaching of mathematics and very many teachers are making considerable efforts to satisfy a public expectation of achievement in mathematics which is by no means always realistic. Nevertheless, as we hope our report will have made clear, there are at present many pupils who are being offered mathematics courses which are not suited to their needs and many teachers of mathematics who lack suitable qualifications. For these and other reasons, the mathematical education which many pupils are receiving is not satisfactory. We therefore believe that major changes are essential.

801   Those who teach mathematics must take into account the great variation which exists between pupils both in their rate of learning and also in their level of attainment at any given age. It follows that mathematics courses must be matched both in level and pace to the needs of pupils; and therefore that a 'differentiated curriculum' must be provided so that pupils will be enabled to develop to the full their mathematical skill and understanding, a positive attitude towards mathematics, and confidence in making use of it. Examinations at secondary level must also be better suited to the needs of pupils than is the case at present.

802   We believe, too, that it is essential to improve the overall quality of the mathematics teaching force. This means that active efforts must be made to attract more well qualified mathematicians into the teaching profession, to retain those mathematics teachers who are well qualified and effective, and to provide increased levels of in-service support and training. Good support at classroom level is essential for all who teach mathematics; this requires not only the provision of adequate teaching facilities and equipment, but also good leadership by mathematics co-ordinators and heads of department who have been trained for their task.

803   Among the suggestions and recommendations which are included in our report are many to which teachers themselves can respond and, indeed, for whose implementation teachers themselves must accept the main responsibility. We therefore hope that teachers will read with particular care those parts of the report, notably Chapters 5 to 11, in which these suggestions are concentrated and consider how best they can help to bring about the changes we are recommending. For example, computational skills should be related to practical situations and applied to problems. Mathematics teaching for pupils of all ages should include exposition, discussion, appropriate practical work, problem solving, investigation, consolidation and practice, as well as mental and oral work. Assessment should be both diagnostic and supportive, and teaching should be based on a scheme of work which is appraised and revised regularly. All of this and more is necessary if mathematics teaching is to be effective; all of this is in the hands of teachers.

804   On the other hand, we have made several suggestions and recommendations which teachers cannot implement by themselves and which require action by others. For example, in Chapters 6 and 9 we have stressed the importance of the roles of the mathematics co-ordinator and head of department; it

is essential that these teachers should be given training for their task and continuing support to enable them to carry it out. Those who teach mathematics require the support of effective advisory services as well as opportunities for in-service training of various kinds; they also need suitable facilities and equipment in their schools. Responsibility for provision of these kinds must rest primarily with local education authorities. We discuss the facilities required to teach mathematics in Chapter 12, and in-service support for teachers in Chapter 15.

805 Several of our recommendations relate to examinations and tests. In Chapter 10 we argue that examinations at all levels should provide suitable targets and reflect suitable curricula. They should enable candidates to demonstrate what they know and should not undermine the confidence of those who attempt them. We recommend that in the new single system of examination at 16+ there should be a range of papers in mathematics focussed at different levels. We also recommend that a study should be commissioned to consider whether it is possible to devise a means of providing evidence of achievement for lower attaining pupils in ways which will assist, and not conflict with, the provision of suitable courses for these pupils; the outcome of such a study could have a bearing on the need for employers' tests, which are discussed in Chapter 3. We recommend in Chapter 8 that in the near future an overall appraisal should be prepared of the educational implications of the mathematics testing which has been carried out by the Assessment of Performance Unit. Action in these matters must rest primarily with examination boards and, in certain respects, with central and local government. In studying our recommendations they should bear in mind the needs of pupils in later life, which we discuss in Chapters 2, 3 and 4.

806 No efforts to improve the quality of mathematics teaching are likely to succeed unless there is an adequate supply of suitably qualified mathematics teachers. In Chapter 13 we discuss the supply and recruitment of teachers and include proposals for greater flexibility within the salary system, for financial incentives to suitably qualified students and for a guarantee of employment for newly trained mathematics teachers. In many of these matters we believe that primary responsibility rests with central government; local education authorities as employers of teachers also have considerable responsibility.

807 All who teach mathematics require initial training which fits them both mathematically and professionally to start teaching. We hope that those who are engaged in initial training in mathematics will review the content of their courses in the light of our discussion of mathematics teaching. We would draw the attention of all who are responsible for teacher training to our discussion in Chapter 14, in which we recommend a review and evaluation of the initial training of all who will teach mathematics.

808 Finally, we have drawn attention in the course of our report to the need for curriculum development of various kinds leading to the preparation of materials for use in the classroom. For example, work is required to develop materials which will reflect the content of the 'foundation list' we have set out

in Chapter 9 and also to develop the range of materials which are written in Welsh. We suggest in Chapter 7 that materials are required which will enable teachers to exploit the use of calculators as an aid to mathematics teaching and that software of good quality is required for use with micro-computers. We hope that those public and private bodies which fund curriculum development and educational research will bear our suggestions in mind when considering proposals which are submitted to them.

809 In this chapter we have set out our recommendations in a way which many will not have expected. Our report includes many suggestions which we could have set out in a list, but we have preferred instead to draw attention to them by the use of heavier type in the main body of our report. We have done this because we believe that the teaching of mathematics must be approached as a whole and that the significance of many of the points which we could have set out separately becomes more apparent when they are read within the context of our argument. We have therefore identified instead, in paragraphs 803 to 808, six agencies whose active response to our report we believe to be essential if the changes in mathematical education which we are recommending are to be brought about. These agencies are teachers, local education authorities, examination boards, central government, training institutions and those who fund and carry out curriculum development and educational research. It will not be sufficient if only some of these agencies respond, since the contribution of all will be necessary if progress is to be made. Educational change cannot be accomplished overnight and the full implementation of our proposals will take time. For this reason we believe it to be essential that action should not be delayed and that the necessary resources for this action to begin must be made available.

810 But even the active co-operation of the six agencies we have identified is not likely to lead to lasting results unless there is support from a seventh, the public at large, including especially parents, employers and those engaged in public work. The setting up of our Committee demonstrated a widespread view that action was needed in order to meet the perceived national need for a numerate population. During the past three years we have received many expressions of support for our work which indicate a widespread belief that every boy and every girl needs to develop, while at school, an understanding of mathematics and confidence in its use. In our view this can only come about as the result of good mathematics teaching by teachers who have been trained for their work and who receive continuing in-service support. It must therefore be the task of all who share this belief to support and encourage the implementation of the changes which we believe to be necessary and to make it clear that, as part of the education which our children receive, mathematics counts.

Appendix 1   Statistical information

A1 This appendix contains a selection of the considerable amount of statistical information which has been made available to the Committee from a number of sources. We are most grateful to all who have helped us. We acknowledge in particular the assistance we have received from DES Statistics Branch and the Universities Statistical Record; both have, at our request, devised and carried out additional investigations. Some reordering, re-classification and amalgamation of the data has been undertaken by members of the Committee and its secretariat.

List of tables

Pupil numbers

A2 The figures in Table 1 are based on age in September. In broad terms, 'primary' refers to pupils aged 5 to 11 and 'secondary' to pupils aged 11 to 16, together with those who elect to remain in full-time education in schools after the age of 16. The secondary projections therefore make assumptions about staying-on rates. Schools include sixth form colleges but not tertiary colleges, which operate under FE regulations.

Table 1   School population

	Maintained schools: England											Thousands
	Actual numbers			Projected numbers								
	1970	1975	1980	1981	1982	1983	1984	1985	1990	1995	2000
Type of school:											
Primary	4187	4335	3850	3691	3531	3338	3202	3154	3471	3938	4140
Secondary	2848	3558	3818	3779	3712	3660	3573	3457	2760	2953	3370
All schools	7035	7893	7668	7470	7243	6998	6775	6611	6231	6891	7510

Source: DES Statistics Branch

**Examination
performance of
pupils in schools**

A3 The only information which is available about the examination performance of pupils in schools on a national scale comes from the annual 10 per cent survey of school leavers in both maintained and independent schools which is carried out by the DES and Welsh Office. This survey collects information about all school leavers whose birthdays fall on the 5th, 15th or 25th day of each calendar month and includes details of their examination entries and results.

A4 Because the information relates to pupils who *leave* school in any one year, the examination results which are recorded may have been obtained over a period of years before leaving school. The figures do not therefore relate to a complete year group. However, the relativities between the number of leavers in any year, the number of examination entries in that year and the size of the year group have remained reasonably constant for the years 1977 to 1979. We therefore believe that the picture displayed in the tables which follow is unlikely to differ significantly from the picture which would have emerged if it had been possible to obtain comparable information for a complete year group.

A5 The information for all three years relates to both maintained and independent schools in England and Wales. It refers to pupils in schools (including sixth form colleges) but not to students in FE or tertiary colleges. The examination results of this latter group are therefore included only up to the time at which they left school. In very general terms this means that most of their O-level and CSE results are included but that their A-level results are not. In cases in which a school leaver has entered for the same examination more than once he is credited with the highest grade obtained. 'Mathematics' includes Statistics and Computer Studies when these are taken as separate subjects.

A6 The information displayed is for the whole population, derived by multiplying up from the 10 per cent sample surveyed, and is therefore subject to sampling error. In cases where the numbers are large the true figures are unlikely to differ from those given by more than half of one percentage point. In tables which relate to smaller numbers, the error may be greater, up to 3 per cent.

Table 2   Numbers of school leavers

England and Wales

	Number of leavers		
	Boys	Girls	All
Year of leaving:			
1977	384 020	367 070	751 070
1978	394 890	373 570	768 460
1979	399 630	381 610	781 240

Source: DES 10% Leavers' Survey

Table 3   O-level grades in Mathematics: as percentages of all leavers

England and Wales   Percentages

	1977			1978			1979		
Year of leaving	Boys	Girls	All	Boys	Girls	All	Boys	Girls	All
O-level grades:									
A	3.5	2.1	2.8	4.9	2.4	3.7	5.5	2.6	4.1
B	6.7	4.9	5.9	8.2	5.8	7.0	8.7	6.2	7.5
C	8.7	7.9	8.3	9.9	8.4	9.2	10.2	8.8	9.5
Pass[1]	3.8	1.2	2.5	0.6	0.1	0.3	0	0	0
A, B, C or Pass	22.8	16.1	19.5	23.7	16.7	20.3	24.5	17.6	21.1
D	2.3	2.5	2.4	2.3	2.5	2.4	2.4	2.5	2.5
E	2.6	2.8	2.7	2.7	3.0	2.9	3.0	3.3	3.1
U or Fail	4.2	4.6	4.4	4.2	4.8	4.5	4.3	5.1	4.7
D, E, U or Fail	9.1	9.8	9.5	9.2	10.2	9.7	9.7	10.9	10.3
% of all leavers	31.9	25.9	29.0	32.9	26.9	30.0	34.2	28.5	31.4
Numbers entered	122 580	94 980	217 560	130 050	100 610	230 660	136 770	108 730	245 500

Source: DES 10% Leavers' Survey

[1] Grades at O-level were not awarded until the Summer of 1975 (see note to paragraph 448).

Table 4   CSE grades in Mathematics: as percentages of all leavers

England and Wales   Percentages

	1977			1978			1979		
Year of leaving	Boys	Girls	All	Boys	Girls	All	Boys	Girls	All
CSE grades:[1]									
1	7.4	5.8	6.6	7.4	6.1	6.8	7.7	6.6	7.2
2	7.4	6.8	7.1	7.2	7.4	7.3	7.5	7.5	7.5
3	9.3	9.6	9.4	9.8	9.9	9.8	10.3	10.8	10.5
4	11.9	13.4	12.6	12.6	13.9	13.2	13.5	15.3	14.4
5	8.1	9.5	8.8	8.5	10.3	9.4	8.2	10.5	9.3
U	6.8	8.5	7.6	6.8	8.3	7.5	6.4	8.1	7.2
% of all leavers	50.9	53.6	52.2	52.3	55.8	54.0	53.5	58.8	56.1
Numbers entered	195 400	196 620	392 020	206 350	208 600	414 950	213 850	224 370	438 220

Source: DES 10% Leavers' Survey

[1] Some CSE boards offer a separate examination in Arithmetic; the table shows the combined results for Mathematics and Arithmetic.

A7 Some pupils attempt both O-level and CSE examinations in mathematics, either in the same year or in different years. An aggregation of the O-level results given in Table 3 and the CSE results given in Table 4 does not therefore give a picture of the overall situation. Table 5 amalgamates the O-level and CSE results obtained by school leavers in such a way that each pupil appears only once, according to his 'best' grade. In the case of leavers whose highest grades included both O-level grade A, B, C or O-level pass and CSE grade 1 we have taken account of the O-level result and discounted the CSE result. We have grouped together those whose 'best' grade was either O-level grade D or CSE grade 2 and those whose 'best' grade was either O-level grade E or CSE grade 3. 'Best' grades have been defined in the order presented in Table 5; for example, if a pupil obtained O-level grade E and CSE grade 2 it is the CSE grade which has been included, whereas if a pupil obtained O-level grade E and CSE grade 4 it is the O-level grade which has been included. We emphasise that these steps have been taken solely for the purpose of drawing up the table and should not be interpreted in any other sense. In the table, 'higher grades' include O-level grades A, B or C, O-level pass and CSE grade 1; 'lower grades' include O-level grades D and E and CSE grades 2 to 5.

Table 5   'Best'[1] grades in mathematics: as percentages of all leavers

	England and Wales								Percentages
	Year of leaving								
	1977			1978			1979		
	Boys	Girls	All	Boys	Girls	All	Boys	Girls	All
O-level or CSE grades:									
A	3.5	2.1	2.8	4.9	2.4	3.7	5.5	2.6	4.1
B	6.7	4.9	5.9	8.2	5.8	7.0	8.7	6.2	7.5
C	8.7	7.9	8.3	9.9	8.4	9.2	10.2	8.8	9.5
Pass	3.8	1.2	2.5	0.6	0.1	0.3	0	0	0
1	4.7	4.2	4.4	4.4	4.3	4.4	4.4	4.5	4.5
All 'higher grades'	27.4	20.2	23.9	28.1	21.0	24.6	28.9	22.1	25.6
D or 2	7.9	7.8	7.8	7.8	8.3	8.0	8.0	8.3	8.2
E or 3	10.2	10.8	10.5	10.6	11.1	10.8	11.2	12.0	11.6
4	11.5	13.0	12.2	12.2	13.5	12.8	13.0	14.8	13.9
5	8.0	9.3	8.6	8.3	10.1	9.2	8.0	10.2	9.1
All 'lower grades'	37.5	40.9	39.1	38.9	43.0	40.9	40.2	45.4	42.7
U or failed	8.3	10.3	9.3	8.0	10.0	9.0	7.8	9.8	8.8
Total entry	73.2	71.4	72.3	75.0	74.1	74.5	76.8	77.3	77.1

Source: DES 10% Leavers' Survey

[1] See paragraph A7.

A8 The table which follows analyses examination results in Additional Mathematics at O-level or AO-level according to the age at which the best result was obtained. 'By age 16' means during or before the school year in which the pupil reached the age of 16; 'at age 17', during the school year in which the pupil reached the age of 17; 'at age 18 or over', during or after the school year in which the pupil reached the age of 18. Thus, in general, 'by age 16' refers to results obtained in the fifth form, or earlier; 'at age 17' to results obtained in the first-year sixth; 'at age 18 or over' to results obtained in the second year sixth or later. The proportion of leavers who had attempted the examination at some stage increased from 3.5 per cent in 1977 to 3.7 per cent in 1979 but the overall pattern of entry and success was similar for all three years. We therefore give the figures for leavers in 1979 only.

Table 6   O-level grades in Additional Mathematics: by age

England and Wales

| | Leavers in 1979 | | | | | |
| | By age 16[1] | | At age 17 | | At age 18 or over | |
	Boys	Girls	Boys	Girls	Boys	Girls
O-level grade:						
A	2660	220	1290	570	130	50
B	3580	530	2030	800	210	100
C	3870	790	2140	900	600	160
A, B or C	10 110	1540	5460	2270	950	300
D, E or U	3380	1040	2120	580	730	260

Source: DES 10% Leavers' Survey

[1] See paragraph A8.

Table 7   Sixth form pupils studying Mathematics at A-level

England[1]

	First year sixth			Second year sixth[2]		
	Pupils on A-level courses	Pupils[3] taking mathematics	% taking mathematics	Pupils on A-level courses	Pupils[3] taking mathematics	% taking mathematics
School year:[4]						
1976–77:						
Boys	77 704	35 283	*45.4*	72 434	30 768	*42.5*
Girls	68 887	12 702	*18.4*	60 529	10 173	*16.8*
All	146 593	47 985	*32.7*	132 963	40 441	*30.8*
1977–78:						
Boys	78 313	37 721	*48.2*	71 932	32 135	*44.7*
Girls	65 940	14 374	*20.9*	60 606	11 165	*18.4*
All	147 253	52 095	*35.4*	132 538	43 300	*32.7*
1978–79:						
Boys	79 428	40 046	*50.4*	72 872	35 149	*48.2*
Girls	72 444	15 693	*21.7*	62 117	12 901	*20.8*
All	151 872	55 739	*36.7*	134 989	48 050	*35.6*
1979–80:[5]						
Boys	81 050	41 250	*50.9*	73 750	36 450	*49.4*
Girls	75 150	17 300	*23.0*	65 050	13 900	*21.4*
All	156 200	58 550	*37.5*	138 800	50 400	*36.3*

Source: DES Statistics Branch (Form 7)

[1] Figures for England only because comparable figures for Wales have not been collected since 1977.
[2] Includes third year pupils who are taking A-level courses.
[3] Mathematics includes Statistics and Computer Studies if studied as separate A-level subjects.
[4] Numbers at end of January in each school year.
[5] These figures include a small amount of estimation in respect of some independent schools for which no information was available.

Table 8   A-level grades in Mathematics: sixth form pupils

England and Wales   Percentages

	Year of leaving								
	1977			1978			1979		
	Boys	Girls	All	Boys	Girls	All	Boys	Girls	All
A-level grades[1]:									
A	*14.4*	*10.3*	*13.4*	*13.6*	*9.5*	*12.6*	*15.4*	*10.1*	*14.0*
B	*16.1*	*18.7*	*16.7*	*15.4*	*13.9*	*15.0*	*14.6*	*16.1*	*15.0*
C	*11.9*	*12.1*	*12.0*	*12.7*	*13.0*	*12.8*	*12.2*	*11.3*	*12.0*
D	*14.1*	*15.9*	*14.6*	*15.9*	*19.3*	*16.8*	*15.0*	*16.0*	*15.3*
E	*17.2*	*18.5*	*17.5*	*16.2*	*19.3*	*17.0*	*16.2*	*19.5*	*17.1*
Grades A-E	*73.7*	*75.4*	*74.1*	*73.8*	*74.9*	*74.2*	*73.4*	*73.0*	*73.3*
Numbers entered[2]	30 420	10 080	40 500	31 660	11 010	42 670	34 670	12 350	47 020

Source: DES 10% Leavers' Survey

[1] If a pupil has attempted more than one mathematical subject at A-level, the best result is recorded.
[2] Entry numbers in this table are lower than the numbers of pupils in the second year of the sixth form shown in Table 7. The numbers in this table are actual entries, those in Table 7 are for pupils studying the course when schools made their returns in January.

Table 9   A-level Mathematics grades related to O-level grades of the same pupils[1]

England and Wales

Year of leaving	O-level grade	Number of pupils	A-level grade (% of each row)					
			A	B	C	D	E	O/Fail
	A	10 760	20.2	26.6	15.0	14.5	13.5	10.4
1977	B	10 420	2.4	9.3	11.2	17.5	24.1	35.4
	C	3840	0.0	2.3	6.5	11.5	18.2	61.7
	A	15 650	22.9	24.2	15.5	16.0	12.6	8.9
1978	B	14 680	3.1	10.7	12.7	21.2	21.7	30.6
	C	5290	0.4	3.2	6.4	12.9	19.1	58.0
	A	17 940	28.1	23.6	14.9	13.5	11.3	8.5
1979	B	16 460	3.8	12.1	11.9	18.3	22.4	31.5
	C	5710	1.4	1.4	5.1	11.2	17.5	63.4

Source: DES 10% Leavers' Survey

[1] These tables include only those pupils who had obtained O-level grade A, B or C (as their highest grade) by age 16 (see paragraph A8). The total numbers of A-level candidates to whom the table refers are therefore smaller than those shown in Table 8.

Table 10   A-level passes in Mathematics: as percentages of all leavers

England and Wales

	Year of leaving						
	1972	1973	1976	1977	1978	1979	1980
Mathematical subjects passed at A-level[1]							
1 subject only	20 130	20 330	23 440	24 900	26 070	28 090	30 410
% of all leavers	3.1	3.1	3.3	3.3	3.4	3.6	3.8
2 or more subjects	7040	6210	5740	5120	5570	6370	6770
% of all leavers	1.1	0.9	0.8	0.7	0.7	0.8	0.9
At least 1 subject	27 170	26 540	29 180	30 020	31 640	34 460	37 180
% of all leavers	4.2	4.0	4.1	4.0	4.1	4.4	4.7
Leavers with 2 or more passes in mathematics as % of leavers with at least 1 pass in mathematics	25.9	23.4	19.7	17.1	17.6	18.5	18.2

Source: DES 10% Leavers' Survey

[1] Includes passes at grades A to E in any mathematical subjects, including Statistics and Computer Science.

Comparison of examination results in mathematics and English

A9 Tables 11, 12 and 13 compare the O-level and CSE results obtained by pupils in mathematics/arithmetic with those obtained in English. The 'best' grades in English have been defined in the way which is explained in paragraph A7. If more than one English subject was attempted, the overall 'best' grade has been used.

Table 11   Comparison of O-level and CSE results in Mathematics and English: as percentages of all leavers

England and Wales   Percentages

| | Year of leaving | | | | | |
| | 1977 | | 1978 | | 1979 | |
	Mathematics or Arithmetic	English	Mathematics or Arithmetic	English	Mathematics or Arithmetic	English
O-level or CSE grades:						
A, B, C or 1	23.9	34.7	24.6	34.5	25.6	34.5
D or 2	7.8	11.8	8.0	12.1	8.2	13.1
E or 3	10.5	12.9	10.8	13.3	11.6	14.1
4	12.2	10.7	12.8	11.5	13.9	11.1
5	8.6	4.3	9.2	4.5	9.1	4.7
U or Fail	9.3	3.7	9.0	3.8	8.8	3.9
Not entered	27.7	21.9	25.6	20.3	22.8	18.6
Total	100.0	100.0	100.0	100.0	100.0	100.0

Source: DES 10% Leavers' Survey

Table 12   Comparison of O-level and CSE results in Mathematics and English: cumulative percentages

England and Wales   Cumulative percentages

| | Year of leaving | | | | | |
| | 1977 | | 1978 | | 1979 | |
	Mathematics or Arithmetic	English	Mathematics or Arithmetic	English	Mathematics or Arithmetic	English
O-level or CSE grades:						
A, B, C or 1	23.9	34.7	24.6	34.5	25.6	34.5
and D or 2	31.7	46.5	32.6	46.6	33.8	47.6
and E or 3	42.2	59.4	43.4	59.9	45.4	61.7
and 4	54.4	70.1	56.2	71.4	59.3	72.8
and 5	63.0	74.4	65.4	75.9	68.4	77.5
and U or Fail	72.3	78.1	74.4	79.7	77.2	81.4
and Not entered	100.0	100.0	100.0	100.0	100.0	100.0

Source: DES 10% Leavers' Survey

Table 13   Comparison of grades in Mathematics and English: by sex, as percentages of all leavers

England and Wales   Percentages

| | Leavers in 1979[1] | | | |
| | Boys | | Girls | |
	Mathematics	English	Mathematics	English
O-Level/CSE grades:				
'Higher grades'[2]	28.9	30.1	22.1	39.1
'Lower grades'	40.2	44.3	45.4	41.8
Ungraded	7.8	4.7	9.8	3.0
Total entry	76.8	79.0	77.3	84.0

Source: DES 10% Leavers' Survey

[1] Results in 1977 and 1978 followed a similar pattern.
[2] See paragraph A7.

Table 14   Average class sizes: selected subjects[1]

England and Wales   Average class size

| | Subject | | | | |
	Mathematics	English	History	French	Physics
Secondary schools:					
Year 1	25.7	25.6	27.0	27.4	26.9
Year 2	26.1	26.3	27.5	27.9	27.6
Year 3	26.2	26.3	27.1	27.8	27.1
Year 4	24.5	24.2	22.3	21.7	22.8
Year 5	24.1	24.1	21.8	20.7	22.1

Source: DES 1977 Survey of Secondary School Staffing

[1] These figures are averages for all pupils across all schools. Mathematics and English are taken by the great majority of pupils throughout years 1 to 5, whereas the other subjects are often optional at some stage.

Qualifications of those who teach mathematics

A10 Tables 15 to 18 are based on information obtained from the Survey of Secondary School Staffing carried out by the DES in 1977.

Table 15   DES 1977 Survey of Secondary School Staffing: composition of sample

England and Wales

Type of school:	Number of each type of school
Middle 9 to 13[1]	56
Comprehensive 13/14 to 18	62
Comprehensive 11/12 to 18: Large[2]	39 ⎫
Comprehensive 11/12 to 18: Medium[2]	70 ⎬ 153
Comprehensive 11/12 to 18: Small[2]	44 ⎭
Comprehensive 11/12 to 16	96
Comprehensive 10/11 to 14	18
Modern	42
Grammar	32
Sixth form college[3]	29
All schools	488

Source: DES 1977 Survey of Secondary School Staffing

[1] The middle schools were all for pupils aged from 9 to 13 and "deemed secondary". For middle schools and 10/11 to 14 schools, information refers only to the teaching of pupils aged 11 or over.
[2] See paragraph A11.
[3] The sixth form colleges did not include any which still contained a residual element of younger pupils.

A11 Some of the analysis displayed in this section is restricted to the schools in the sample. In other cases, estimated national totals have been derived by using "weighted multiplers" which take account of the percentage of schools of each type in the sample in relation to the total numbers of such schools in the country. In categorising the 11/12 to 18 schools by size, those with up to 800 pupils were regarded as small, those with 801 to 1200 pupils as medium and those with more than 1200 pupils as large. Information about the staffing of middle schools is given in paragraph A15 and is not included in Tables 16 to 18.

A12 The four categories of qualification of teachers to which we referred in paragraph 625 were defined as follows:

'Good'
- Trained graduates, or equivalent, with mathematics as the first, main or only subject of a degree course.
- Bachelors of Education (BEd) with mathematics as a main specialist subject.
- Teachers whose general qualifications were of either of these types with mathematics as a subsidiary subject provided their main specialism was in a related* subject.

'Acceptable'
- Trained graduates, graduate equivalents, or BEd with mathematics as a second or subsidiary specialism if their first subject was not related.
- Untrained graduates with mathematics as first, main or only subject.
- Teachers holding the Certificate in Education, having followed a secondary course in which mathematics was their first, main or only specialism.
- Teachers with no initial mathematical qualifications who had a further qualification resulting from a course of at least one year in which mathematics was the main subject.

'Weak'
- Teachers holding the Certificate in Education having followed a secondary course with mathematics as a second or subsidiary subject provided their first or main subject was related.
- Teachers holding the Certificate in Education having followed a Junior or Junior/Secondary course with mathematics as their first or main subject.
- Teachers in the immediately preceding category with subsidiary mathematics provided their main subject was related.
- Graduates in any subject provided their course included a related subject.

'Nil'
- Qualified teachers without any recorded mathematics and not covered by any previous specification.
- Teachers holding the Certificate in Education with mathematics subsidiary to an unrelated subject.
- Teachers without any initial qualification who possessed a further qualification which did not lead to graduate status and in which mathematics was not the main subject.

A13 Tables 16 to 19 examine the distribution of all mathematics teaching between teachers in the qualification categories defined above. Numbers of periods have been standardised to a 40 period week.

*'Mathematics' includes statistics. Related subjects, at any level, include Computer Studies, Physics, Engineering (or Engineering Science) and Combined Physical Sciences at graduate level.

Table 16   Mathematics teaching in maintained secondary schools: by levels of qualification of teachers

Maintained schools in England and Wales

| | No of maths periods taught[1] | | | | % of mathematics teaching | | | |
| | Qualifications of staff[2] | | | | Qualifications of staff | | | |
	Good	Accept.	Weak	Nil	Good	Accept.	Weak	Nil
432 schools in sample:								
Comprehensive								
13/14−18	5 774	3 532	1 569	1 860	*45*	*28*	*12*	*15*
11/12−18 Large	4 905	3 711	2 245	2 078	*38*	*29*	*17*	*16*
Medium	5 754	3 858	2 739	3 534	*36*	*24*	*17*	*22*
Small	2 298	2 039	1 042	1 828	*32*	*28*	*14*	*25*
11/12−16	4 037	5 865	3 410	4 015	*23*	*34*	*20*	*23*
10/11−14	364	438	519	821	*17*	*20*	*24*	*38*
Modern	889	1 762	873	1 860	*17*	*33*	*16*	*35*
Grammar	2 937	1 408	383	283	*59*	*28*	*8*	*6*
Sixth form college	2 892	1 269	312	234	*61*	*27*	*7*	*5*
Estimated total for all secondary schools (excluding middle)	278 000	243 800	136 900	178 000	*33*	*29*	*17*	*21*

Source: DES 1977 Survey of Secondary School Staffing

[1] The numbers of periods shown are those recorded for the schools in the sample, adjusted to a 40 period week.
[2] As defined in paragraph A12.

Table 17   Estimated percentages of mathematics curriculum 'suitably' staffed

Maintained schools in England and Wales

	Estimated number of schools[1]		Cumulative totals
% of curriculum 'suitably' staffed:[2]		Cumulative % of curriculum 'suitably' staffed:	
0	152	*0*	152
1−10	27	*0− 10*	179
11−20	49	*0− 20*	228
21−30	260	*0− 30*	488
31−40	250	*0− 40*	738
41−50	543	*0− 50*	1281
51−60	796	*0− 60*	2077
61−70	780	*0− 70*	2857
71−80	548	*0− 80*	3405
81−90	528	*0− 90*	3933
91−99	199	*0− 99*	4132
100	241	*0−100*	4373

Source: DES 1977 Survey of Secondary School Staffing

[1] Middle schools are not included.
[2] 'Suitably' covers the aggregate of teaching by teachers with 'good' and 'acceptable' qualifications.

A14 The percentages and totals in Table 16 conceal the extensive variation between individual schools of the same type. One aspect of this is shown in Table 18.

Table 18   Percentage of mathematics curriculum 'suitably' staffed: schools included in the sample

	England and Wales						Schools in sample
	Type of school						
	Comprehensive				Modern	Grammar	Sixth form college
	13/14−18	11/12−18	11/12−16	10/11−14			
% of mathematics curriculum 'suitably' staffed:[1]							
0	0	1	1	1	8	0	0
1−10	0	0	1	2	0	0	0
11−20	0	3	0	0	1	0	0
21−30	0	9	6	4	3	1	0
31−40	3	8	7	4	2	0	0
41−50	6	17	20	0	6	0	0
51−60	4	31	21	3	9	0	2
61−70	10	34	16	1	9	0	1
71−80	16	16	13	1	1	9	3
81−90	12	26	4	1	1	10	7
91−99	10	3	3	1	1	3	5
100	1	5	4	0	1	9	11
Number of schools:	62	153	96	18	42	32	29

Source: DES 1977 Survey of Secondary School Staffing

[1] See note 2 to Table 17.

Table 19   Deployment of 'unsuitably' qualified teachers in comprehensive schools: by year groups

	England and Wales		Percentages
	Comprehensive schools		
	11/12−18	11/12−16	all
% of mathematics curriculum taught by 'unsuitably' qualified teachers in:[1]			
Year 1	48	46	47
Year 2	45	47	46
Year 3	39	42	40
Year 4	37	40	38
Year 5	36	40	37

Source: DES 1977 Survey of Secondary School Staffing

[1] 'Unsuitably' covers the aggregate of teaching by teachers with 'weak' and 'nil' qualifications.

A15 Analysis of the qualifications of those teaching mathematics to pupils aged 11 or over in the middle schools for pupils aged 9 to 13 which were included in the sample shows that, according to the categories set out in paragraph A12, 6 per cent of the teaching was by teachers whose qualification was 'good', 17 per cent by teachers whose qualification was 'acceptable', 15 per cent by teachers whose qualification was 'weak' and 62 per cent by teachers whose qualification was 'nil'. As we point out in paragraph 629, it could be argued that, for teachers in middle schools, some of the qualifications classified as 'weak' on the ground that training had been for a younger age group should be regarded as 'acceptable'. If all the qualifications classified as 'weak' were to be regarded as 'acceptable', it would be the case that 38 per cent of the teaching of mathematics to pupils aged 11 or more would have been by 'suitably' qualified teachers, and 62 per cent by teachers whose qualification was 'unsuitable'.

Supply of teachers in maintained schools

A16 Precise definition of the term 'mathematics graduate' is not possible; for example, the Universities Statistical Record lists 63 different degree courses under the main heading of 'mathematics'. (It is this group of courses which we have called 'mathematical studies' in the main body of the report; see note to paragraph 150.) Published information about numbers of teachers who are mathematics graduates is therefore difficult to reconcile because of the differing definitions which have been used to compile the information. For this reason Tables 20 to 23 which follow do not all show the same total of mathematics graduates. When recording the qualifications of graduates, DES records show education as the first subject of qualification of all holders of the BEd degree. For holders of BEd who have taken mathematics as a main subject, mathematics appears as the second subject of qualification. In Tables 20 to 23, 'all graduates' include holders of the BEd degree, but references to graduates who have mathematics as the only or first subject of their degree do not include holders of BEd.

Table 20   Full-time teachers in maintained nursery, primary and secondary schools

England and Wales

	Year ending 31 March				
	1970	1976	1977	1978	1979
Qualification: Mathematics graduates[1]	5 340	7 221	7 467	7 748	7 876
Other graduates	46 561	107 573	116 180	124 389	134 662
All graduates	51 901	114 794	123 647	132 137	142 538
Non-graduates	260 762	316 183	312 649	306 734	300 490
All teachers	312 663	430 977	436 296	438 871	443 028
Mathematics graduates as % of all graduates	*10.3*	*6.3*	*6.0*	*5.9*	*5.5*

Source: DES *Statistics of Education* Vol. 4

[1] Restricted to graduates in mathematics only or in mathematics and physics.

A17 Table 21 analyses mathematics graduates by the sector in which they are teaching and displays the number of men and women who make up the total. The figures include those who have mathematics named as the only subject of their degree or who have mathematics named as the first of two subjects in their degree. Consequently the numbers shown in Table 21 are somewhat larger than those in Table 20.

Table 21   Full-time teachers in maintained schools who are mathematics graduates: by sex and phase

England and Wales

	Year ending 31 March				
	1975	1976	1977	1978	1979
Mathematics graduates:[1]					
Primary:[2]					
Men	131	121	117	113	106
Women	174	195	179	183	165
All	305	316	296	296	271
Secondary:					
Men	4804	5033	5222	5364	5455
Women	2296	2387	2460	2619	2668
All	7100	7420	7682	7983	8123
All schools:					
Men	4935	5154	5339	5477	5561
Women	2470	2582	2639	2802	2833
All	7405	7756	7978	8279	8394

Source: DES *Statistics of Education* Vol. 4

[1] Includes graduates who have mathematics as the only or the first of two subjects in their degree.
[2] Includes nursery.

A18 Table 22 gives details by age and sex of the rates at which mathematics graduates leave teaching, and compares these with the rates of leaving for all graduate teachers. Table 23 compares entry and wastage rates for trained and untrained mathematics graduates. Both tables relate to graduates who have mathematics as the only subject of their degree or as the first subject of a degree which does not include science. The tables have been compiled from information which is supplied to the DES each year by local education authorities.

Appendix 1   Statistical information

Table 22   Rates of leaving teaching: by age and sex

(i) Men

	Age at 31 March				
	Under 25	25-29	30-34	35-39	40 & over
1974−75[1]					
Maths graduates:					
Stock[2]	290	1233	582	413	1081
Leavers[3]	44	113	30	12	93
Percentage leaving:					
Maths graduates	*15.2*	*9.2*	*5.2*	*2.9*	*8.6*
All graduates	*16.2*	*9.6*	*6.0*	*4.3*	*7.2*
1975−76					
Maths graduates:					
Stock	302	1367	680	454	1067
Leavers	38	111	30	12	86
Percentage leaving:					
Maths graduates	*12.6*	*8.1*	*4.4*	*2.6*	*8.1*
All graduates	*9.5*	*7.5*	*4.1*	*2.8*	*6.0*
1976−77					
Maths graduates:					
Stock	322	1448	831	497	1081
Leavers	37	69	20	13	85
Percentage leaving:					
Maths graduates	*11.5*	*4.8*	*2.4*	*2.6*	*7.9*
All graduates	*7.1*	*5.4*	*3.0*	*2.4*	*6.0*
1977−78					
Maths graduates:					
Stock	295	1524	1028	528	1112
Leavers	27	99	36	10	83
Percentage leaving:					
Maths graduates	*9.2*	*6.5*	*3.5*	*1.9*	*7.5*
All graduates	*6.5*	*5.6*	*3.5*	*2.2*	*5.5*
1978−79					
Maths graduates:					
Stock	271	1547	1202	599	1153
Leavers	22	121	46	17	60
Percentage leaving:					
Maths graduates	*8.1*	*7.8*	*3.8*	*2.8*	*5.2*
All graduates	*5.9*	*5.9*	*3.5*	*2.6*	*5.0*

Source: DES Statistics Branch

[1] Year from 1 April to 31 March
[2] Stock at 1 April.
[3] Leavers by 31 March.

Table 22 (continued)

(ii) Women

| | Age at 31 March | | | | |
	Under 25	25-29	30-34	35-39	40 & over
1974—75 [1]					
Maths graduates:					
Stock [2]	386	802	216	153	634
Leavers [3]	54	171	38	13	56
Percentage leaving:					
Maths graduates	14.0	21.3	17.6	8.5	8.8
All graduates	13.9	20.9	17.9	9.2	8.8
1975—76					
Maths graduates:					
Stock	397	948	234	183	628
Leavers	41	188	31	14	50
Percentage leaving:					
Maths graduates	10.3	19.8	13.2	7.7	8.0
All graduates	11.6	18.9	15.1	8.8	7.6
1976—77					
Maths graduates:					
Stock	428	1007	328	188	661
Leavers	31	181	44	13	53
Percentage leaving:					
Maths graduates	7.2	18.0	13.4	6.9	8.0
All graduates	8.5	15.9	13.9	6.5	6.7
1977—78					
Maths graduates:					
Stock	409	1131	362	239	685
Leavers	26	176	44	17	44
Percentage leaving:					
Maths graduates	6.4	15.6	12.2	7.1	6.4
All graduates	7.3	15.0	13.5	6.6	6.5
1978—79					
Maths graduates:					
Stock	408	1240	384	286	747
Leavers	48	216	55	17	81
Percentage leaving:					
Maths graduates	11.8	17.4	14.3	5.9	10.8
All graduates	7.5	15.8	15.0	7.2	6.8

Source: DES Statistics Branch

[1] Year from 1 April to 31 March
[2] Stock at 1 April.
[3] Leavers by 31 March.

Table 23   Trained and untrained mathematics graduates: entry and wastage, by age

Men and women

	Ages at 31 March							
	Trained				Untrained			
	Under 25	25–29	30–34	all ages	Under 25	25–29	30–34	All ages
1974–75[1]								
In service 1 April	565	1592	535	4275	111	443	263	1515
Wastage[2]	76	222	51	494	29	101	32	215
Entrants[3]	561	121	21	727	112	66	13	194
Re-entrants[4] etc	15	70	56	195	2	14	13	68
In service 31 March	1065	1561	561	4703	196	422	257	1562
1975–76								
In service 1 April	593	1860	642	4703	106	455	272	1557
Wastage	63	237	56	462	23	89	21	207
Entrants	568	128	32	756	91	78	17	200
Re-entrants etc	11	63	35	172	3	19	18	86
In service 31 March	1109	1814	653	5169	177	463	286	1636
1976–77								
In service 1 April	642	2009	848	5153	105	456	314	1638
Wastage	42	197	45	405	28	69	22	172
Entrants	559	120	28	729	47	32	13	102
Re-entrants etc	13	68	44	168	0	17	12	64
In service 31 March	1172	2000	875	5645	124	436	317	1632
1977–78								
In service 1 April	645	2242	1043	5680	59	413	347	1633
Wastage	47	228	54	444	8	65	30	154
Entrants	557	124	25	743	49	32	11	102
Re-entrants etc	8	47	38	150	4	15	20	73
In service 31 March	1163	2185	1052	6129	104	395	348	1654
1978–79								
In service 1 April	624	2421	1231	6170	55	366	355	1667
Wastage	60	296	84	565	14	67	35	175
Entrants	529	126	53	756	35	35	4	84
Re-entrants etc	6	45	60	190	3	16	13	69
In service 31 March	1099	2296	1260	6551	79	350	337	1645

Source: DES Statistics Branch

[1] Year from 1 April to 31 March.

[2] Wastage includes those leaving teaching and those transferring to other sectors of education e.g. FE or teacher training.

[3] Entrants denote those entering teaching for the first time.

[4] This line includes those re-entering teaching, and those transferring from other sectors of education, together with those who change status e.g. from non-graduate to graduate.

Entry to initial training

A19 Tables 24 to 26 provide information about entry to initial training courses.

Table 24   Entry to Postgraduate Certificate in Education (PGCE) courses

England and Wales   Men and Women

| | Year of entry | | | | | |
	1975	1976	1977	1978	1979	1980
University Departments of Education:						
Total entry[1]	4959	4808	4935	4923	4947	5523
No taking main method courses in mathematics[2]	445.5	458.3	495.5	436	401.5	518
% taking mathematics	9.0	9.5	10.0	8.9	8.1	9.4
Non-university institutions:						
Total entry	5609	5415	4438	4431	4514	5198
No taking main method courses in mathematics	320.5	342	393.3	291.2	234.7	310
% taking mathematics	5.7	6.3	8.9	6.6	5.2	6.2
All institutions:						
Total entry	10568	10223	9373	9354	9461	10721
No taking main method courses in mathematics	766	800.3	888.8	727.2	636.2	828
% taking mathematics	7.2	7.8	9.5	7.8	6.7	7.7

Source: Graduate Teacher Training Registry

[1] Numbers in October of each year..
[2] If a student is taking main method courses in two subjects, $\frac{1}{2}$ is recorded against each subject; for main method courses in three subjects, $\frac{1}{3}$ is recorded. Subsidiary subjects are not recorded.

Table 25   Qualifications of students entering PGCE courses who are taking main method courses in mathematics

England and Wales   Men and Women

| | Year of entry | | |
	1978	1979	1980
PGCE students taking main method courses in mathematics:			
University Departments of Education:			
Total number	436	401.5	518
Graduates in mathematics or in in mathematics/physics	323	305	343
Non-university institutions:			
Total number	291.2	234.7	310
Graduates in mathematics or in in mathematics/physics	123	98	139
All institutions:			
Total number	727.2	636.2	828
Graduates in mathematics or in in mathematics/physics	446	403	482
Mathematics or mathematics/physics graduates as % of all PGCE students	4.8	4.3	4.5

Source: Graduate Teacher Training Registry

Table 26   Entry to BEd courses in 1980

England and Wales   1980[1]

	Men	Women	Total
Entry to BEd courses:[2]			
Students taking one main subject only:			
Mathematics	97	176	273
Other subjects	583	2520	3103
Total	680	2696	3376
Students taking more than one main subject:			
Including mathematics	85	127	212
Not including mathematics	278	963	1241
Total	363	1090	1453
All students following subject-oriented courses:			
Including mathematics	182	303	485
Not including mathematics	861	3483	4344
Total	1043	3786	4829
Students not following subject-oriented courses	93	390	483
Total entry to BEd courses	1136	4176	5312

Source: Central Register and Clearing House and DES

[1] Comparable figures for earlier years were not collected.
[2] Numbers include full-time home students only and exclude those transferring to BEd courses after a 3-year Certificate in Education course.

Entrants to universities in England and Wales

A20 Tables 27 to 30 which follow are derived from information supplied by the Universities Statistical Record. They refer to entrants to full-time and sandwich first degree or first degree and first diploma courses at universities in England and Wales who had home fee-paying status and whose entry qualification was based on A-levels. Over the years 1973–1979 these constituted about 92 per cent of all entrants to these courses at universities in England and Wales.

Table 27   Entrants to first degree or first degree and first diploma courses: numbers with A-level mathematics

Universities in England and Wales

	Year of entry to first degree course						
	1973	1974	1975	1976	1977	1978	1979
Men:							
Entrants with A-levels	29 173	30 532	32 456	33 430	34 793	34 974	35 097
Any maths A-level	14 690	14 514	15 694	16 360	17 173	17 891	18 903
% with A-level maths	50.4	47.5	48.3	48.9	49.4	51.2	53.9
Women:							
Entrants with A-levels	16 227	17 479	18 462	19 126	20 531	21 308	22 928
Any maths A-level	4 230	4 269	4 395	4 560	4 887	5 331	6 244
% with A-level maths	26.1	24.4	23.8	23.8	23.8	25.0	27.2

Source: Universities Statistical Record

A21   Tables 28 to 30 provide details of the distribution of the entrants defined in paragraph A20 between categories of degree courses, and also details of their mathematical qualifications. The categories used are

● Engineering and technology, including the various types of engineering, mining, metallurgy, surveying, other technologies and combinations of these

● Physical sciences, including mathematics/physics, physics, chemistry, general and combined physical sciences.

● Mathematical studies, including mathematics, statistics, computer science and combinations of these; and combinations of mathematics with other subjects.

● Medical and dental, including pre-clinical medicine and dentistry, and para-medical courses such as pharmacy.

● Biological sciences, including veterinary and agricultural studies and forestry.

● Other sciences, including combinations of physical and biological sciences and of various sciences with other subjects.

● Business studies, including business and management studies, economics and accountancy.

● Geography.

● All other subjects.

Geography was examined separately in order to investigate whether its allegedly increasing mathematical emphasis was reflected in the mathematical qualifications of those enrolling.

A-level qualification in Mathematics denotes a pass in at least one of the subjects Pure Mathematics, Applied Mathematics, Pure and Applied Mathematics (including Mathematics with Statistics), Further Mathematics; double-subject Mathematics denotes a pass in Applied Mathematics or Further Mathematics.

Appendix 1   Statistical information

Table 28   Entrants to first degree or first degree and first diploma courses:[1] qualifications in mathematics

Universities in England and Wales

	Year of entry to first degree course						
	1973	1974	1975	1976	1977	1978	1979
Subject group[2]							
Engineering and technology							
Entrants with A-levels	5459	5499	5864	6295	6704	7100	7451
of whom Any maths A-level	5122	4896	5429	5810	6235	6666	7045
of whom Double subject maths	1617	1416	1534	1662	1535	1703	1714
Physical sciences							
Entrants with A-levels	4166	3950	4021	4148	4350	4519	4852
Any maths A-level	3717	3356	3411	3505	3641	3871	4185
Double subject maths	1048	826	860	865	817	877	968
Mathematical studies							
Entrants with A-levels	2649	2637	2539	2590	2796	2926	3143
Any maths A-level	2643	2604	2534	2572	2759	2903	3107
Double subject maths	2103	1944	1871	1859	1688	1669	1728
Medical and dental							
Entrants with A-levels	3852	4152	4386	4533	4635	4696	4764
Any maths A-level	1285	1468	1499	1628	1693	1885	2095
Double subject maths	139	129	142	147	117	138	161
Biological sciences							
Entrants with A-levels	3770	4189	4441	4676	4819	4847	4914
Any maths A-level	1097	1124	1236	1331	1469	1554	1656
Double subject maths	61	51	60	62	45	52	47
Other sciences							
Entrants with A-levels	2897	2931	3021	2945	3052	3123	3108
Any maths A-level	1506	1434	1512	1458	1457	1469	1646
Double subject maths	422	324	327	298	222	202	281
Business studies							
Entrants with A-levels	2062	2343	2828	3060	2884	2873	2986
Any maths A-level	863	1064	1342	1364	1446	1447	1621
Double subject maths	185	189	260	232	163	155	177
Geography							
Entrants with A-levels	1506	1657	1716	1795	1805	1804	1823
Any maths A-level	372	364	436	459	474	413	454
Double subject maths	34	30	32	20	23	16	19
Other subjects							
Entrants with A-levels	19 089	20 653	22 102	22 514	24 279	24 394	24 984
Any maths A-level	2315	2473	2689	2793	2886	3014	3338
Double subject maths	397	354	360	356	282	244	246
All subject groups							
Entrants with A-levels	45 400	48 011	50 918	52 566	55 324	56 282	58 025
Of whom Any maths A-level	18 920	18 783	20 089	20 920	22 060	23 222	25 147
Of whom Double subject maths	6006	5263	5446	5501	4892	5056	5341

Source: Universities Statistical Record

[1] With home fee-paying status.
[2] See paragraph A21.

Table 29 Entrants to first degree or first degree and first diploma courses: [1]
as percentages of all entrants with stated qualifications

Universities in England and Wales   Percentages

	Year of entry to first degree course						
	1973	1974	1975	1976	1977	1978	1979
Subject group[2]							
Engineering and technology							
Entrants with A-levels[3]	12.0	11.5	11.5	12.0	12.1	12.6	12.8
Any maths A-level[4]	27.1	26.1	27.0	27.8	28.3	28.7	28.0
Double subject maths[5]	26.9	26.9	28.2	30.2	31.4	33.7	32.1
Physical sciences							
Entrants with A-levels	9.2	8.2	7.9	7.9	7.9	8.0	8.4
Any maths A-level	19.6	17.9	17.0	16.8	16.5	16.7	16.6
Double subject maths	17.5	15.7	15.8	15.7	16.7	17.3	18.1
Mathematical studies							
Entrants with A-levels	5.8	5.5	5.0	4.9	5.1	5.2	5.4
Any maths A-level	14.0	13.9	12.6	12.3	12.7	12.5	12.4
Double subject maths	35.0	36.9	34.4	33.8	34.5	33.0	32.4
Medical and dental							
Entrants with A-levels	8.5	8.6	8.6	8.6	8.4	8.3	8.2
Any maths A-level	6.8	7.8	7.5	7.8	7.7	8.1	8.3
Double subject maths	2.3	2.5	2.6	2.7	2.4	2.7	3.0
Biological sciences							
Entrants with A-levels	8.3	8.7	8.7	8.9	8.7	8.6	8.5
Any maths A-level	5.8	6.0	6.2	6.4	6.7	6.7	6.6
Double subject maths	1.0	1.0	1.1	1.1	0.9	1.0	0.9
Other sciences							
Entrants with A-levels	6.3	6.1	5.9	5.6	5.5	5.5	5.4
Any maths A-level	8.0	7.6	7.5	7.0	6.6	6.3	6.5
Double subject maths	7.0	6.2	6.0	5.4	4.5	4.0	5.3
Business studies							
Entrants with A-levels	4.5	4.9	5.6	5.8	5.2	5.1	5.1
Any maths A-level	4.6	5.7	6.7	6.5	6.6	6.2	6.5
Double subject maths	3.1	3.6	4.8	4.2	3.3	3.1	3.3
Geography							
Entrants with A-levels	3.3	3.5	3.4	3.4	3.3	3.2	3.1
Any maths A-level	2.0	1.9	2.2	2.2	2.1	1.8	1.8
Double subject maths	0.6	0.6	0.6	0.4	0.5	0.3	0.4
Other subjects							
Entrants with A-levels	42.0	43.0	43.4	42.8	43.9	43.3	43.1
Any maths A-level	12.2	13.2	13.4	13.4	13.1	13.0	13.3
Double subject maths	6.6	6.7	6.6	6.5	5.8	4.8	4.6
All subject groups							
Entrants with A-levels	100	100	100	100	100	100	100
Any maths A-level	100	100	100	100	100	100	100
Double subject maths	100	100	100	100	100	100	100

Source: Universities Statistical Record

[1] With home fee-paying stuatus.
[2] See paragraph A21.
[3] As percentage of all entrants with A-levels.
[4] As percentage of all entrants with any maths A-level.
[5] As percentage of all entrants with double-subject maths.

Table 30 Entrants to first degree or first degree and first diploma courses: as percentages of entrants to each subject group with stated qualification

Universities in England and Wales　　　　　　　　　Percentages

			Year of entry to first degree courses				
	1973	1974	1975	1976	1977	1978	1979
Subject group[1]							
Engineering and technology							
Single-subject maths[2]	64.2	63.3	66.4	65.9	70.1	69.9	71.5
Double-subject maths[2]	29.6	25.8	26.2	26.4	22.9	24.0	23.0
Any maths A-level[2]	93.8	89.0	92.6	92.3	93.0	93.9	94.6
Physical sciences							
Single-subject maths	64.0	64.1	63.4	63.6	64.9	66.3	66.3
Double-subject maths	25.2	20.9	21.4	20.9	18.8	19.4	20.0
Any maths A-level	89.2	85.0	84.8	84.5	83.7	85.7	86.3
Mathematical studies							
Single-subject maths	20.4	25.0	26.1	27.5	38.3	42.2	43.9
Double-subject maths	79.4	73.7	73.7	71.8	60.4	57.0	55.0
Any maths A-level	99.8	98.7	99.8	99.3	98.7	99.2	98.9
Medical and dental							
Single-subject maths	29.8	32.2	30.9	32.7	34.0	37.2	40.6
Double-subject maths	3.6	3.1	3.2	3.2	2.5	2.9	3.4
Any maths A-level	33.4	35.4	34.2	35.9	36.5	40.1	44.0
Biological sciences							
Single-subject maths	27.5	25.6	26.5	27.1	29.5	31.0	32.7
Double-subject maths	1.6	1.2	1.4	1.3	0.9	1.1	1.0
Any maths A-level	29.1	26.8	27.8	28.5	30.5	32.1	33.7
Other Sciences							
Single-subject maths	38.1	37.9	39.2	39.4	40.5	40.6	43.9
Double-subject maths	14.8	11.1	10.8	10.1	7.3	6.5	9.0
Any maths A-level	52.9	48.9	50.0	49.5	47.7	47.0	53.0
Business studies							
Single-subject maths	32.9	37.3	38.3	37.0	44.5	45.0	48.4
Double-subject maths	9.0	8.1	9.2	7.6	5.7	5.4	5.9
Any maths A-level	41.9	45.4	47.5	44.6	50.1	50.4	54.3
Geography							
Single-subject maths	22.4	20.2	23.5	24.5	25.0	22.0	23.9
Double-subject maths	2.3	1.8	1.9	1.1	1.3	0.9	1.0
Any maths A-level	24.7	22.0	25.4	25.6	26.3	22.9	24.9
Other subjects							
Single-subject maths	10.0	10.3	10.5	10.8	10.7	11.4	12.4
Double-subject maths	2.1	1.7	1.6	1.6	1.2	1.0	1.0
Any maths A-level	12.2	12.0	12.2	12.4	11.9	12.4	13.4
All subject groups							
Single-subject maths	28.4	28.2	28.8	29.3	31.0	32.3	34.1
Double-subject maths	13.2	11.0	10.7	10.5	8.8	9.0	9.2
Any maths A-level	41.7	39.1	39.5	39.8	39.9	41.3	43.3

Source: Universities Statistical Record

[1] See paragraph A21.
[2] Percentage of entrants to subject group who have this qualification.

Destinations of graduates in mathematical studies

A22 Tables 31 to 33 provide information about the destination of those obtaining first degrees in mathematical studies (see paragraph A16 and note to paragraph 150) from universities in England and Wales in the years 1977, 1978 and 1979. They are not restricted to those with home fee-paying status. In each of these years the overall proportion of those obtaining first degrees who had home fee-paying status was more than 90 per cent of men and more than 95 per cent of women.

Table 31 Destinations of graduates in mathematical studies[1]

Universities in England and Wales

	Year of completing degree course					
	1977		1978		1979	
	Men	Women	Men	Women	Men	Women
Destination:						
Further study or training (other than for teaching)	280	71	309	68	299	69
Teacher training	156	165	111	100	109	108
Total (study or training)	436	236	420	168	408	177
Permanent home employment[2]	934	406	1080	404	1102	419
Already employed or not available[3]	134	67	139	71	143	93
Temporary home employment or unemployed	117	26	72	22	89	19
Not known	152	30	135	34	192	48
Total	1773	765	1846	699	1934	756

[1] See paragraph A22.   Source: Universities Statistical Record
[2] Analysed further in Table 32.
[3] Includes return overseas or employment overseas.

Table 32 Type of permanent home employment: Graduates in mathematical studies[1]   Universities in England and Wales

	Year of completing degree course					
	1977		1978		1979	
	Men	Women	Men	Women	Men	Women
Type of work:						
Research, design in engineering/technology	78	26	101	27	121	43
Services to management[2]	462	182	560	191	589	192
Financial work[2]	255	108	269	118	265	105
Teaching (not only in schools)	19	18	20	14	11	15
Other	120	72	130	54	116	64
Total	934	406	1080	404	1102	419

[2] Analysed further in Table 33.   Source: Universities Statistical Record
See paragraph A22

Table 33   Services to management and financial work: Graduates in mathematical studies[1]

Universities in England and Wales

	Year of completing degree course					
	1977		1978		1979	
	Men	Women	Men	Women	Men	Women
Services to management:						
Computer programming	318	127	420	134	440	144
Systems analysis	50	21	53	17	58	18
Operational research	20	6	22	15	27	13
Other	74	28	65	25	64	17
Total	462	182	560	191	589	192
Financial work:						
Accountancy articles	110	56	146	80	123	59
Other accountancy	43	25	15	15	39	15
Actuarial	68	7	74	13	75	15
Banking	15	8	12	1	18	9
Other	19	12	22	9	10	7
Total	255	108	269	118	265	105

Source: Universities Statistical Record

[1] See paragraph A22.

A23. Table 34 compares the number of graduates in mathematical studies who gained employment in computer programming with the number of graduates from other subject groups who were similarly employed.

Table 34   Graduate employment in computer programming: by subject group

Universities in England and Wales

	Year of completing degree course					
	1977		1978		1979	
	Men	Women	Men	Women	Men	Women
Subject group:[1]						
Engineering and technology	50	5	74	6	95	6
Physical sciences	98	21	115	24	151	35
Mathematical studies	318	127	420	134	440	144
Medical and dental	1	0	4	1	1	2
Biological sciences	22	17	47	39	56	39
Other sciences	53	38	75	31	73	39
Business studies	24	8	42	18	41	11
Geography	11	7	18	18	27	17
Other subjects	39	33	71	71	91	74
All	616	256	866	342	975	367
% who were graduates in mathematical studies	*51.6*	*49.6*	*48.5*	*39.2*	*45.1*	*39.2*

[1] See paragraph A21.

Source: Universities Statistical Record

Appendix 2   Differences in mathematical performance between girls and boys

By Miss H B Shuard

B1 Pupils in schools are often classified according to their sex, and discussion of their educational programme may take their sex into account, either directly or because of social custom. Until fairly recently in Britain, it was commonplace to discuss separately the education of boys and girls, including the mathematics courses they should follow, and expectations of their mathematical attainment were different. In this appendix, differences in mathematical performance are described, and possible causes of these differences are discussed. Historical and statistical evidence, and the considerable volume of research on educational differences between the sexes, are drawn upon. Suggestions are made for measures which might help to improve the mathematical performance of girls.

Historical Background

B2 Mathematics established itself in the curriculum of boys' public and secondary schools in the first half of the nineteenth century, and when girls' secondary schools began to be founded later in the century, the pioneers of girls' education wished to introduce its study into their schools. Lecturing in 1848, Professor F D Maurice discussed the curriculum of the newly opened Queen's College; his remarks about mathematics have often been quoted out of context as a belief "that women students were unlikely to advance far in mathematics"[1]. However, the belief which he actually expressed was that although "we are aware that our pupils are not likely to advance far in Mathematics", there were positive benefits which girls would gain from its study, and "the least bit of knowledge ... must be good"[2]. At another new girls' school, Cheltenham Ladies' College, Miss Dorothea Beale acknowledged that although she wished to introduce mathematics into the curriculum, it would spell financial ruin for the school, because parents did not wish their daughters to study it. Even arithmetic was suspect; around 1860, a father wrote to Miss Beale, on deciding to send his daughters to another school, "My dear lady, if my daughters were going to be bankers, it would be very well to teach arithmetic as you do, but really there is no need"[3]. However, the opening of the Cambridge Local Examinations to girls in 1863 gave impetus to the teaching of arithmetic and mathematics in girls' secondary schools. Of the first 25 candidates from the North London Collegiate School, 10 failed in arithmetic; the headmistress, Miss Buss, was horrified, and the teaching of arithmetic at once became a matter of extreme importance in her school[3]. Only three years later, girls were doing as well in arithmetic as in other subjects in the Cambridge Local Examinations, and when substantial numbers of girls' secondary schools were founded after 1873, mathematics became a regular subject of the curriculum. At first there were great difficulties because of the inadequate supply of teachers; this slowly improved as more women graduated from universities, but in 1912 it was estimated that, out of some 900 women teaching mathematics in secondary schools, only about one-third had themselves studied as far as the calculus. However, the boys' schools were no better off, for about the same proportion of men teachers of mathematics in secondary schools had studied calculus. [1]

B3 In the public elementary schools, both boys and girls learned arithmetic, but as the Royal Commission on the Elementary Education Acts[4] reported in 1888, "as the time of the girls is largely taken up with needlework, the time they can give to arithmetic is less than that which can be given by boys". They therefore recommended that the arithmetical requirements of the curriculum should be modified in the case of girls.

B4 In the 1912 Report on the teaching of mathematics in Britain[1], Miss E R Gwatkin, headmistress of Queen Mary High School, Liverpool, noted that mathematics occupied a good position in the curriculum of most girls' secondary schools, but the pressure caused by the introduction of a wider range of subjects into the curriculum was causing this position to be questioned. Among the objections to mathematics as an important subject for girls were that the subject was uninteresting to most girls, that its utilitarian value to them was negligible, which could explain the lack of interest, and that its difficulty put a strain on pupils out of all proportion to the benefit received. Miss Gwatkin presented a clear case against these objections, using such arguments as the following:

> All girls ought to grow up reasonable beings, and many of them do not; all girls ought to acquire a knowledge of the meaning of language, and a power of using it accurately, and many of them do not. Mathematics, properly taught, will help to both ends. In all branches of Mathematics, though perhaps more especially in Geometry, it is necessary to be clear-headed.

> Women are said to be inconsistent, and, moreover, to be quite unable to recognise their inconsistency. There is no place where the penalties of inconsistency are more striking than in the mathematical classroom.

> Mathematics . . . offers unique opportunities to the teacher for recognising and encouraging independent thought. . . . Girls, perhaps, need a greater stimulus to independent thought than boys do.

> Much harm is done by rating too highly a girl's power to produce—somehow, anyhow—the correct answer to a given sum, and more attention should be paid to style, and more credit given for connected and intelligible explanations, both oral and written.

> A teacher of girls is, perhaps, too easily satisfied when her pupils are working steadily and conscientiously along the lines which she has laid down for them; a boy is almost certain to go off at a tangent . . . routine does him less harm, because he is less susceptible to its influence. Probably one of the weaknesses of girls is that they will submit to so much dullness without resentment. . . . Many girls who are apparently good workers are really mentally lazy, they reproduce, but they do not produce. A teacher needs to be alive to this danger, and to realise that it is her business to stimulate intellectual curiosity, the desire to know, and not only to know, but to

find out at first hand; she has every opportunity in Mathematics, where hardly any of the work should be reproductive.

The evidence cited later in this appendix shows that some of the reasons given for girls' need for mathematics as part of their education in 1912 are now advanced as reasons for their failure to perform better at the subject.

B5 In 1923, when the Consultative Committee of the Board of Education reported on differentiation of the curriculum for boys and girls[5], the Regulations for Secondary Schools provided that:

> an approved course in a combination of these (domestic) subjects may for girls over 15 years of age be substituted partially or wholly for Science and for Mathematics other than Arithmetic.

Moreover, out of 230 Advanced Courses in Mathematics and Science provided in secondary schools, less than one quarter were in girls' schools. The Committee, however, noted that:

> The present degree of girls' inferiority in this subject should not be regarded as permanent, being due partly to unskilful teaching of an old-fashioned kind and partly to an impression among parents, which has influenced the timetable, that Mathematics is unsuitable for girls.

B6 At intervals between 1905 and 1937 the Board of Eduction issued handbooks of Suggestions for the Guidance of Teachers[6]. These suggestions were addressed to teachers in elementary schools, including in the 1937 edition the senior schools which were the forerunners of the secondary modern schools. In the early issues there was little suggestion that the arithmetic curriculum should differ for boys and girls, but as standards rose and children stayed longer at school, comment on curriculum difference for the sexes appeared. From 1918 onwards it was suggested that "difference of sex must affect to some extent the treatment of many ordinary subjects in the curriculum", so that in arithmetic, "a course suitable for boys often requires considerable modification if it is to serve the needs and interest of girls". Girls should deal with "detailed accounts accompanying shopping and housekeeping", while boys "established by experimental methods some of the more important theorems of elementary geometry". In 1927 it was noted that in the senior classes, girls spent less time on mathematics and were less likely to be using it in other subjects. However, this threw more responsibility on the teacher for providing a basis of reality for girls' mathematical work:

> the fact that the girls miss the scale-drawing and "plan and elevation" work of the handicraft course should be a reason not for doing less but for doing more of such work in mathematics, and the same remarks apply to the practical measurement which girls miss through not studying elementary physics.

By the next edition of the Handbook in 1937, the education of girls and boys had come closer together, and this edition stated that:

> In mental capacity and intellectual interests they have much in common, the range of difference in either sex being greater than the difference between the sexes. But in early adolescence the thoughts of boys and girls are turning so strongly towards their future roles as men and women that it would be entirely inappropriate to base their education solely on their intellectual similarity.

Thus by 1937 educational opinion had come to regard girls and boys as in-
tellectually similar, although different in interests.

**Present-day differences
in the examination
performance of
boys and girls**

B7 Today, almost all girls as well as boys study mathematics up to the age of
16, but girls are still not as successful in mathematics examinations as are
boys. The proportion of entries for public examinations in mathematics by
girls decreases as the level of the examination increases. At CSE, almost equal
numbers of boys and girls enter[7], although the same was not true in earlier
years. In O-level mathematics, only 44 per cent of the 1979 entry was from
girls, while in A-level, 26 per cent of the 1979 entry was from girls. However,
the position is improving; in 1968, 37 per cent of the mathematics O-level entry was from girls, and only 17 per cent of the mathematics A-level entry.

B8 Information about the relative performance of boys and girls has also
been obtained from the DES Survey of 10 per cent of school leavers, discussed
in Appendix 1. This data relates to pupils who left school in 1979, so that not
all their examination results were obtained in the same year. At O-level, 24.5
per cent of boys who left school in 1979 had a pass in mathematics at grade A,
B or C, but only 17.6 per cent of girls, and at the highest grade, grade A, more
than twice as many boys (5.5 per cent) as girls (2.6 per cent) obtained this
grade. This is significant for their future careers, as amongst the pupils who
went on to take A-level mathematics, 91 per cent of both boys and girls who
had obtained a grade A pass at O-level passed at A-level, while only 48 per cent
of boys and 36 per cent of girls who had a grade C O-level pass obtained an
A-level pass. However, a smaller proportion of the girls who had a grade A
pass in O-level mathematics continued to A-level than did boys; the percentages were 48.6 per cent of girls as against 66.9 per cent of boys. In summary, a
smaller proportion of girls than boys enter for O-level mathematics; of those
who do enter, a smaller proportion of girls achieve high grades, and of those
with high grades, a smaller proportion of girls than boys proceed to A-level.
The result of this is that nearly three times as many boys as girls entered for
A-level mathematics in 1979: 12 350 girls as against 34 670 boys. Although
the overall pass rates for boys and girls were very similar (boys 73.4 per cent,
girls 73.0 per cent), at the highest grade boys were more successful; 10.1 per
cent of girls obtained a grade A pass, as against 15.4 per cent of boys.

B9 Although boys are more successful than girls in public examinations in
mathematics, the opposite holds in English, as Appendix 1 shows. The picture
which emerges is that girls under-achieve in public examinations in
mathematics compared with boys, while in English the position is reversed.
This pattern continues into the teaching careers of men and women, so that in
1978, among the 258 811 women teachers and 180 060 men teachers in maintained schools, there were 2484 women mathematics graduates, as against
5264 men mathematics graduates[7]. The same pattern holds among students
who follow the BEd route into teaching; the percentages of 1980 entrants to
subject-oriented BEd courses who took mathematics either as a single subject
or in combination were: men 17.4 per cent, women 8 per cent.

Details of differences in mathematical performance

B10 In order to see how girls' performance in mathematics might be improved, it is necessary to look at details of the differences. This information can be gathered from a number of sources at different age levels. More work has been done in the USA than in Britain on sex differences in mathematics; the position in the two countries seems to be roughly comparable, and American as well as British work is therefore referred to.

B11 In the project reported in *Mathematics and the 10-year-old,* Ward[8] tested 2296 children in England and Wales; each item was administered to more than 550 children. Girls performed significantly better than boys on 11 items out of 91. These items were on computation with whole numbers and money, and on entirely verbal items involving naming geometric shapes and making a deduction from given verbal (non-numerical) information. Girls also did significantly better on one logic item involving mappings. Boys performed significantly better on 14 items out of 91. Four of these items were on place-value, and of the other five place-value items, boys did better on four. The other items on which boys did significantly better involved measurement and visual items, word problems and reversing an operation, as in $105 \div \Delta = 21$. The items on which girls did significantly better were easier, with average success rate 64 per cent, as against an average success rate of 49 per cent for the items on which boys did significantly better. The items on which girls did better were also thought to be more important by their teachers, in a survey carried out as part of the project. The types of items on which boys did better at the age of 10 become more important as children get older. An understanding of place-value, an ability to reason about a word-problem and to reverse an operation become more important as pupils proceed towards O-level, when routine computation is less important than problem-solving ability.

B12 The first APU Primary Survey, in 1978[9], found some differences between boys and girls in the results of its written tests at the age of eleven:

> The girls' mean score is significantly higher statistically in computation (whole numbers and decimals). The boys' mean score is significantly higher statistically in three sub-categories: length, area, volume and capacity; applications of number; and rate and ratio.

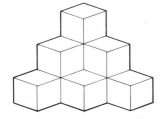

However, in the 1979 APU Primary Survey[10], girls were not significantly in advance of boys in computation, while there were two additional categories in which boys scored significantly higher: the measurement of money, time, weight and temperature; and concepts of decimals and fractions. Both the 1978 and 1979 Primary Surveys also found differences between the sexes in the practical testing. Boys were significantly better at building from a diagram a model which needed four hidden blocks to make it stand; in 1979, 82 per cent of boys succeeded, but only 63 per cent of girls. Although 71 per cent of the girls in the first survey could halve a given piece of string and then cut off one quarter of this piece as against 60 per cent of boys, the percentages who knew what fraction of the whole string they now had were: boys 43 per cent and girls 40 per cent; a similar result occurred the next year. There were also two significant differences in favour of boys among the questions on the topic of weighing. Girls, on the other hand, were slightly better at giving change from a sum of money; there was a difference of between 6 per cent and 9 per cent on each of the items on this topic.

B13   The first APU Secondary Survey[11] found that by the age of 15/16, the composition of the top 10 per cent of achievers was: boys 61.5 per cent and girls 38.5 per cent. However, the difference between proportions of boys and girls in the middle 10 per cent of achievers and in the bottom 10 per cent was small. Over all the written tests, boys had higher scores than girls in every sub-category, and were furthest ahead in descriptive geometry (a success rate of 7 per cent more than that for girls), rate and ratio (6 per cent) and mensuration (5 per cent). There were also considerable differences between boys and girls in the practical tests, particularly in the mass/weight topic, in which 19 per cent more boys than girls succeeded in finding the mass of one peg from a bag of equal small pegs when given only a 20 gram mass and a balance. This topic was an extension of the task on weighing used in the Primary Surveys.

B14   In 1973 and 1974, Wood[12] analysed performance on the London Board O-level Syllabus C papers. On the multiple-choice papers he found that, in ad-dition to the fact that girls generally performed worse than boys, they were particularly weak on items on scale, pie-charts and probability, and had a poor grasp of size and of distance-time graphs. Girls were strong on sets, Venn diagrams, matrices, the real number-line or line-segments, and on straightforward vector addition. In the free-response questions, probability and geometry were much more popular among boys, and girls' relative per-formance was worst on a question about size and estimation. Girls in single-sex schools did rather better than girls in mixed schools, but the items on which girls from single-sex schools out-performed girls from mixed schools were on substitution, intersection of sets, reflection and matrix definition. The performance on the "boys' items" remained low. The comparison bet-ween single-sex and mixed schools needs to be treated with some caution, as in 1973-4 many of the single-sex schools were selective, and it is therefore not clear how far like was compared with like. Wood summarises

> None of the items on which girls out-performed boys required what could be termed problem-solving behaviour; instead they call for recognition or classifica-tion, the supplying of definitions, application of techniques, substitution of numbers into an algebraic expression and so forth, just the kind of operations which are most susceptible to drilling.

Wood puts forward two hypotheses to explain the differences. The first is the difference between the sexes in ability in spatial visualisation, which is well documented, and which is discussed in paragraph B18. The second is what he calls in another article[13] Cable's Comparison Factor. He uses the term 'com-parison factor' for a number which is used to state how one quantity compares with another, and he sees the existence of a comparison factor as a common thread linking the difficult (for girls) items of fractions, proportion and measures.

B15   A number of similar studies have been carried out in the USA. Many of those which dealt with children of ages between 10 and 14 were surveyed by Fennema in 1974[14].

She concluded that

> Girls performed slightly better than did boys in the least complex skill (computation). ... In the 77 tests of more complex cognitive skills (comprehension, application and analysis) five tests had results that favoured girls, while 54 tests showed significant differences in favour of boys. The conclusion is inescapable that the boys of these populations learned the mathematics measured by these tests better than did the girls ...

> ... in overall performance on tests measuring mathematics learning ... there are no significant differences that *consistently* appear between the learning of boys and girls in the fourth to ninth grade. ... if a difference does exist, girls tend to perform better in tests of mathematical computation, and boys tend to perform better in tests of mathematical reasoning.

B16 Fennema also analysed the 1978 mathematics test of the USA National Assessment of Educational Progress (NAEP)[15]. Testing was carried out at ages 9, 13 and 17, and the test was analysed into scores on knowledge, skills, understanding and applications. It was found that with the exception of the skills scores of the 9 and 13 year olds, boys did better than girls in all cases, and the higher the cognitive level, the greater the difference between the sexes. Moreover, in items related to geometry, such as measurement skills, geometric manipulations and items on perimeter, area and volume, the differences were particularly large.

B17 The testing carried out in 1964 as part of the International Study of Achievement in Mathematics[16] showed a similar pattern. In all twelve developed countries which took part in the study, the performance of boys was higher than that of girls at the age of 13, and the performance of boys was further ahead on verbal mathematical problems than on computational problems. However, there was a considerable difference between countries, the sex differences in performance being greatest in Belgium and Japan, and least in the USA and Sweden.

B18 There has been much research on the subject of differential performance between the sexes on non-mathematical tests of various types, and it is well established that males tend to perform better than females on tests of spatial visualisation, which include the ability to rotate objects in the mind and to orient oneself or other objects in space[17]. Evidently, some aspects of mathematics make use of this ability, but it is not yet clear how far spatial visualisation is directly related to the learning of mathematics. In tests of verbal ability, on the other hand, the average ability of girls is greater than that of boys, but it is not known how important this ability is in the learning of mathematics. On all tests, however, the overlap between the sexes is very large, and it would be a gross distortion to expect that most boys would be better than most girls at, for example, tasks involving spatial visualisation.

Reasons suggested for differences in performance between boys and girls

B19 A number of biological theories have been put forward to explain differences between the sexes in spatial visualisation. Three of these theories concern a recessive gene on the x-chromosome, the role of sex hormones, and differences in brain lateralisation[18]. The question of biological differences in either spatial visualisation or mathematical ability is not yet fully understood, but in view of the fact that differences between the sexes in mathematical attainment are more marked in some countries than in others[16], there would seem to be factors other than differences in ability in spatial visualisation which influence differences in mathematical attainment. These factors may be classified as follows: first, patterns of socialisation may be produced by child-rearing practices and peer-group pressures; secondly, the expectations of schools and individual teachers may affect pupils' performance; finally, the pupils' own motivation may have significant effects on their attainment. All these factors may be expected to interact with one another in influencing the different mathematical attainments of boys and girls. These factors are now surveyed, and reference is made to research studies concerned with them.

Socialisation patterns

B20 A child-rearing practice which may have an effect on mathematical attainment is the fact that boys are given significantly more spatial and scientific toys, rather than the dolls which girls receive[17]. A recent British study[19] found significant differences between the spontaneous play of boys and girls in a nursery school; girls engaged in more fantasy and creative play, while boys chose more construction play and play with sand and water. Throughout childhood, boys play more with constructional toys and take part in more physical games, both of which promote spatial awareness and problem-solving activity. Boys are encouraged to be more independent, a valuable characteristic for problem-solving, while girls are expected to be more passive and conformist, and to spend time helping mother around the house, rather than helping father with 'do-it-yourself' and with the car, both of which are more directly related to measurement, shape and calculation than are washing-up and simple cooking and sewing. Boys seem to receive more attention, more punishment and more praise from adults, and adults respond to boys as if they find them more interesting and more attention-provoking than girls.[20] Thus, boys' ideas seem to be more valued by adults, and so boys put their ideas forward. This may have some consequences for later mathematics learning in school.

B21 Peer-group pressures centre on sex-stereotyping, and on the view of mathematics taken by children's contemporaries. Peer-group pressure increases in adolescence, and is enhanced by the influence of pop culture, the media and the teenage magazines, some of which still put forward stereotypes which confirm the restricted image of the girl who is engaged at the age of 16 or 17 and married at 20[20]. Thus, the study of mathematics may seem pointless to girls who are influenced by these pressures.

B22 Recent studies suggest that the sex-typing of mathematics is decreasing, but that male prejudice against girls' involvement in mathematics still exists, or at least girls believe it does[21]. It has also been shown that mathematically gifted girls fear that their achievement will have negative consequences for their relationships with boys[22] and that girls who under-achieve in mathematics see intellectual achievement as only appropriate for men[21]. The evidence as to whether boys do in fact stereotype mathematics as a male domain is somewhat conflicting [23],[21]. In providing opportunities for mathematically gifted pupils in Baltimore, Fox[24] found that more 13 year old boys than girls were eager to enrol in a special mathematics course, and that many girls dropped out of the course. They appeared to be afraid that their participation would have negative social consequences for them.

Factors within the school

B23 It is widely believed in the USA that the most important influence on learning and achievement in mathematics is how great an opportunity the pupil has to learn mathematics, and that boys take more advanced mathematics courses than girls do, so that they have more opportunity to learn mathematics. However, the 1978 National Assessment of Educational Progress in the USA[15] shows only very small differences in course-taking in mathematics between the sexes. This is not the case in England and Wales, where more boys than girls take O-level as against CSE courses, and where this pattern is greatly intensified at A-level. Hence, in the APU's testing of mathematical attainment in England and Wales at the age of 16, boys might be expected to score rather higher than girls, because more boys than girls take O-level courses.

B24 Mathematics is not only learned in mathematics lessons; a great deal of mathematics is used and learned in science, especially in the physical sciences, and in technical drawing. In these subjects, not only is mathematics learned and practised in a practical context, but the message is conveyed that mathematics is useful in the technical, scientific and employment worlds, particularly in the traditional worlds of boy's employment and interests. In the physical sciences and technical drawing there is a great difference between numbers of boys and girls who take these subjects.

B25 In mixed schools, in groups in which boys and girls are following the same course, there is some evidence that boys still have more opportunity to learn that do girls. Secondary school teachers have been shown to interact more with boys than they do with girls[25] and to give more serious consideration to boys' ideas[20]; they also give boys more opportunity to respond to higher cognitive level questions[26]. High-achieving girls have been found to receive significantly less attention in mathematics classes than do high-achieving boys[27].

B26 Even in the primary school, there are differences between the treatment of boys and girls. In a 1973 survey[28] HMI found that

> ... Boys engage in a wider range of crafts involving the use of a range of tools and materials leading to three-dimensional modelling and construction and the use of measurement ... the boys' experience further helps to familiarise them with geometrical ideas ...

However, HMI also found that, in middle schools, opportunities for mixed classes in the crafts were on the increase. This trend towards common experience for boys and girls in the crafts has no doubt continued since 1973.

B27 In primary schools, too, children are expected to behave in ways regarded as appropriate to their sex. Boys and girls are asked to help in different ways in the classroom; boys move PE apparatus and milk crates, while girls tidy up and arrange displays[29]. Primary teachers see girls as sensible, obedient, hard-working and co-operative, while boys are excitable and talkative, and need more supervision and attention[20]. Thus, no doubt, teachers find themselves interacting more with boys than with girls.

B28 Dweck has investigated a phenomenon she calls 'learned helplessness'[30] 'Learned helplessness' exists when a person believes that failure at a task is insurmountable, and it is accompanied by a deterioration in performance. Dweck found that some children become incompetent following a failure, while others rise to the challenge, persist and improve their performance. She found that the child's interpretation of the failure, what he thinks caused it and whether he thinks he can overcome it, predicts his response. She was able to train children to attribute their failures to insufficient effort rather than to innate inability; most of them then showed improvement in a problem-solving task following a failure. In contrast, children whom she arranged to have experiences only of successful problem-solving during her experiment did not improve their ability to cope with failure when it occurred. She then investigated by classroom observation (in the USA), the ways in which elementary school teachers commented on pupils' failures. She found that more of the negative feedback received by boys referred to conduct than to intellectual ability, and even feedback about their work was largely about neatness, instruction-following or style of presentation. On the other hand, much of the negative feedback girls received for their work related to its intellectual aspects. However, Fennema[31] made similar observations of pupils aged 12 in mathematics classes, and found relatively few instances in which the teacher provided feedback to either boys or girls which attributed success or failure to such causes. Fennema, however, noted that boys who were high in confidence received a great deal more attention from their mathematics teachers than did any of the girls, whether they were confident or not, or than the less confident boys. In another recent study[32], Fennema has found significant differences between boys and girls in the ways in which they themselves attribute their successes and failures in mathematics learning. People may attribute their successes or failures to relatively stable and unchanging causes such as their own ability (or lack of it), or to relatively unstable and changeable causes such as lack of effort, the difficulty or ease of the task, the unfairness of the teacher, or luck. Fennema's work confirmed, for the learning of mathematics, findings obtained by workers in other areas, that boys are more likely to attribute their successes to stable causes such as ability and their failures to unstable causes such as lack of effort, while girls attribute their successes to unstable causes such as the effort they put into their work, and their failures to stable causes such as their lack of ability.

Motivation and attitudes towards mathematics

B29 There is some evidence that boys see more clearly than do girls that mathematics will be useful in their future lives and work. Fox, surveying American research in 1977[22], found many studies which suggested that girls are less oriented towards careers outside the home than are boys, and that the usefulness of mathematics in the traditional women's careers in business, nursing, teaching and the social services is less plain than is its usefulness in traditional men's careers. Fennema, however, has reported that in the 1978 NAEP survey[15], no significant differences were found between boys and girls in the perceived usefulness of mathematics. This suggests that American attitudes may be changing. The recent British evidence on this point is inconclusive. A survey of the attitudes of 13 year old Sheffield schoolchildren[33], found a small but significant difference in the mean scores for boys and girls on the attitude scales, with boys showing a greater liking for mathematics. It was also found that the most common reason for liking mathematics is its perceived usefulness in getting a job, or in doing a job, or in general everyday use; one boy is quoted as saying "It is helping you to get a better job with good pay even though the lessons may be boring and confusing".

B30 The APU has surveyed the attitudes of 11 year old children to mathematics. In the first survey[9], it was found that most children agreed with statements about the usefulness of mathematics, and that there were no sex differences. However, primary mathematics is fairly clearly related to everyday life, while the mathematics of the secondary school is less clearly oriented towards daily living, and more towards qualifications, further study, and use in science and technology. In the second survey, in 1979[10], it was found that significantly more boys than girls believed that they usually understood a new mathematical idea quickly, that they were usually correct in their work and that mathematics was one of their better subjects. However, significantly more girls than boys confirmed that they often got into difficulty with mathematics and were surprised when they succeeded. Thus, 11 year old girls were already showing a tendency to attribute failure to stable causes such as lack of ability, while boys showed greater self-confidence in their mathematics. In their comments on individual mathematical topics, significantly more boys than girls liked and found easy topics such as measurement and geometry, while girls preferred numerical topics such as factors and multiplication tables.

B31 Many people also regard mathematics as a male domain. Weiner[20] maintains that
> ... mathematics is regarded by pupils of all ages both primary and secondary (and by teachers) as a subject at which boys excel.

Mathematics teachers in secondary schools tend to be men, and in primary schools men teachers tend to teach the older classes, where the mathematics is more advanced, while women teachers are concentrated in younger age-groups where the emphasis on language and reading is greater.

B32 Children acquire many of their attitudes from their parents. Parents still often hold lower educational aspirations for girls than for boys, and it has been found in the USA that low levels of mathematical achievement are more easily accepted by parents of girls than by parents of boys[34]. Teachers, too, may unconsciously play a part in the sex-role stereotyping which reinforces children's attitudes to mathematics. It is well-known that children's books often reinforce traditional pictures of sex-roles; mathematics textbooks and examination papers sometimes do the same. It has been said that in one well-known British secondary mathematics series, all the shopping is done by the female sex, while in another example, the girls knit while the boys make concrete or play football[35].

B33 Pupils' self-confidence and self-concept may affect their achievement in mathematics, and sex differences in self-confidence in mathematics have been found in several studies[21]. In one study[36] most students gave lack of effort as the reason why they failed to be successful in most subjects of the curriculum. In mathematics, however, 26 per cent of girls gave their lack of ability as the reason, but only 15 per cent of boys. Fennema[32] states that females often feel inadequate about intellectual, problem-solving activities, and under-estimate their ability to solve mathematical problems. The APU Primary Surveys confirm that there is a comparative lack of mathematical self-confidence among girls in England at an age as early as 11. High anxiety is also associated with lower achievement in mathematics, and as schooling progresses, American girls have been found to display greater anxiety of a debilitating type. Recent work by Buxton[37], as well as the report of the study undertaken by ACACE for our inquiry, suggest that anxiety about mathematics also occurs frequently in Britain.

Evidence submitted to the Committee

B34 In view of the evidence cited earlier which shows the differences in examination success in mathematics between boys and girls in England and Wales, and the fact that there is a large volume of research evidence on sex differences in mathematics, it is surprising that little evidence was submitted to the Committee relating to this topic. However, the Committee for Girls and Mathematics pointed out to us that

> The debate on standards, particularly in relation to the mathematical needs of industry, has often seemed to be directed very largely towards boys. Schools, the careers service and industry appear to have shown very little initiative in encouraging or attracting girls with ability in mathematics into some of the fields where there are shortfalls of good applicants.

Strategies for improvement

B35 The evidence reviewed in this appendix suggests a number of strategies which might improve the attitudes and achievement of girls in mathematics. Girls should be helped to realise that mathematics is as important for their daily lives, and in their future careers, as it is for boys. Investigations into the uses of mathematics in employment have shown the importance of mathematical knowledge in, for instance, the traditional women's career of nursing, while failure to obtain O-level mathematics at grade C or higher now provides a barrier to entry to teaching. In less traditional women's careers, public awareness of equal opportunities for women is increasing, and more girls are entering these careers. It will be a consequence of this development that greater demands will be made on girls' mathematics. Boys as well as girls need to be aware of the importance of mathematics for women's careers and daily lives, so that boys do not unconsciously emphasise outdated stereotypes in their expectations of girls.

B36 Teachers should ensure that girls receive additional help and encouragement in the areas of measuring, spatial and diagrammatic work and problem-
solving, and should ensure that girls attain a good grasp of the principles of
place-value. Girls should be encouraged to tackle higher cognitive level tasks,
and not be content with success at low-level tasks such as routine computation.

B37 Teachers need to become aware of the fact that they may unconsciously
give cues to both boys and girls, and that these cues may affect not only attitudes to mathematics but also the learning of mathematics. If a teacher
responds to pupils in a way which conveys the message to a boy that
mathematics is important for him, that he is expected to succeed and that lack
of success is due to his lack of effort, while a girl receives the message that her
lack of success is due to lack of ability and that lack of mathematical ability is
common and unimportant in girls, then it is not surprising if the girl gives up
trying while the boy tries harder. Thus, teachers need to be consciously aware
of the importance of helping girls to see their successes in mathematics as the
result of their good mathematical ability, and not solely due to their hard
work. Teachers also need to be well informed of the specific areas of
mathematics in which girls may need additional experience and help if they are
to achieve well, and of the importance of ensuring that girls are given the
necessary opportunities.

B38 Authors, publishers, examiners and teachers should ensure that written
material in mathematics does not reinforce the stereotyping of boys as active,
exploratory problem-solvers while girls are portrayed as passive helpers
whose interests do not extend beyond fashion and the home. Applications of
mathematics should encompass those with which girls as well as boys can
identify. Teachers also need to ensure that mathematics is not presented as a
male domain in the daily oral work of the classroom, as well as in written
materials.

B39 In choosing their options in the secondary school, girls should be encouraged to take more subjects in which the uses of mathematics are made
plain. As well as the traditional subjects of physics and technical drawing, the
newer Craft, Design and Technology and computer studies can encourage a
problem-solving approach which is relevant to today's world.

B40 Careers guidance should make plain to girls, early in the secondary
school, the qualifications which they will need for entry to various occupations, and the importance of mathematics among those qualifications.
Mathematics often acts as a 'filter', whose absence as a qualification can exclude girls from many fields of employment, training and further education.

B41 Research is needed in Britain into the causes of girls' comparative under-
achievement in mathematics. A good deal of the research quoted above was
undertaken in the USA, and it is not clear how accurate is its application in the
British educational system, and in a society whose expectations are not exactly
the same as those of society in the USA. It is clear, however, from the statistics

of public examinations in England and Wales that, even in the last few years, many fewer girls than boys were studying mathematics at the higher levels, and that those who continued the study of mathematics did not perform as well as did boys at the highest grades.

References

[1] Special Reports on Educational Subjects: Volume 26, *The teaching of mathematics in the United Kingdom*, HMSO, 1912.

[2] Maurice F.D., *Queen's College, London: its objects and methods*, London, 1848.

[3] Kamm, J., *How Different From Us*, Bodley Head, 1958.

[4] *Final Report of the Commissioners appointed to Inquire into the Elementary Education Acts, England and Wales*, (the Cross Report), HMSO, 1880.

[5] Board of Education, *Report of the Consultative Committee on Differentiation of the Curriculum for Boys and Girls respectively in Secondary Schools*, HMSO, 1923.

[6] Board of Education, *Suggestions for the Consideration of Teachers and others concerned in the Work of Public Elementary Schools*, HMSO. First issue 1905. The handbook was re-issued in 1909, 1912, 1915, 1918, 1922, 1923, 1927 and 1937. There were substantial changes to the section on mathematics in 1915, 1927 and 1937.

[7] DES, *Statistics of education*, HMSO, 1978.

[8] Ward, M., *Mathematics and the 10-year-old*, Schools Council Working Paper 61, Evans/Methuen Educational, 1979.

[9] Assessment of Performance Unit, *Mathematical development. Primary survey report No.1*, HMSO, 1980.

[10] Assessment of Performance Unit, *Mathematical development. Primary survey report No.2*, HMSO, 1981.

[11] Assessment of Performance Unit, *Mathematical development. Secondary survey report No.1*, HMSO, 1980.

[12] Wood, R., Sex differences in mathematics attainment at GCE Ordinary Level, *Educational Studies*, **2**, 2, 1976, 141–160.

[13] Wood, R., Cable's Comparison Factor: Is this where girls' troubles start? *Mathematics in School*, **6**, 4, Sept. 1977, 18–21.

[14] Fennema, E., Mathematics learning and the sexes: a review, *Journal for Research in Mathematics Education*, 5, 1974, 126.

[15] Fennema, E., Carpenter, T., Sex-related differences in mathematics: Results from national assessment, *Mathematics Teacher*, **74**, 7, Oct. 1981, 555–559.

[16] Husen, T., (ed.), *International Study of Achievement in Mathematics, Vols I and II*, John Wiley and Sons, 1967.

[17] Maccoby, E.E., Jacklin, C.N., *The psychology of sex differences*, OUP, 1975.

[18] Sherman, J., Effects of Biological Factors on Sex-Related Differences in Mathematics Attainment, in *NIE Papers in Education and Work: No.8, Women and mathematics: research perspectives for change*, National Institute of Education, Washington, D.C., 1977.

[19] Eynard, R., Walkerdine, V., *The practice of reason, Vol. 2: Girls and mathematics*, Univ. of London, Institute of Education, 1981.

[20] Weiner, G., in Deem, R., *Schooling for women's work*, Routledge and Kegan Paul, 1980.

[21] Schildkamp-Kündiger, E., Mathematics and gender, in *Comparative Studies of Mathematics Curricula—Change and Stability 1960–80*, Materielien und Studien, Band 19, Institut für Didaktik der Mathematik der Universität Bielefeld, 1980.

[22] Fox, L.H., The effects of sex role socialization on Mathematics participation and achievement, in *NIE Papers in Education and Work No. 8* (see ref. 18 above.)

[23] Fennema, E., Women and girls in mathematics—Equity in mathematical education, *Educational Studies in Mathematics*, **10**, 1979, 384–401.

[24] Fox, L.H., Brody, L., Tobin, D., (eds)., *Women and the mathematical mystique*, Johns Hopkins, U.P., 1980.

[25] Becker, J., *A study of differential treatment of females and males in mathematics classes*, unpublished doctoral dissertation, University of Maryland, 1979.

[26] Sears, J., Development of Gender Role, in Beach, F., *Sex and behaviour*, John Wiley, 1965.

[27] Good, T.L., Sykes, J.N., Brophy, J.E., Effects of teacher sex and student sex on classroom interaction, *Journal of Educational Psychology*, **65**, 1, 1973, 74–87.

[28] Department of Education and Science, *Curricular differences for boys and girls in mixed and single-sex schools*, Education Survey 21, HMSO, 1975.

[29] Lobban, G., Sexism in primary schools, *Women Speaking*, **4**, 1975.

[30] Dweck, C.S., Bush, E.S., Sex differences in learned helplessness I, *Developmental Psychology*, **12**, 2, 1976, 147–156.

Dweck, C.S., Davidson, W., Nelson, S., Enna, B., Sex differences in learned helplessness, II and III, *Developmental Psychology*, **14**, 3, 1978, 268–276.

[31] Fennema, E., Attribution theory and achievement in mathematics, in Yussen, S.R. (ed), *The Development of Reflection*, New York, Academic Press, 1981.

[32] Fennema, E., The sex factor: real or not in Mathematics Education, in Fennema, E. (ed.), *Mathematics education research: Implications for the 80s*. Association for Supervision and Curriculum Development, Washington, D.C., 1981.

[33] Preece, M., Sturgeon, S., *Mathematics education and girls*, draft report of BP project, Sheffield City Polytechnic (unpublished), 1981.

[34] Levine, M., *Identification of reasons why qualified women do not pursue mathematical careers*, report to the National Science Foundation, USA, 1976.

[35] Berrill, R., Wallis, P., Sex roles in mathematics, *Mathematics in School*, **5**, 2, March 1976, 28.

[36] Dornbusch, S., To try or not to try, *Stanford Magazine*, **2**, 2, 1974, 50–54.

[37] Buxton, L., *Do you panic about maths?*, Heinemann, 1981.

Appendix 3 List of organisations and individuals who have made submissions to the Committee

denotes those who have met members of the Committee for discussion.

Mr J Abramsky
Mr G Adlam
Advisory Unit For Computer Based Education
Professor M Aitken
Mr A G Aitken & Mrs N E Hughes
Staff of Albemarle Primary School, London SW19
Rotary Club of Aldridge
Mrs E F Allan
Mrs A Allen
Mr J H Allen
Amersham (Old Town) Women's Institute
Ms M L Andison
Mr D Armer
Professor A M Arthurs
Ash & Lacy Limited
Mrs S Ashby
Mr E R Ashley
*Assistant Masters & Mistresses Association
Associated Examining Board for the General Certificate of Education
Association for Science Education, Education (Co-ordination) Committee
Association of British Chambers of Commerce
Association of Career Teachers
Association of Educational Psychologists
Association of Graduate Careers Advisory Services
Association of Headmistresses of Preparatory Schools
*Associated Lancashire Schools Examining Board
Association of Polytechnic Teachers

Association of Teachers of Domestic Science Limited
Association of Teachers of Geology
*Association of Teachers of Mathematics
Association of University Teachers
Mr R Atherton
Mr G M Austin
Miss M E Austin
Avery Hill College

Mr J K Backhouse
Mr C W Baker
Mr D Ball
Rotary Club of Bangor
Banking Information Service
Mr B Banks
*Barbers of Fulham
Mr J K Barnes
Barnet LEA
Rotary Club of Barnet and East Barnet
Rotary Club of Barnstaple
Miss Julia Barrett
Barton Conduits Limited
Bath College of Higher Education
Rotary Club of Bath
University of Bath, School of Education
Mr E Bathgate
Miss J S Batty
Professor Dr H Bauersfeld
Mr S R Beaumont
Mr T Beck
Bedford College of Higher Education
Rotary Club of Bedford
Bedfordshire LEA
Mr K M Bedwell

*Dr A W Bell
Mr J A Bell
Mr B D Bennett
Mr D H Bennett
Mr D Bent
Berkshire LEA
Miss B J Berry
Rotary Club of Beverley
J Bibby & Sons Limited
Mr R G Biddlecombe
Bifurcated Engineering Limited
*Miss E Biggs
Rotary Club of Billingham
Mr D Bird
Ms M H Bird
Mr J G Birkett
City of Birmingham LEA
Rotary Club of Birmingham
Bishop Burton College of
    Agriculture
Bishop Grosseteste College
Rotary Club of Bishop's Stortford
*Dr A J Bishop
Rotary Club of Blackpool
Bolton LEA
Mr C L Boltz
Boots Company Limited
*Mr J W G Boucher
Rotary Club of Bournemouth
Mrs C Bowler
Boxfoldia Limited
Mrs V R Bradbury
Bradford College
City of Bradford LEA
University of Bradford, Under-
    graduate School of Studies in
    Mathematical Sciences
Rotary Club of Braintree & Bocking
Mr F T Brawn and Mr G Johns
Mr S E Bray
Mrs S Brennard
Brent LEA
Bretton Hall College
Brighton Polytechnic
Rotary Club of Brighton
University of Bristol, School of
    Education
Bristol Polytechnic Faculty of
    Education

Rotary Club of Bristol
British Airports Authority
British Airways
British Association for Early Child-
    hood Education
British Broadcasting Corporation,
    BBC Education
*British Broadcasting Corporation,
    Open University Productions
British Council
British Petroleum Company
    Limited
British Rail
British Society for the History of
    Mathematics
British Society for the Psychology
    of Learning Mathematics
British Steel Corporation
Mr W R Broderick
*W Brooks and Sons
Bromley LEA
Mr J Brown
*Mrs M L Brown
Mr R V G Brown
Professor M Bruckheimer
Brunel University Department of
    Education
Rotary Club of Buckingham
Buckinghamshire College of Higher
    Education
Buckinghamshire LEA
Professor H Burkhardt
Bury LEA
Business Education Council
Mr L G Buxton

C & A Modes
Mr J Cable
*Mr J Cain
Calderdale LEA
Cambridge Institute of Education,
    Academic Board
Cambridgeshire LEA
University of Cambridge,
    Department of Education
University of Cambridge, General
    Board of the Faculties
University of Cambridge Local
    Examinations Syndicate

University College, Cardiff, Department of Education

Careers Service Advisory Council for Wales

Mr J B Carmel

Carpet Industry Training Board

Mrs J E Carrick

Centre for Science Education, Chelsea College

Centre for Statistical Education

Ceramics, Glass & Mineral Products Industry Training Board

Mrs M Chadwick

Mr B R Chapman

Chartered Institute of Public Finance and Accountancy

Chartered Institute of Transport

Rotary Club of Chatham

Chelmer Institute of Higher Education, Faculty of Education, Arts and Humanities

Rotary Club of Chelmsford

Rotary Club of Cheltenham

Chemical and Allied Products Industry Training Board

Chemical Society

Cheshire LEA

*Rotary Club of Chester

Staff of the Mathematics Department, Chipping Sodbury School, Bristol

Christ's College, Liverpool

City and Guilds of London Institute

City University

Civil Service Commission

Staff of Clapthorne Church of England Primary School, Peterborough

Ms S I Clark

Mr C G Clayton

Mr H F Cleaves

Cleveland LEA

Mr C Cleveland

Rotary Club of Clitheroe

Clothing and Allied Products Industry Training Board

Clwyd County Council

College of Preceptors

Mr A P W Collins

Commercial Union Assurance Company Limited

Committee for Girls and Mathematics

Committee of Heads of University Geological Departments

Committee of Professors of Statistics

Committee of Vice-Chancellors and Principals of the Universities of the United Kingdom

*Confederation of British Industry

*Confederation of British Industry (Wales)

Confederation for the Advancement of State Education

Conference of Professors of Applied Mathematics

Construction Industry Training Board

Professor D E Conway

Mrs F Conway

Mr E C Cooper

Mrs J Cooper

Dr M G Cooper

Rotary Club of Corby

Mr M L Cornelius

Cornwall LEA

Mr G B Corston

Cotton and Allied Textiles Industry Training Board

Council for Educational Technology for the United Kingdom

*Council for National Academic Awards

*Council of Engineering Institutions

*Council of Local Education Authorities

Council of Science & Technology Institutes Limited

Council of Subject Teaching Associations

Courtaulds Limited

Coventry & District Engineering Employers' Association

Coventry LEA

Rotary Club of Coventry

Professor J Crank

Ms A Crawshaw

Mr M B Cresswell
Crewe & Alsager College of Higher
    Education
Rotary Club of Crompton & Royton
Mr G T Cross
Croydon LEA
Rotary Club of Croydon
Cumbria LEA
Mr D M Cundy
Mr H J Curtis

Dacorum Education and Training
    Forum
Mrs E Daincey
Rotary Club of Darlington
Rotary Club of Dartford
Mr J R M Davies
*Mr J D Dawson
Mrs D M Dean
Debenhams Limited
*Mr J A Deft
Delta Metal Company Limited
Staff of the Mathematics Depart-
    ment, Denstone College,
    Uttoxeter
Derby College of Further Educa-
    tion, Mathematics Study Group
Derby Lonsdale College of Higher
    Education, Division of
    Mathematics Education
Derbyshire LEA
Development of Ideas in Mathe-
    matical Education (DIME)
    Project
Rotary Club of Devizes
Ms D M Diamond
Mrs A Disney
*Distributive Industry Training
    Board
Mr C Dixon
Dr M H Dodson
Dr M M Dodson
Doncaster LEA
Rotary Club of Dorchester
Dorset   Institute   of   Higher
    Education
Dorset LEA
Ms E Drury
Rotary Club of Dudley

Miss J M Duffin
Mr J Dunford
Mrs A Dunn
Durham Industry Commerce
    Education Association
University of Durham Board of
    Studies in Engineering Science
Dyfed LEA

Ms M R Eagle
University of East Anglia
East Anglian Examinations Board
East Anglian Regional Advisory
    Council for Further Education
East Midland Educational Union
*East Midland Regional
    Examinations Board
Rotary Club of Eastbourne
Eaton Hall College of Education
Edge Hill College of Higher
    Education
Mr N A Edney
Educational Publishers Council
Mr M N Edwards
Dr R Edwards
Professor A S C Ehrenberg
Mrs E R Ehrhardt
Electricity Supply Industry
    Training Committee
Rotary Club of Ellesmere Port
Mr R C Ellis
Mr H Elston
Professor L R B Elton
EMI Limited
Enfield LEA
*Engineering Industry Training
    Board
*Engineering Professors' Conference
English China Clays Limited
Equal Opportunities Commission
Rotary Club of Erith
Rotary Club of Esher
Essex LEA
Mr D M Esterson
Mr I Evans
Rotary Club of Exeter
University of Exeter, School of
    Education

Appendix 3   List of organisations and individuals who have made submissions to the Committee

Mr J C F Fair
Rotary Club of Falmouth
Mrs P Farndell
Mr M D Fellows
Mr D S Fielker
Dr C R Fletcher and Dr I Danicic
Food, Drink & Tobacco Industry
    Training Board
Mrs B Foord
Mr A G Foot
Mr S Forbes
Ford Motor Company Limited
Mrs M A Fowler
Mrs R E Fraser
*Mr D W French
Professor M J French
Staff   of   Friarswood   County
    Primary School,
    Newcastle-under-Lyme
Mr C Frisby
Furniture & Timber Industry
    Training Board

*Gaines and Gaines (Overseas)
    Limited
Mrs D Gale
Miss S Gale
Mr A Gallant
Mrs T Gant
Mr A Gardiner
Garnett College
Rotary Club of Garston
Staff of Gastrells County Primary
    School, Stroud
Gateshead LEA
General Electric Company Limited
GEC – Marconi Electronics Limited
General   Nursing   Council   for
    England and Wales
Geographical Association
Miss R F Gibbons
Mr F D Gibson
Dr W G Gilchrist
Mr G Giles
Rotary Club of Gillingham
Mid Glamorgan LEA
South Glamorgan LEA
South Glamorgan Institute of
    Higher Education
West Glamorgan LEA

West Glamorgan Institute of
    Higher Education
Mr J A Glenn
Gloucestershire LEA
Ms R Goldhawk
Mrs S J Gowar
Mr J D Graham
Mr M J Graham
*Dr M A Grant
Mrs O J Gray
Ms A D Green
Mr L J Green
*Mr F Gregory
Mr B Griffiths
Mr D Griffiths
Mr E N Griffiths
*Professor H B Griffiths
*Mr M Griffiths
Mr P L Griffiths
*Mr A G Gronow
Gwent College of Higher Education
Gwent LEA
Gwynedd LEA

Mr G Haig
Rotary Club of Hailsham
Rotary Club of Halesworth
Staff of Halifax Primary School,
    Ipswich
Mr F Hall
Mr P J D Halpenny
Hampshire, Sixth Form Colleges,
    Heads of Mathematics
Mr A Hanley
Ms T Hardy
Haringey LEA
Mrs M Harris
Rotary Club of Harrogate
*Dr K M Hart
Mr J R Hartley
Mrs J Harvey
Rotary Club of Haverfordwest
Havering LEA
Mrs A Haworth
Mrs M Hayman
Mr J Hayter & Miss J Yates
Mr S Hayward
*Headmasters' Conference
Mrs S Heaney
Rotary Club of Hemel Hempstead

Mr J M Hepplewhite
Dr J Herbert
Hereford & Worcester LEA
*HM Inspectorate of Schools
Hertfordshire College of Higher
  Education
Hertfordshire LEA
Professor J Heywood
Master Callum Hickey
Mrs P Hickey
Hillingdon LEA
Dr S Hilsum
Dr B Hilton
Dr D Hilton
Mr D Hindson
Mr M Hiscox
Mr G Hitch
Mr B E Holley
Mr K Holling
Miss B Hollinshead
Mrs M Holloway
Rotary Club of Honiton
Homerton College
Mr R Hooper
Hotel & Catering Industry Training
  Board
Hotel, Catering & Institutional
  Management Association
Hounslow LEA
Dr A G Howson
Mr G Hubbard
Rotary Club of Hucknall
Hull College of Higher Education
Hull, Senior High Schools, Heads
  of Mathematics
Rotary Club of Hull
University of Hull Department of
  Educational Studies
Humberside LEA
Humberside, East Riding Division,
  Heads of Mathematics
Humberside, Grimsby Division,
  Heads of Mathematics

IBM United Kingdom Ltd
Ilkley College
*Imperial   Chemical   Industries
  Limited
Incorporated Association of
  Preparatory Schools

Independent Broadcasting
  Authority
Mrs D M Ingham
Inner London Education Authority
ILEA Division 10, Heads of
  Mathematics
Institute of Actuaries
Institute of Bankers
Institute of Biology
Institute of Careers Officers
Institute of Chartered Accountants
  in England and Wales
Institute of Chartered Secretaries
  and Administrators
Institute of Geologists
*Institute of Mathematics and its
  Applications
Institute of Personnel Management
Institute of Purchasing and Supply
Institute of Quantity Surveyors
Institution of Civil Engineers
Institution of Mechanical
  Engineers, North Midlands
  Branch
Institution of Mining Engineers
Institution of Mining and
  Metallurgy
Institution of Structural Engineers
International Association of Lions
  Clubs
International Computers Limited
Rotary Club of Ipswich
Dr I C Irmler
Staff of Irthlingborough Infant
  School, Northants
Isle of Wight LEA

Mr H Jackson
Dr H L W Jackson
Mr D Jacobs
Mr P E Jaynes
Mr P G Jenkins
Ms L Joffe
Mrs A Johncock
Staff of John Ruskin School
  (Infants Department), London
  SE5
Professor D C Johnson
Mr F T Johnson
Mr J R Johnson

Mr H T Joint
Joint Mathematical Council of the United Kingdom
Joint Matriculation Board
*Miss A D Jones
Mr A P Jones
*Mr E Jones
Mr E G Jones

Mr P Kaner
Mr C Keal & Mr J D Warren
Mr D A Keane
Kent LEA
Kent Mathematics Project
*Mr J M Kenyon
Mr J K Kerley
Miss D Kerslake
Rotary Club of Kidderminster
Ms P Kilmister
King Alfred's College, Winchester
Rotary Club of King's Lynn
Dr B M Kingston
Mr C Kiralfy
Mr John Kirkham
Kirklees LEA
Professor Dr U Knauer
Knitting, Lace and Net Industry Training Board
Mrs E Knott
Knowsley LEA

Mr C W Lamble
M. Mathieu Lambrecht
Lancashire LEA
Rotary Club of Lancaster
University of Lancaster Department of Environmental Sciences
Professor F W Land
Mr N Langdon
Mr D Laycock
Leapfrogs Group
Leeds LEA
University of Leeds, School of Education
Mr D J Lee
Rotary Club of Leicester
University of Leicester, School of Education
Leicestershire LEA
Rotary Club of Leighton Buzzard

Mr J M Letchford
Rotary Club of Letchworth Garden City
Mr A Levy
Mr R Lewis
Mr W H Lewis
Rotary Club of Lichfield
*Sir James Lighthill
Mr G H Littler
Littlewoods Organisation Limited
Lincolnshire LEA
City of Liverpool College of Higher Education, Department of Mathematics
City of Liverpool LEA
Rotary Club of Liverpool
University of Liverpool, Department of Applied Mathematics and Theoretical Physics
Mr A H Livingstone
Rotary Club of Llandudno
Rotary Club of Llangollen
Mr A S Llewellyn
Mr J Lockett
London Mathematical Society
*London Chamber of Commerce and Industry (Commercial Education Scheme)
Rotary Club of London
London Transport Executive
University of London Entrance and School Examinations Council
West London Institute of Higher Education
Loughborough University of Technology, Department of Education
Miss M Louis
Mr E Love
Mr A W Lupton

Miss S M Macaskill
Miss M I McClure
Mrs J McConachie
Miss M J McConnell
Mr G McFarlane
Mr H McMahon
Mr P J McVey
Mr K Madgett
Rotary Club of Maidstone
Mr A F Makinson

Man-Made Fibres Producing
  Industry Training Board
City of Manchester LEA
Manchester Polytechnic, Didsbury
  School of Education
Rotary Club of Manchester
University of Manchester,
  Faculty of Technology
Mr M J Mapleton
Mr R J Margetts
*Marks and Spencer Limited
Mr J Marshall
Mrs M M Massey
*Mathematical Association
'Mathematical Association, Diploma'
  (Mathematical Education) Board
Mathematical Association, Schools
  and Industry Committee
Mathematics Advisory Unit
Mathematics Applicable Group
Mathematics Education on Mersey-
  side
Mathematics Teachers' Group
  (Coventry)
Matlock College of Education
Professor G Matthews and
  Mrs J Matthews
Mr P Matts
*Miss S Maughan
Mr D J Maxwell
Ms J M Maxwell
Merchant Navy Training Board
Professor J E Merritt
Rotary Club of Merthyr Tydfil
Merton LEA
Metal Box Limited
Metrication Board
Metroplitan Regional Examinations
  Board
Rotary Club of Middlesbrough
Middlesex Polytechnic Mathematical
  Education Group
Middlesex Regional Examining
  Board
Milton Keynes College of Education
Mr B Molloy
Mr R Moon
Mr D I Morgan
*Professor A O Morris

Mr B Morris
Mr R W Morris
Mullard Limited
Mr A Murphy

National Association for Gifted
  Children
National Association for Remedial
  Education
National Association for the
  Teaching of English
*National Association of Head
  Teachers
National Association of Inspectors
  and Educational Advisers
*National Association of
  Mathematics Advisers
*National Association of School-
  masters/Union of Women
  Teachers
National Association of Teachers
  in Further and Higher Education
*National Association of Teachers
  in Further and Higher
  Education, Mathematical
  Education Section
National Coal Board
*National Coal Board (Wales)
National Conference of Teachers'
  Centre Leaders
National Extension College
National Federation of Women's
  Institutes
National Institute of Adult
  Education
National Institute for Careers
  Education and Counselling
National Scientific Society of Wales
*National Union of Teachers
Mrs P Neal
Nene College
City of Newcastle upon Tyne LEA
Newcastle upon Tyne Polytechnic
  Faculty of Education and
  Librarianship and St Mary's
  College, Fenham
University of Newcastle upon Tyne
  School of Education

University of Newcastle upon Tyne Faculties of Engineering and Science
Newham LEA
Rotary Club of Newport (Mon)
*Mr M D Noddings
Norfolk College of Arts and Technology
Norfolk LEA
Normal College of Education, Bangor
North East London Polytechnic
North Oxfordshire Technical College and School of Art
North Regional Examinations Board
University College of North Wales, School of Mathematics and Computer Science
North Yorkshire LEA
North Western Regional Advisory Council for Further Education
Northamptonshire LEA
Northern Counties Technical Examinations Council
Northumberland College of Higher Education
Northumberland LEA
Staff of Northumberland Park Secondary School, London
Nottinghamshire LEA
University of Nottingham
University of Nottingham School of Education
Nuffield Mathematics National Committee
*Rotary Club of Nuneaton

Rotary Club of Oadby
Mrs P Odogwn
Oldham LEA
Mr A J Oldknow
*Open University
Mr C Ormell
Oxford and Cambridge Schools Examination Board
Oxford Delegacy of Local Examinations
University of Oxford Department of Educational Studies

University of Oxford Department of Physics
University of Oxford Board of the Faculty of Mathematics
Oxfordshire LEA

Parke-Davis and Company
Mr A R Parr
Miss D M Parry
Mrs P Parsonson
Mrs M E Payne
Mr A Peace
Mrs M Peach
Mrs E Pearson
Mr J L Pearson
*Mr J Peel
Staff of Mathematics Department, Peers School, Oxford
Mr G Pemberton
Rotary Club of Pembroke
Rotary Club of Penarth
Mr D J Pentecost
Rotary Club of Penzance
Miss D M Perry
Peter Brotherhood Limited
Mr H W Pettman
Philips Industries Limited
Pilkington Brothers Limited
Mr B S Pinfield
Sister Mary Timothy Pinner
Pirelli Limited
Mr D Pitman
Mrs P D Playle-Mitchell
Plenty Group Limited
Mr S P O Plunkett
Plymouth Rotary Club
Mr G E Pohu
Rotary Club of Porthcawl
Portsmouth Polytechnic, Faculty of Educational Studies
Rotary Club of Portsmouth & Southsea
Post Office
Powys LEA
Rotary Club of Preston
Professional Association of Teachers
Professors of Mechanical Engineering in the Northern Universities
*Professors of Statistics

Provident Mutual Life Assurance
  Association
Prudential Assurance Company
  Limited
Mr D W Purdy
Dr C W Puritz
Mr F J Purkiss

*Mr D A Quadling
Miss E Quartly
Miss V Quin

Mr N Radcliffe
Mr N F Radford
Staff of Ravenstone Infant School,
  London SW12
Mr J W M Read
University of Reading School of
  Education
Redbridge LEA
Rotary Club of Redcar
Mr D L Rees
Professor C M Reeves
Regional Advisory Council for the
  Organisation of Further Educa-
  tion in the East Midlands
Regional Advisory Council for Tech-
  nological Education, London and
  Home Counties
Regional Council for Further
  Education for the South West
Mr S J Relf
Mrs A I G Renton
Staff of La Retraite High School,
  London SW12
*Mr P Reynolds
Mr F Rhodes
Ms K Rich
Mr A Richards
Mr F G Richards
Sir Alan Richmond
Staff of Ripley County Junior
  School, Derby
Miss D Roake
Mr P T Robbins
Mr A Roberts
Dr H G F Roberts

Rochdale Borough Council
Rotary Club of Rochester
Staff of Rodford Junior School,
  Bristol
Mr A Rodgman
Mr G W Rodda
Roehampton Institute of Higher
  Education
Mrs L Roland
Rolls Royce Limited, Aero Division
Mr D J Rooke
Professor A M Ross
Mr T D Ross
Rotary International in Great
  Britain and Ireland
Rotherham LEA
Royal Institution of Chartered
  Surveyors
Royal Institution of Naval
  Architects
Royal Navy
Royal Society
Royal Society of Arts
  Examinations Board
Royal Statistical Society/Institute
  of Statisticians Education
  Committee
Rugby District Chamber of
  Commerce
Rotary Club of Ruislip &
  Northwood
Staff of Ruskin County Junior
  School, Wellingborough
Professor W R Russell

Mrs M J Sadler
J Sainsbury Limited
Rotary Club of St Austell
St Helens LEA
Staff of St John's Church of
  England School, Bristol
St Katharine's College, Liverpool
St Martin's College, Lancaster
Colleges of St Mary and St Paul,
  Cheltenham
St Mary's College, Strawberry Hill
Rotary Club of St Pancras
City of Salford LEA

University of Salford
Rotary Club of Salisbury
Mrs B Sanders
Sandwell LEA
Mr G J Sasse
School Broadcasting Council for the United Kingdom
*School Mathematics Project
Schools Calculators Working Party
*Schools Council
*Schools Council Project *Learning through science*
Schools Council Project *Low attainment in mathematics*
Schools Council Project on Statistical Education
*Professor R L E Schwarzenberger
Scottish Mathematical Council
Rotary Club of Scunthorpe
*Secondary Heads Association
Sentra Engineering Training Limited
Mr K E Selkirk
Miss B E Severs
Mr T R Sharma
Dr M J R Shave
Mr R W Shaw
Mrs Sheaney
City of Sheffield LEA
Sheffield City Polytechnic
Sheffield and District Engineering Group Training Association
University of Sheffield (Careers Advisory Service)
Shell UK Limited
Shipbuilding Industry Training Board
Mrs J Shoenberg
Shoreditch College
Mr G Short
Miss H B Shuard, Mr A Rothery and Mr M Holt
Mr A R Shuttleworth
Mrs E Shuttleworth
Rotary Club of Silloth on Solway
Professor R R Skemp
Miss S L Skuba
Mr D K Sledge
Rotary Club of Slough
Professor C A B Smith

Mr C H Smith
Mr J H Smith
Mr J J Smith
W H Smith & Son Limited
Society of Company and Commercial Accountants
Society of Education Officers
Society of Headmasters of Independent Schools
Solihull LEA
Rotary Club of Solihull
Somerset County Council
*South-East Hampshire Chamber of Commerce and Industry
South-East Regional Examinations Board
South Nottinghamshire Project
South Western Examinations Board
Rotary Club of Southampton
Southern Regional Council for Further Education
*Southern Regional Examinations Board
Southern Universities Joint Board for School Examinations
Stafford College of Further Education
Rotary Club of Stafford
Staffordshire LEA
Mr M A Stamp
Standard Telephone and Cables Limited
Standing Conference of Heads of Department of Mathematics in Polytechnics and Major Colleges
Standing Conference on University Entrance
Staveley Industries Limited
*Mrs J Stephens
Rotary Club of Stevenage
Ms S Stewart
Stockport LEA
Rotary Club of Stockport
Mr E L Stocks
Major (Rtd) A R Stockton
Mr R J Stone
Rotary Club of Stratford-upon-Avon
Mr R T Street
Mr D Strong

Mrs S B Sturgeon
Mr D A Sturgess
Suffolk LEA
Dr R Sumner
Sunderland Polytechnic Faculty
    of Education
Rotary Club of Sunderland
Surrey LEA
East Sussex LEA
West Sussex LEA
Sutton Coldfield Teachers' Centre,
    Mathematics School/Employment
    Interface Working Party
Swan Hunter Shipbuilders Limited
Rotary Club of Swanage & Purbeck
Rotary Club of Swansea
University College of Swansea
    Department of Education
Mr J C B Sweeten
Mr K Swinbourne
Rotary Club of Swindon
Sir Peter Swinnerton-Dyer

Dr W Tagg
Mr D G Tahta
Tameside LEA
Rotary Club of Taunton
Mrs G Taylor
Mr G V Taylor
Mr H Taylor
Mrs J Taylor
Mr P Taylor
Mr Peter Taylor
Mr R M Taylor
Technician Education Council
Tesco Stores Limited
Thames Polytechnic
Mr M Thirlby
Dr B Thwaites
Mr B Tillbrook
Mrs B A Tilling
Mr J D Tinsley
Mr C Todd
Mr D Toeman
Rotary Club of Tonbridge
Mr J Towers
Trades Union Congress
Trafford LEA
Mr D Traviss

Trent Polytechnic Department of
    Education
Trent Polytechnic, Department of
    Mathematical Sciences
Trinity College, Carmarthen
Trinity and All Saints' Colleges,
    Division of Mathematics and
    Science
Rotary Club of Truro
North Tyneside LEA
South Tyneside LEA

*Understanding British Industry
Unilever UK Holdings Limited
United Kingdom Atomic Energy
    Authority
Universities Council for the
    Education of Teachers
*University Departments of
    Education Mathematics Study
    Group
University Grants Committee
Mr A Unsworth

*Mr P Vallom
Staff of Vicarage Infant School,
    London E6

Mr F Wade
Mr G T Wain
Wakefield LEA
Mr A A G Walbridge
University College of Wales,
    Aberystwyth, Faculty of
    Education
Mr M C Wales
Mr T W Wales
Miss R Walker
*Professor C T C Wall
Walsall LEA
Waltham Forest LEA
Ms M Warden
Mrs C Warm
University of Warwick, Depart-
    ment of Science Education
University of Warwick, Mathemat-
    ics Institute
Warwickshire LEA
Mr N C C Webb
Mr J M Weseen

Mr D Wells
*Welsh Joint Education Committee
Welsh Secondary Schools
  Association
Rotary Club of Welshpool
Mr P M West
West Cumbria Careers Association,
  Working Party on Mathematics
  and Industry
West London Institute of
Higher Education
West Midlands College
West Midlands Examinations Board
West Midlands Mathematics
  Advisers/HMI Group
Westhill College
Rotary Club of Whitchurch, Salop
Mrs E R B White
Mr G M Whitehead
Mr M N Whitley
Rotary Club of Whitley Bay
Wigan LEA
Staff of William Shrewsbury
  Primary School,
  Burton-on-Trent
*Mrs B Williams
Mr D L Williams
Mrs J Williams
Mr M Williams
Mr R S Williams
Mr B J Wilson

Wiltshire LEA
Wirral LEA
Rotary Club of Wisbech
Wolverhampton LEA
Wolverhampton Polytechnic
Mr P G Wood
Dr D Woodrow
Mrs M D Woods & Mr P A Rodgers
Wool, Jute and Flax Industry
  Training Board
Worcester College of Higher
  Education
Working Mathematics Group
Mr R Wort
Mr A E Wright
Mr D Wright
Professor J Wrigley and
  Mr D D Malvern

Ms J Yates
Rotary Club of York
University of York Department of
  Education
Yorkshire & Humberside Council
  for Further Education
Yorkshire Regional Examinations
  Board
Miss J O Young

*Professor E C Zeeman

Appendix 4   Visits to schools, industry and commerce: meetings with teachers

Schools

Albury Manor Comprehensive
School, Surrey

Bishop Perowne Church of England
Secondary School, Hereford and
Worcester

Bodedern School, Gwynedd

Cavendish School, Hertfordshire

Chipping Norton Comprehensive
School, Oxfordshire

City Road Junior and Infants
School, Birmingham

Crescent Junior School,
Leicestershire

Daubeney CP Junior School, ILEA

Daynecourt Comprehensive School,
Nottinghamshire

Downsend School, Leatherhead

Great Cornard Middle School,
Suffolk

Haberdashers' Aske's School,
Elstree

Hammersmith County Girls'
School, ILEA

Hermitage Comprehensive School,
Durham

Highbury Secondary School,
Wiltshire

Hipperholme Grammar School,
Calderdale

Holloway Boys' School, ILEA

Holyhead County Secondary
School, Gwynedd

Houghton County Primary School,
Cambridgeshire

Howden Clough High School,
Kirklees

Hylton Red House Junior School,
Sunderland

Islington Green Comprehensive
School, ILEA

Light Hall School, Solihull

Liverpool Parkhill Primary School,
Liverpool

Manor Field First School, Norfolk

Mead Junior School, Havering

Okehampton School and Community College, Devon

Oulder Hill Community School,
Rochdale

Page Hill County Middle School,
Buckinghamshire

Parsloes Manor School, Barking

Peasedown St John Primary School,
Avon

Peter Lea Junior School,
South Glamorgan

Prior Western County Primary
School, ILEA

Rhodes Avenue Junior Mixed and
Infants School, Haringey

Ripley County Junior School,
Derbyshire

Riverside School, Bexley

St Asaph VC Infants School, Clwyd

St Oswald's Roman Catholic Primary School, Newcastle upon Tyne

St Wilfrid's Roman Catholic High
School, Sefton

Sandylands County Primary School,
Lancashire

Shaftesbury Church of England
Primary School, Dorset

Stirling Junior School, Doncaster

Stokesley Comprehensive School,
Cleveland

Thomas Telford High School,
Sandwell

Tinsley Junior School, Sheffield
Upwood County Primary School,
   Cambridge
William Shrewsbury School,
   Staffordshire
Winstanley County Primary School,
   Wigan
Wombwell High School, Barnsley
Wyche County Primary School,
   Cheshire
Wyndham School, Cumbria
Ysgol Cymerau Pwllheli, Gwynedd
Ysgol Esgob Morgan Junior, Clwyd
Ysgol Gyfun Llanhari,
   Mid Glamorgan

Industry and Commerce
Abbey Panels
Barclays Bank Limited
Baugh and Weedon Limited
British Airways
British Rail
British Steel Corporation
Clarks Limited
Coral Leisure Group Limited
Debenhams Limited
Delta Metal Company Limited
English China Clays Limited
Ford Motor Company Limited
John Williams of Cardiff Limited

Littlewoods Organisation Limited
Marks and Spencer Limited
Metal Box Limited
Mid Glamorgan Health Authority
Mullards Mitcham
Porth Textiles Limited
Post Office
Prudential Assurance Company
   Limited
Sheffield Industrial Training
   Company
Standard Life Insurance Company
Swan-Hunter Shipbuilders Limited
Vickers Limited
Wales Gas (Nwy Cymru)
William Walkerdine Limited

Meetings with teachers
The Lamorbey Park Teachers'
   Centre, Bexley
Mathematics Teachers' Centre,
   Brunel University
William Tyndale Mathematics
   Centre, ILEA
Ogmore Residential Education
   Centre, Mid Glamorgan
Coxlodge Teachers' Centre,
   Newcastle upon Tyne
The Reading Centre, University of
   Reading

Appendix 5   Abbreviations used within the text of the report

ACACE Advisory Council for Adult and Continuing Education
AFE Advanced Further Education
AGCAS Association of Graduate Careers Advisory Services
APU Assessment of Performance Unit

BBC British Broadcasting Corporation
BEC Business Education Council
BEd Bachelor of Education

CBI Confederation of British Industry
CEE Certificate of Extended Education
CGLI City and Guilds of London Institute
CNAA Council for National Academic Awards
CSE Certificate of Secondary Education
CSMS Concepts in Secondary Mathematics and Science

DES Department of Education and Science

EITB Engineering Industry Training Board

FE Further Education

GCE General Certificate of Education

HMI Her Majesty's Inspector(ate)
HNC Higher National Certificate
HND Higher National Diploma

IMA Institute of Mathematics and its Applications
IBA Independent Broadcasting Authority

LEA Local Education Authority

MEI Mathematics in Education and Industry

NFER National Foundation for Educational Research

ONC Ordinary National Certificate
OND Ordinary National Diploma

PGCE Postgraduate Certificate in Education
POSE Schools Council Project on Statistical Education

RSA Royal Society of Arts

SCUE Standing Conference on University Entrance
SEN State Enrolled Nurse

SI	Système International d'Unités
SLAPONS	School Leaver's Attainment Profile of Numerical Skills
SMP	School Mathematics Project
SRN	State Registered Nurse
TEC	Technician Education Council
UCCA	Universities Central Council on Admissions
USR	Universities Statistical Record
VAT	Value Added Tax

Index

References in the index are to paragraphs in the main report unless otherwise stated. An outline of the contents of each chapter will be found in the index under the entry for the chapter heading.

Printed in England for Her Majesty's Stationery Office
by Commercial Colour Press, London E.7.
Dd.736065 C50 1/83